IB
STUDY
GUIDES

BIOLOGY
for the IB Diploma
STANDARD AND HIGHER LEVEL

Andrew Allott

OXFORD
UNIVERSITY PRESS

OXFORD
UNIVERSITY PRESS

Great Clarendon Street, Oxford OX2 6DP

Oxford University Press is a department of the University of Oxford.
It furthers the University's objective of excellence in research, scholarship,
and education by publishing worldwide in

Oxford New York

Auckland Bangkok Buenos Aires Cape Town Chennai
Dar es Salaam Delhi Hong Kong Istanbul Karachi Kolkata
Kuala Lumpur Madrid Melbourne Mexico City Mumbai Nairobi
São Paulo Shanghai Taipei Tokyo Toronto

Oxford is a registered trade mark of Oxford University Press
in the UK and in certain other countries

British Library Cataloguing in Publication Data

Data available

ISBN 0 19 914818 X

10 9 8 7 6 5 4

The publisher would like to thank the following for their permission to
reproduce photographs:
Biophoto Associates for all except:
Alison Allott p.141, Andrew Allott pp.3, 38 and 155, Dr. E.Signer © Nature p.31
(left), Professor T. Burke p.31 (right), Professor G. Leedale/Biophoto Associates
p.49, Microscopix pp.2 and 90, Quest/Science Photo Library p.101 (left), Science
Photo Library/Susan Leavines P.121, Science Photo Library/Steve Grand p.137

Design and illustration by Hardlines Ltd, Charlbury, Oxon.

Printed in Great Britain

Introduction and acknowledgements

The IB Biology Programme includes many recent developments and much innovative material, which can be hard to find, even in bulky textbooks. This book is intended to help students obtain necessary information quickly and easily. It follows the programme that will be examined from 2003 onwards.

All topics needed for Higher Level (HL) and Standard Level (SL) Biology are covered, including all eight options. The topics are covered in the same sequence as in the syllabus, but within some topics the sequence of sub-topics has been slightly altered to give a more coherent progression of ideas. An index of assessment statements in the syllabus has been included to show where each one is covered.

- Topics 1–5 are core topics studied at both HL and SL.
- Topics 6–13 are additional topics studied only at HL.
- Options A–C are studied only at SL (each student takes two of the eight options).
- Options D–G can be studied at HL or SL, with the extra material needed at HL separated on clearly marked pages at the end of the option.
- Option H is studied only at HL.

Practice questions are included at the end of topics. Almost all of these are taken, with permission, from past IB examination papers. Answers to each question are given, though students and teachers may be able to find other valid answers!

Guidance is given for students working on internal assessment, extended essays or preparing for final exams.

For biologists, these are the best of times and the worst of times. There are unprecedented opportunities for using recently developed techniques in beneficial ways, but there are also greater threats to the natural world than for millions of years. A thorough understanding of the principles of biology is essential if we are to counter the threats and make the most of the opportunities. Biology teachers worldwide should be commended for the work that they do in promoting this understanding. Teachers of IB Biology often take on an additional challenge – to promote internationalism and international understanding. There are many opportunities for this in Biology. Apart from humans, living organisms do not recognize national frontiers. Living organisms throughout the biosphere, including humans, are inter-dependent. Human activities have national impacts, so international co-operation is essential to protect the biosphere and its treasure-house of biodiversity.

I am very grateful for the help that many fellow teachers have given me during the writing of this book. I am indebted to Mary Sanders for her invaluable work as Editor. I would also like to thank my wife Alison for her support and understanding during the period of intense work and for the drawings of plants and animals that she contributed. The work that I did on the book is dedicated to my sister Rachel who was bravely fighting illness at the time.

Andrew Allott

CONTENTS

(Italics denote Higher level pages)

1 CELLS (IB TOPIC 1)

Cells, tissues and organs	1
Light and electron microscopes	2
Size in cell biology	3
The cell theory	4
Prokaryotic cells	5
Eukaryotic cells	6
Membrane structure and membrane proteins	7
Transport across membranes	8
Cell division	9
Exam questions on Topic 1	10

2 THE CHEMISTRY OF LIFE (IB TOPIC 2)

Water	11
Organic and inorganic compounds	12
Building macromolecules	13
Enzymes and substrates	14
Enzymes in action	15
Cell respiration and energy	16
Photosynthesis	17
Introducing DNA	18
Genes to polypeptides	19
Exam questions on Topic 2	20

3 GENETICS (IB TOPIC 3)

Genes and chromosomes	21
Meiosis	22
Mendel's law of segregation	23
Inheritance of blood groups	24
Genes and gender	25
Pedigree analysis	26
Genetic screening	27
Genetic disease and gene therapy	28
Genetic modification	29
Cloning	30
DNA profiling	31
Exam questions on Topic 3	32

4 ECOLOGY AND EVOLUTION (IB TOPIC 4)

Classification	33
Identifying living organisms	34
Measuring populations	35
Variation in populations	36
Population dynamics	37
Natural selection	38
Evolution in action	39
Trophic levels	40
Energy flow	41
Food webs and energy pyramids	42
Nutrient recycling	43
Global impacts	44
Local impacts	45
Exam questions on Topic 4	46

5 HUMAN HEALTH AND PHYSIOLOGY (IB TOPIC 5)

Digestion	47
Blood and the blood system	48
Pathogens and disease	49
Defence against infectious disease	50
Gas exchange	51
Maintaining the internal environment	52
Negative feedback and homeostasis	53
Reproductive systems	54
Pregnancy and childbirth	55
Menstrual cycle	56
Controlling fertility	57
Exam questions on Topic 5	58

6 NUCLEIC ACIDS AND PROTEIN (IB TOPIC 6)

DNA structure and replication	59
DNA in eukaryotes and prokaryotes	60
Gene expression in eukaryotes	61
Transcription and reverse transcription	62
Translating the genetic code	63
Polysomes and polypeptide elongation	64
Starting and stopping translation	65
Intramolecular bonding in proteins	66
Protein structure	67
Protein functions	68
Enzymes and activation energy	69
Enzyme inhibition	70
Controlling metabolic pathways	71
Exam questions on Topic 6	72

7 CELL RESPIRATION AND PHOTOSYNTHESIS (IB TOPIC 7)

Glycolysis	73
Krebs cycle	74
Oxidative phosphorylation	75
Mitochondria	76
Light and photosynthesis	77
Light-dependent reactions	78
Light-independent reactions	79
Chloroplasts	80
Limiting factors on photosynthesis	81
Exam questions on Topic 7	82

8 GENETICS (IB TOPIC 8)

Mendel's law of independent assortment	83
Dihybrid crosses	84
Polygenic inheritance	85
Statistical testing	86
Gene linkage and recombination	87
Crossing-over	88
Phases of meiosis	89

9 HUMAN REPRODUCTION (IB TOPIC 9)

Spermatogenesis	90
Oogenesis	91
Gametes	92
Fertilization	93
Pregnancy and the placenta	94
Exam questions on Topics 8 and 9	95

10 DEFENCE AGAINST INFECTIOUS DISEASE (IB TOPIC 10)

Types of defence 96
Antibody production 97
Helping to defend the body 98

11 NERVES, MUSCLES AND MOVEMENT (IB TOPIC 11)

Nerve impulses 99
Neurones and synapses 100
Muscle contraction 101
Muscles, joints and locomotion 102
Exam questions on Topics 10 and 11 103

12 EXCRETION (IB TOPIC 12)

Excretory products 104
Kidney structure and ultrafiltration 105
Urine production and osmoregulation 106

13 PLANT SCIENCE (IB TOPIC 13)

Plant diversity 107
Leaf structure and function 108
Stems and roots 109
Reproduction of flowering plants 110
Exam questions on Topics 12 and 13 111

14 DIET AND HUMAN NUTRITION (IB OPTION A)

Human diets 112
Nutritional needs 113
Malnutrition 114
Diet and disease 115
Food and health 116
Exam questions on Option A 117

15 PHYSIOLOGY OF EXERCISE (IB OPTION B)

Skeleton, joints and muscle 118
Co-ordination of muscle activity 119
Muscles and energy 120
Fitness, training and injuries 121
Exam questions on Option B 122

16 CELLS AND ENERGY (IB OPTION C)

Exam questions on Option C 123

17 EVOLUTION (IB OPTION D)

Origin of life on Earth 124
Theories of evolution 125
Evidence for evolution 126
Evolutionary history 127
Human origins 128
Human evolution 129
Population genetics 130
Mutations and polymorphisms 131
Species and speciation 132
Exam questions on Option D 133

18 NEUROBIOLOGY AND BEHAVIOUR (IB OPTION E)

Examples of behaviour 134
Perception of stimuli 135
Innate behaviour 136
Reflexes in humans 137
Learned behaviour 138
Social behaviour 139
Autonomic nervous system 140
Neurotransmitters and synapses 141
Psychoactive drugs 142
Exam questions on Option E 143

19 APPLIED PLANT AND ANIMAL SCIENCE (IB OPTION F)

Crop production 144
Plant productivity 145
Livestock production 146
Controversies in agriculture 147
Plant growth hormones 148
Plant and animal breeding 149
Genetic engineering in agriculture 150
Sexual reproduction in plants 151
Asexual reproduction in plants 152
Exam questions on Option F 153

20 ECOLOGY AND CONSERVATION (IB OPTION G)

Constructing pyramids of energy 154
Distribution of plants and animals and succession 155
Ecological niches 156
Statistics in ecology 157
Biodiversity 158
Conservation 159
Nitrogen cycle 160
Ozone depletion and acid rain 161
Eutrophication and biological fuels 162
Exam questions on Option G 163

21 FURTHER HUMAN PHYSIOLOGY (IB OPTION H)

Hormonal control 164
Secretion of digestive juices 165
Digestive enzymes 166
Absorption of digested foods 167
Liver 168
Cardiac cycle 169
Lymph, lipoproteins and CHD 170
Oxygen transport 171
Carbon dioxide transport 172
Exam questions on Option H 173

GUIDANCE FOR STUDENTS WORKING ON INTERNAL ASSESSMENT 174

GUIDANCE FOR STUDENTS WORKING ON EXTENDED ESSAYS IN BIOLOGY 176

GUIDANCE FOR STUDENTS PREPARING FOR FINAL EXAMS 178

ANSWERS TO QUESTIONS 179

INDEX TO ASSESSMENT STATEMENTS 182

INDEX 183

ORIGIN OF INDIVIDUAL QUESTIONS

The questions detailed below are all taken from past IB examination papers. The questions are from May (M) or November (N), Sample, 1998 (98), 1999 (99) and 2000 (00) paper 2 (P2) or paper 3 (P3) with question number in brackets. All other questions are IB style questions written by the author of this book.

TOPIC 1 CELLS
1. MOOSLP2(1)   2. M99SLP2(1)

TOPIC 2 THE CHEMISTRY OF LIFE
2. SAMPLE SLP2(2)   3. N99SLP2(3)

TOPIC 3 GENETICS
1. SAMPLE HLP2(3)   2. MOOSLP2(2)   3. M99SLP2(3)

TOPIC 4 ECOLOGY AND EVOLUTION
1. M99SLP2(1)   2. M98HLP2(3)

TOPIC 5 HUMAN HEALTH AND PHYSIOLOGY
1. N99SLP2(1)   2. N99SLP2(2)   3. SAMPLE SLP2(3)

TOPIC 6 NUCLEIC ACIDS AND PROTEINS
2. M98HLP2(2)   3. N98HL2(3)

TOPIC 7 CELL RESPIRATION AND PHOTOSYNTHESIS
1. N98SLP3(C1)   2. N99SLP3(N99)   3. M98SLP3(C3)

TOPICS 8 AND 9 GENETICS AND HUMAN REPRODUCTION
1. M99HL2(3)   2. N98HLP2(2)   3(c) N99HLP2(3)

TOPICS 10 AND 11 DEFENCE AGAINST INFECTIOUS DISEASE
AND NERVES MUSCLES AND MOVEMENT
1(b) M99HL2(2)   2. M99SL3(B1)   3. MOOHLP2(1)

TOPICS 12 AND 13 EXCRETION AND PLANT SCIENCE
1. MOOHL2(3)   2. N99HLP2(2)   3. M98SLP3(C1)

OPTION A – DIET AND HUMAN NUTRITION
1. M99SLP3(A1);   2. NOOSLP3(A2);   3. M99SL3(A3).

OPTION B – PHYSIOLOGY OF EXERCISE
1. N98SLP3(B1);   2. N98SLP3(B2);   3. M99SLP3(B3).

OPTION C – CELLS AND ENERGY
1. SAMPLE SLP3(C1);   2. M98SLP3(C2);   3. M99SLP3(C2).

OPTION D – EVOLUTION
1. M98HLP3(D1);   2. N99SLP3(D2);   3. N98HLP3(D2).

OPTION E – NEUROBIOLOGY AND BEHAVIOUR
1. M99HLP3(E2);   2. N99SLP3(E2);   3. N98HLP3(E3).

OPTION F – APPLIED PLANT AND ANIMAL SCIENCE
1. N99SLP3(F1);   2. M98SLP3(F3);   3. M99SLP3(F2).

OPTION G – ECOLOGY AND CONSERVATION
1. NOOSLP3(G1);   2. M98SLP3(G3);   3. N98HLP3(G3).

OPTION H – FURTHER HUMAN PHYSIOLOGY
1. N99HLP3(H1);   2.. N99HLP3(H2);   3. MOOHLP3(H2).

Cells, tissues and organs

INTRODUCING CELLS

Cells consist of **cytoplasm**, enclosed in a **plasma membrane**, usually controlled by a single **nucleus**. Two cell types that can be easily looked at under a light microscope are human cheek cells, scraped from inside the mouth (left) and moss leaf cells (right).

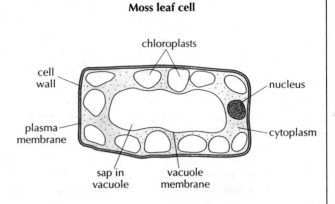

Human cheek cell   **Moss leaf cell**

UNICELLULAR ORGANISMS

Some organisms such as *Amoeba* (below), *Chlorella and Euglena* have only one cell. This single cell has to carry out all the activities essential to living organisms, including obtaining food, excreting waste products and producing offspring.

Amoeba

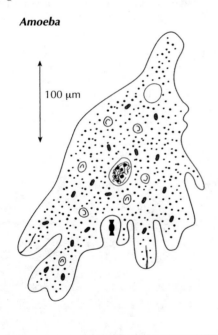

100 μm

ORGANELLES

Cells contain many tiny structures called organelles (little organs). Each one has a specific function in the cell. It is often possible to deduce what the function of a cell is by finding out what organelles it contains. Organelles are discrete structures. This means that they are individually distinct. Many types are enclosed in their own membrane. If the cells of a tissue are burst open (**lysed**), the organelles can be separated using a centrifuge. Mitochondria and chloroplasts are examples of organelles.

TISSUES AND ORGANS

In multicellular organisms the cells are often organized into tissues, organs and organ systems.

- **Tissues** are groups of cells that develop in the same way, with the same structure and function. Heart muscle is an example (below).
- **Organs** are groups of tissues that have combined to form a single structure. In an organ the tissues work together to perform an overall function. The heart is an example.
- **Organ systems** are groups of organs within an organism that together carry out a process. The cardiovascular system is an example.

Heart muscle tissue

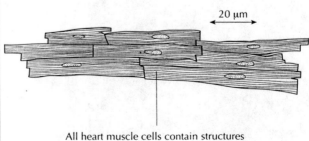

20 μm

All heart muscle cells contain structures made from protein fibres that are used to contract the cell and help to pump blood in the heart.

MULTICELLULAR ORGANISMS

Multicellular organisms consist of many cells. These cells do not have to carry out many different functions. Instead, they can become specialized for one particular function and carry it out very efficiently. Cells in a multicellular organism therefore develop in different ways. This is called **differentiation**. The cells need different genes to develop in different ways. Each cell has all of these genes, so could develop in any way, but it only uses the ones that it needs to follow its pathway of development.

Light and electron microscopes

LIGHT MICROSCOPES

Microscopes are used to study very small structures because they can produce images of them that are larger than the structures themselves. This is called magnification. Light microscopes were the first type to be developed and are still widely used. The figure (below left) shows a light microscope view of leaf cells.

ELECTRON MICROSCOPES

There are different types of electron microscope. In a transmission electron microscope (TEM), an electron beam passes through a very thin section of material. An image is formed because the electrons pass through some parts of the section but not others. In a scanning electron microscope (SEM) a narrow beam of electrons is scanned in a series of lines across the surface of the specimen. The electrons that are reflected or emitted from the surface are collected by a detector and converted into an electrical signal, which is used to build up a three-dimensional image, line by line on a television screen.

In every type of microscope a magnification is eventually reached above which the image can no longer be focused sharply. This is because the resolution of the microscope has been exceeded. The resolution is the ability of the microscope to show two close objects separately in the image. The resolution of a microscope depends on the wavelength of the rays used to form the image – the shorter the wavelength the better the resolution. The figure (below right) shows an electron microscope view of a leaf cell.

ADVANTAGES OF LIGHT AND ELECTRON MICROSCOPES

Biologists use both light and electron microscopes to investigate the structure and activities of living organisms. The two types both have strengths and weaknesses so they are used for different purposes.

Light microscopes

Material can be prepared easily for examination. Often, a sample can simply be placed on a slide with a few drops of water and a cover slip. An image can be obtained within seconds

Living material can be examined, so specimens do not always have to be killed. There is less danger of artificial structures appearing and causing confusion if the specimen is still alive

Movement can be observed if living material is examined, including the flow of blood, streaming of cytoplasm inside cells and the locomotion of microscopic organisms

Colours can be seen – both natural colours and artificial colours caused by staining

The field of view (the area which can be observed at once) is relatively large – 2 mm across at low power with typical microscopes

The resolution of light microscopes is relatively poor – about 0.25 µm so the maximum useful magnification is only about × 600. Many structures within cells cannot be seen clearly

Electron microscopes

Preparation of material for examination always involves a long series of procedures. These take several days to complete and often involve the use of toxic chemicals

Living material cannot survive in the vacuum inside electron microscopes. Tissues therefore have to be killed as the first stage in the preparation of them for examination

No movement can be observed as the material is always dead. Movement can only be deduced indirectly by complex experiments, often involving radioactive tracers

Only monochrome images are produced, with black, white and shades of grey

Only a small field of view can be examined at once – in a TEM the maximum uninterrupted view is about 100 µm across

The short wavelength of electrons gives very good resolution – about 0.25 nm. This allows magnification of up to × 500 000. Very small objects therefore become visible including many of the details of cell structure

Light micrograph of leaf cells

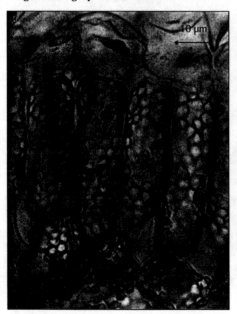

Electron micrograph of a leaf cell

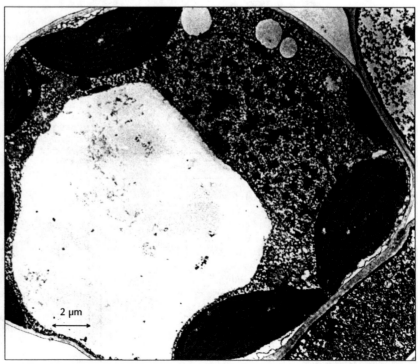

Size in cell biology

LIMITATIONS TO CELL SIZE

Cells do not carry on growing indefinitely. They reach a maximum size and then may divide. If a cell became too large, it would develop problems because its surface area to volume ratio would become too small.

As the size of any object is increased, the ratio between the surface area and the volume decreases. Consider the surface area to volume ratio of cubes of varying size as an example. The rate at which materials enter or leave a cell depends on the surface area of the cell. However, the rate at which materials are used or produced depends on the volume. A cell that becomes too large may not be able to take in essential materials or excrete waste substances quickly enough.

The same principle works for heat. Cells that generate heat may not be able to lose it quickly enough if they grow very large.

Surface area to volume ratios are important in biology. They help to explain many phenomena apart from maximum cell sizes.

UNITS FOR SIZE MEASUREMENTS

Most S.I. units differ from each other by a factor of 1000.
One millimetre is a thousand times smaller than 1 metre.
One micrometre is a thousand times smaller than 1 millimetre.
One nanometre is a thousand times smaller than 1 micrometre.
The most useful units for measuring the sizes of cells and structures within them are nanometres (nm) and micrometres (μm).
The typical sizes of some important structures in biology are shown opposite.

CALCULATING MAGNIFICATION

Photographs or drawings of structures seen under the microscope show them larger than they really are – they magnify them. It is useful to know how much larger the image is than the actual specimen. This factor is called the magnification. It is always helpful to show the magnification on a drawing of a biological structure.

Follow these instructions to calculate magnification.

1. Choose an obvious length, for example the maximum diameter of a cell. Measure it on the drawing.
2. Measure the same length on the actual specimen.
3. If the units used for the two measurements are different, convert one of them into the same units as the other one.
4. Divide the length on the drawing by the length on the actual specimen. The result is the magnification.

$$\text{Magnification} = \frac{\text{size of image}}{\text{size of specimen}}$$

This equation can also be used to calculate the actual size of a specimen if the magnification and size of the image are known.

SCALE BARS

A scale bar is a line added to a micrograph or a drawing to help to show the actual size of the structures.

For example, a 10 μm bar shows how large a 10 μm object would appear.

The figure below shows is a scanning electron micrograph of a leaf with the magnification and a scale bar both shown.

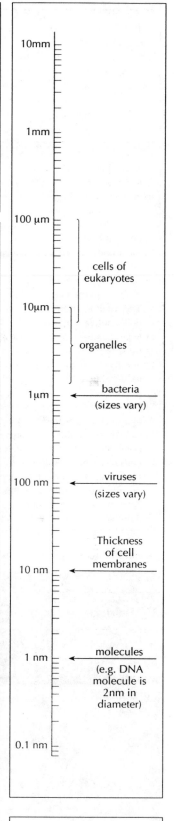

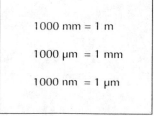

1000 mm = 1 m

1000 μm = 1 mm

1000 nm = 1 μm

Scanning electron micrograph of leaf (× 480)

60 μm

The cell theory

THE ORIGIN OF THE CELL THEORY

When biologists started looking at the structure of animals and plants using microscopes, they found tiny box-like structures making up the tissues (right). They called these cells. More and more living organisms were examined and biologists found that these were also made of cells.

The cell theory was developed, which states that all living organisms are made of cells.

POSSIBLE EXCEPTIONS TO THE CELL THEORY

There are some cases where the idea of living organisms consisting of tiny box-like structures does not seem to fit.

- Skeletal muscle is made up of muscle fibres. These have a membrane around the outside, like a single cell, but contain hundreds of nuclei. They are also much larger than most cells. In humans they can be up to 60 μm in diameter and up to 300 mm long.
- Most fungi consist of thread-like structures called **hyphae**, with a cell membrane and cell wall around the outside. The hyphae often contain many nuclei, without dividing walls between them.
- Many tissues contain **extracellular material** – material outside the cell membrane. In some cases, such as bone and tooth dentine, there is so much of this extracellular material that the cells only make up a very small percentage of the volume of the tissue.
- Some organisms such as *Amoeba* have only one region of cytoplasm, surrounded by a membrane. They are often called unicellular organisms but there are some reasons for considering them to be **acellular**. Instead of having separate cells to carry out different functions, the cytoplasm has to carry out all of the vital functions. These organisms are also usually much larger than typical cells. For example, *Acetabularia* (a giant alga) can be over 70 mm in length (right).

Robert Hooke's drawing of cork cells (1665)

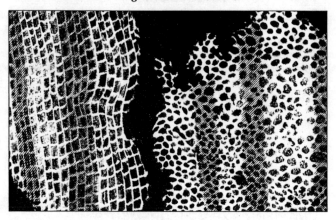

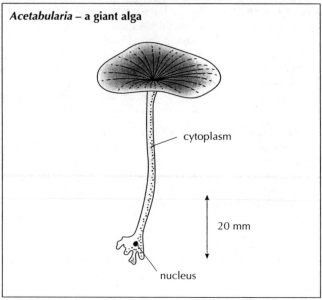

Acetabularia – a giant alga

cytoplasm

20 mm

nucleus

Electron micrograph of adenoviruses (× 120 000)

THE STRUCTURE OF VIRUSES

Viruses are certainly not cells. They are very small, simple particles consisting of some DNA or RNA wrapped up in a protein coat (left and below).

The status of viruses is interesting. They use the same genetic material as living organisms. They can evolve by natural selection. However, they have few other characteristics that biologists expect living organisms to possess. They are therefore not usually considered to be living organisms and they are not named or classified in the same way.

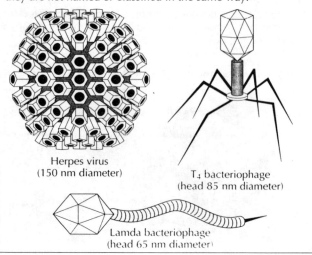

Herpes virus
(150 nm diameter)

T₄ bacteriophage
(head 85 nm diameter)

Lamda bacteriophage
(head 65 nm diameter)

Prokaryotic cells

THE ORIGIN OF CELLS

New cells are formed by the division of a parent cell.
Billions of years ago, when there were no living cells on Earth,
cells presumably developed from non-living chemical
substances. This does not now happen. Cells can only be
formed from other cells.

If a fluid such as some soup in a sealed container is sterilized
to kill all cells present, no cells will ever appear unless they
are allowed to enter from outside (see right).

The first cells had a simple structure which is called
prokaryotic (meaning before the nucleus).

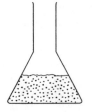

Sterilized soup in
an open container
decays because
bacteria float in

Sterilized soup in a
sealed container does
not decay as no
bacteria are present

STRUCTURE OF PROKARYOTIC CELLS

Figures below show the structure of a prokaryotic cell as seen
in an electron micrograph and a drawing to interpret the
structure.

Electron micrograph and drawing of *Bacillus licheniformis*
(× 45 000)

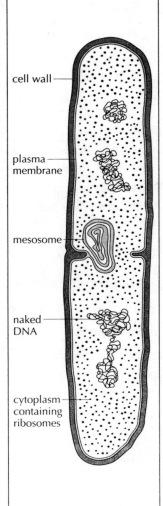

cell wall

plasma membrane

mesosome

naked DNA

cytoplasm containing ribosomes

FUNCTIONS OF THE PARTS OF A PROKARYOTIC CELL

Prokaryotic cell structure	Function
Cell wall	Forms a protective outer layer that prevents damage from outside and bursting if internal pressure is high
Plasma membrane (cell surface membrane)	Controls entry and exit of substances, pumping some of them in by active transport
Mesosome	Increases the area of membrane for ATP production. May move the DNA to the poles during cell division
Cytoplasm	Contains enzymes that catalyse the chemical reactions of metabolism and DNA in a region called the nucleoid
Ribosomes	Synthesize proteins by translating messenger RNA. Some proteins stay in the cell and others are secreted
Naked DNA	Stores the genetic information that controls the cell and is passed on to daughter cells

TYPES OF PROKARYOTE

Prokaryotes are more commonly called bacteria.
Although small and relatively simple in structure, they show a
formidable range of metabolic activity.

- **Photosynthesis** Blue–green bacteria make their own food by
  photosynthesis.
- **Nitrogen fixation** Nitrogen-fixing bacteria convert nitrogen
  from the air into nitrogen compounds.
- **Fermentation** Many bacteria absorb organic substances,
  convert them into other organic substances and release
  them. For example, in yoghurt production, bacteria convert
  lactose into lactic acid. There are many other types of
  fermentation in bacteria.

Eukaryotic cells

EUKARYOTIC CELL STRUCTURE

The figure (below) is an electron micrograph of a liver cell. The figure (right) is an annotated drawing to interpret part of the structure. Liver cells show many typical features of animal cells. Animal and plant cells are eukaryotic.

Electron micrograph of part of a liver cell (× 8000)

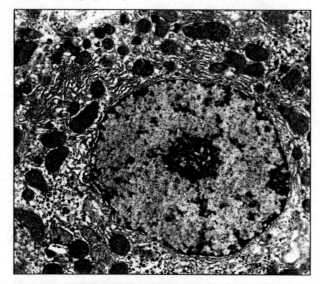

Drawing of part of the electron micrograph

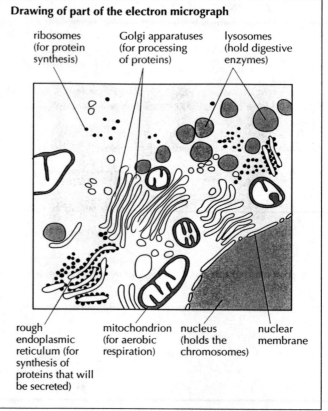

ribosomes (for protein synthesis)

Golgi apparatuses (for processing of proteins)

lysosomes (hold digestive enzymes)

rough endoplasmic reticulum (for synthesis of proteins that will be secreted)

mitochondrion (for aerobic respiration)

nucleus (holds the chromosomes)

nuclear membrane

COMPARING PROKARYOTIC AND EUKARYOTIC CELLS

Feature	Prokaryotic cells	Eukaryotic cells
Type of genetic material	A naked loop of DNA	Chromosomes consisting of strands of DNA associated with protein. Four or more chromosomes
Main location of genetic material	In the cytoplasm in a region called the nucleoid	In the nucleus inside a double nuclear membrane called the nuclear envelope
Mitochondria	Not present. The cell surface membrane and mesosome are used instead	Always present
Ribosomes	Small sized. 70S (S = svedburg units, a measure of the size of organelles)	Larger sized. 80S
Organelles bounded by a single membrane	Few or none are present	Many are present including endoplasmic reticulum, Golgi apparatus and lysosomes

COMPARING PLANT AND ANIMAL CELLS

Plants and animal cells have many similarities because they are both eukaryotic. They also show some differences.

Feature	Animal	Plant
Cell wall	Not present. Animal cells only have a cell surface membrane	Cell wall and cell surface membrane are both present
Chloroplasts	Not present	Present in plant cells that photosynthesize
Carbohydrate storage	Glycogen	Starch
Vacuole	Not usually present. Small or temporary vacuoles are sometimes found	Large fluid-filled vacuole often present
Shape	Able to change shape. Usually rounded	Fixed shape. Usually rather regular

THE PLANT CELL WALL

The main component of plant cell walls is cellulose. Cellulose molecules are arranged in bundles called microfibrils. These give the cell wall great tensile strength and allow high pressures to develop inside the cell.

Membrane structure and membrane proteins

Fluid mosaic model of a biological membrane

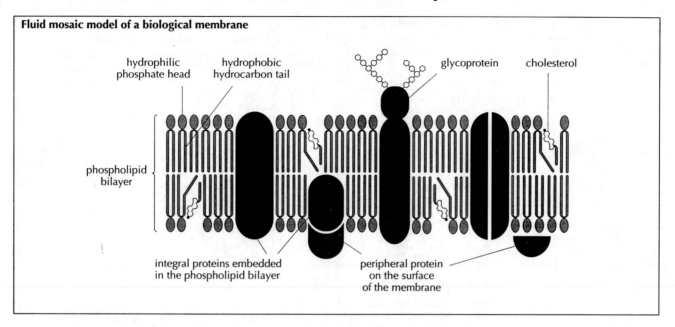

PHOSPHOLIPIDS

Hydrophilic molecules are attracted to water. Hydrophobic molecules are not attracted to water, but are attracted to each other. Phospholipid molecules are unusual because they are partly hydrophilic and partly hydrophobic.

The phosphate head is hydrophilic and the two hydrocarbon tails are hydrophobic. In water, phospholipids form double layers with the hydrophilic heads in contact with water on both sides and the hydrophobic tails away from water in the centre. This arrangement is found in biological membranes. The attraction between the hydrophobic tails in the centre and between the hydrophilic heads and the surrounding water makes membranes very stable.

FLUIDITY OF MEMBRANES

Phospholipids in membranes are in a fluid state. This allows membranes to change shape in a way that would be impossible if they were solid. The fluidity also allows vesicles to be pinched off from membranes or fuse with them.

MEMBRANE PROTEINS

Some electron micrographs show the positions of proteins within membranes. The proteins are seen to be dotted over the membrane. This gives the membrane the appearance of a mosaic. Because the protein molecules float in the fluid phospholipid bilayer, biological membranes are called fluid mosaics. The figure (above) is a diagram showing the fluid mosaic model of a biological membrane. Some of the functions of membrane proteins are shown below.

Functions of membrane proteins

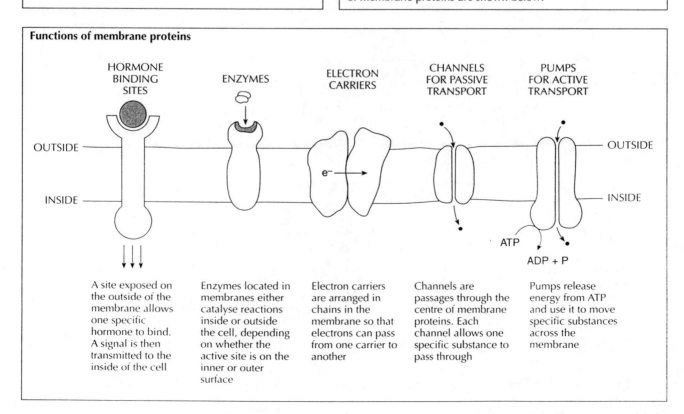

Transport across membranes

PASSIVE TRANSPORT ACROSS MEMBRANES

Diffusion is the passive movement of particles from a region of higher concentration to a region of lower concentration, as a result of the random motion of particles.

In liquids and gases particles are in continual motion. The direction that they move in is random. Particles can diffuse across membranes if the membrane is permeable to them.

higher concentration

lower concentration

Some particles do move from a lower to a higher concentration but more move from a higher to a lower concentration. There is a net movement from the lower to the higher concentration until the concentrations are equal.

Solute unable to diffuse through membrane

Partially permeable membrane

Membranes are partially permeable because they allow some substances to diffuse through but not others.
To allow some substances to diffuse through, channel proteins are needed. This is called facilitated diffusion.

Facilitated diffusion through membrane containing channel proteins.

OSMOSIS

Osmosis is the passive movement of water molecules from a region of lower solute concentration to a region of higher solute concentration, across a partially permeable membrane.

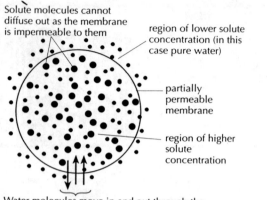

Solute molecules cannot diffuse out as the membrane is impermeable to them

region of lower solute concentration (in this case pure water)

partially permeable membrane

region of higher solute concentration

Water molecules move in and out through the membrane but more move in than out. There is a net movement from the region of lower solute concentration to the region of higher solute concentration

ACTIVE TRANSPORT ACROSS MEMBRANES

Active transport is the movement of substances across membranes using energy from ATP. Active transport can move substances against the concentration gradient – from a region of lower concentration to a region of higher concentration. Protein pumps in the membrane are used for active transport. Each pump only transports particular substances, so cells can control what is absorbed and what is expelled.

Particle enters the pump

Particle binds to a specific site. Other types of particle cannot bind

Energy from ATP is used to change the shape of the pump

Particle is released and the pump then returns to its original shape

TRANSPORT OF MATERIALS BY VESICLES IN THE CYTOPLASM

Proteins are synthesized by ribosomes and then enter the rough endoplasmic reticulum

Vesicles bud off from the rER and carry the proteins to the Golgi apparatus

The Golgi apparatus modifies the proteins

Vesicles bud off from the Golgi apparatus and carry the modified proteins to the plasma membrane

ENDOCYTOSIS

Part of the plasma membrane is pulled inwards

A droplet of fluid becomes enclosed when a vesicle is pinched off

Vesicles can then move through the cytoplasm carrying their contents

EXOCYTOSIS

Vesicles fuse with the plasma membrane

The contents of the vesicle are expelled

The membrane then flattens out again

Cell division

THE CELL DIVISION CYCLE IN EUKARYOTES

New cells are produced by division of existing cells. If many new cells are needed, cells go through a cycle of events again and again. This is called the cell division cycle. The longest phase in this cycle is **interphase**. During interphase the cell carries out many biochemical reactions and grows larger. The DNA molecules in the chromosomes are not coiled up and the genes on them can be transcribed, to allow protein synthesis. If the cell is going to divide again, the DNA is all replicated. These and other processes make interphase a very active period for a cell.

At the end of interphase when DNA replication as been completed, the cell begins **mitosis**. In mitosis, the nucleus divides to form two genetically identical nuclei. Towards the end of mitosis, the cytoplasm of the cell starts to divide and eventually two cells are formed, each containing one nucleus. The process of dividing the cytoplasm to form two cells is **cytokinesis**. The two cells begin interphase when mitosis and cytokinesis have been completed.

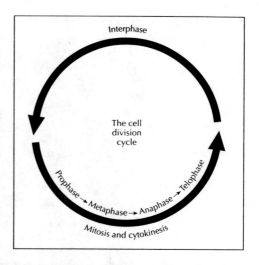

THE PHASES OF MITOSIS

This figure shows how genetically identical nuclei are formed during the four phases of mitosis.

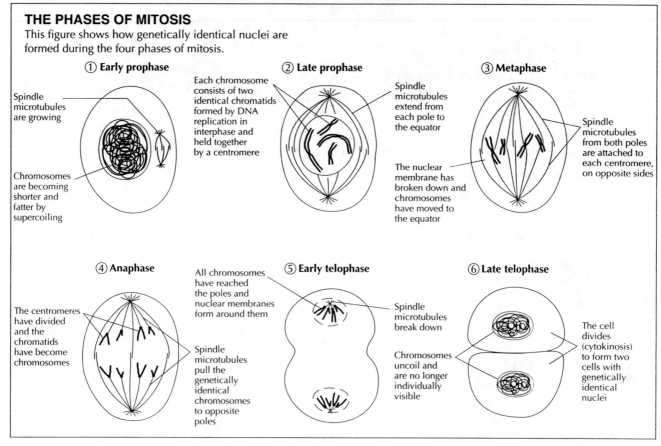

USES OF MITOSIS

Mitosis is used in eukaryotes whenever genetically identical cells are needed:
• during growth
• when tissues have been damaged and need to be repaired
• to reproduce asexually.

TUMOUR FORMATION

Sometimes the normal control of mitosis in a cell fails. This cell divides into two. The two daughter cells divide to form four cells. Repeated divisions soon produce a mass of cells called a tumour. This can happen in any organ. Tumours can grow to a large size and can spread to other parts of the body. The diseases caused by the growth of tumours are called cancer.

DIFFERENCES IN CELL DIVISION BETWEEN PLANT AND ANIMAL CELLS

Plant cells	Animal cells
There are no centrioles in plant cells	Centrioles are found at each pole of animal cells during mitosis
After anaphase, a new cell wall is formed across the equator of the cell, with plasma membrane on both sides. This divides the cell into two	After anaphase, the plasma membrane at the equator is pulled inwards until it meets in the centre of the cell, dividing it into two

EXAM QUESTIONS ON TOPIC 1

1   The photomicrograph below shows a transverse section of part of a liver cell.

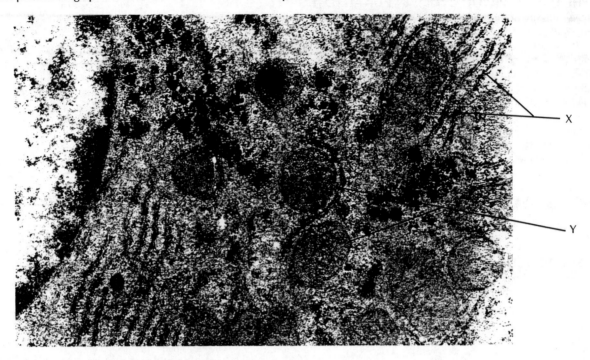

a) Identify the organelles labelled X and Y.   [2]

b) On the photomicrograph, identify the nuclear membrane and show its position with a clear label.   [1]

c) The liver cell shown in the photomicrograph was making large amounts of two substances.

    Deduce what the two substances were, giving reasons for your answer based on the organelles visible in the photomicrograph.   [2]

2   The diagram below represents the fluid model of a cell membrane.

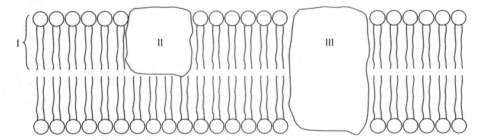

a) (i)   State the name of the molecule labelled I.   [1]

    (ii) Label the diagram to show which part of molecule I is hydrophobic and which part is hydrophilic.   [1]

b) (i)   Identify whether molecule II is an integral or a peripheral protein.   [1]

    (ii) Describe the part played by molecule III in active transport.   [2]

3   a) Distinguish between mitosis and cytokinesis.   [2]

    b) State one difference between plant and animal cells in

        (i)   mitosis   [1]

        (ii) cytokinesis   [1]

    c) Explain the reason for living cells needing to enter a period of interphase after mitosis, before they can carry out mitosis again.   [4]

2 THE CHEMISTRY OF LIFE

Water

THE SIGNIFICANCE OF WATER TO LIVING ORGANISMS

Water is of immense importance to all living organisms. It is used by them in many different ways. These uses can be explained by referring to the properties of water.

Water is used as a **coolant**
　　– refer to thermal properties.

Water is used as a **transport medium**
　　– refer to cohesion, solvent properties and thermal properties.

Water is used as a **habitat**
　　– refer to cohesion and thermal properties.

POLARITY AND HYDROGEN BONDING IN WATER

Water molecules consist of two hydrogen atoms bonded to an oxygen atom. The hydrogen atoms have a slight positive charge and the oxygen atom has a slight negative charge. So, water molecules have two poles – a positive hydrogen pole and a negative oxygen pole (below left). This feature of a molecule is called **polarity**.

A bond can form between the positive pole of one water molecule and the negative pole of another. This is called a hydrogen bond (below centre). In liquid water, many of these bonds form.

Water molecule

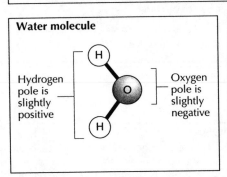

Hydrogen pole is slightly positive

Oxygen pole is slightly negative

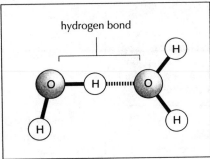

hydrogen bond

Mosquito larva in water

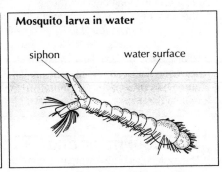

siphon　　　　water surface

THE PROPERTIES OF WATER

Name of the property	Outline of the property	Significance of the property to living organisms
Transparency	Light can pass through water	Light can reach structures inside living organisms such as chloroplasts in plant cells and the retina in the human eye. It can also reach organisms that use freshwater or seawater as their habitat
Cohesion	Water molecules stick to each other because of the hydrogen bonds that form between them.	Strong pulling forces can be exerted to suck columns of water up to the tops of the tallest trees in their transport systems. These columns of water rarely break
	At a surface, the cohesion of water molecules can make it difficult for small objects to break through	Some animals such as mosquito larvae use the surface of water as a habitat (above). Though they are denser than water they remain on the surface and do not sink
Solvent properties	Many different substances dissolve in water because of its polarity. Inorganic particles such as sodium ions and organic substances such as glucose can dissolve	The solvent properties of water allow many substances to be carried dissolved in water in the blood of animals and the sap of plants
Thermal properties: heat capacity	Water has a large heat capacity. This means that large amounts of energy are needed to raise its temperature. The energy is needed to break some of the hydrogen bonds. This heat energy is given out again when the water is cooled	The temperature of water tends to remain quite stable. This is useful for organisms such as fish that use water as a habitat. Blood, which is mainly composed of water, can carry heat from warmer parts of the body to cooler parts
Thermal properties: boiling and freezing points	The boiling point of water (100 °C) is relatively high, because, to change it from a liquid to a gas, all of the hydrogen bonds between the water molecules have to be broken	In natural habitats on Earth, water rarely boils. Living organisms could not survive if the water inside them boiled
	Water also freezes at a relatively high temperature, but because it becomes less dense as it cools to freezing point, ice forms at the surface first	The ice that forms on the surface of lakes or seas insulates the water underneath, so living organisms can survive there
Thermal properties: the cooling effect of evaporation	Water can evaporate at temperatures below boiling point. Hydrogen bonds have to be broken to do this. The heat energy needed to break the bonds is taken from the liquid water, cooling it down	Evaporation of water from plant leaves (transpiration) and from the human skin (sweat) has useful cooling effects

Organic and inorganic compounds

ELEMENTS FOUND IN LIVING ORGANISMS
Living organisms contain many chemical elements, some in large quantities and some in very small amounts. The three commonest chemical elements of life are carbon, hydrogen and oxygen. They are part of all the main organic compounds in living organisms. Examples of other elements that are needed are shown in the table opposite.

ORGANIC AND INORGANIC COMPOUNDS
Living organisms contain many chemical compounds. Some of them are organic and some are inorganic. Organic compounds are defined as compounds containing carbon that are found in living organisms.
There are a few carbon compounds that are inorganic even though they can be found in living organisms. These are all simple carbon compounds that are also widely found in the environment. Carbon dioxide, carbonates and hydrogen carbonates are three examples of inorganic carbon compounds. Three types of organic compound are found in large amounts in living organisms – carbohydrates, lipids and proteins.

THE SUBUNITS OF ORGANIC COMPOUNDS
The molecules of many organic compounds are large and may seem complex, but they are built up using small and relatively simple subunits. Some important subunits are shown below.

EXAMPLES OF CHEMICAL ELEMENTS AND THEIR ROLES

Element	Role in plants or animals
Nitrogen	Part of the amine groups of amino acids and therefore proteins
Calcium	Needed to make the mineral that strengthens bones and teeth.
Phosphorus	Part of the phosphate groups in ATP and DNA molecules
Iron	Needed to make hemoglobin and thus to carry oxygen in blood
Sodium	Used in neurons (nerve cells) for the transmission of nerve impulses

ATOMS AND IONS
An atom is a single particle of a chemical element. If an atom either gains or loses electrons it becomes an ion. Atoms are uncharged particles and ions are charged – they have either positive or negative charges. For example, if a sodium atom (Na) loses an electron, it becomes a sodium ion (Na^+).

Subunits of proteins, carbohydrates and lipids

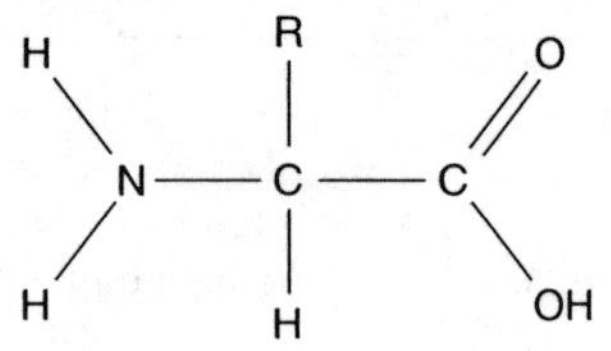

amino acids
(general structure)

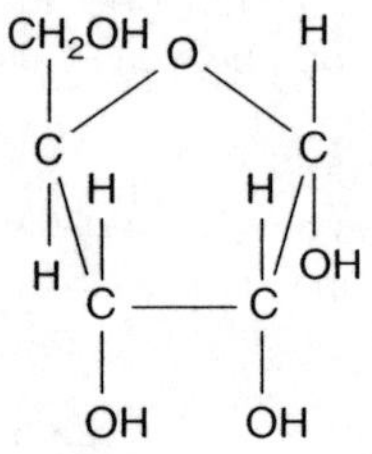

ribose
(a monosaccharide)

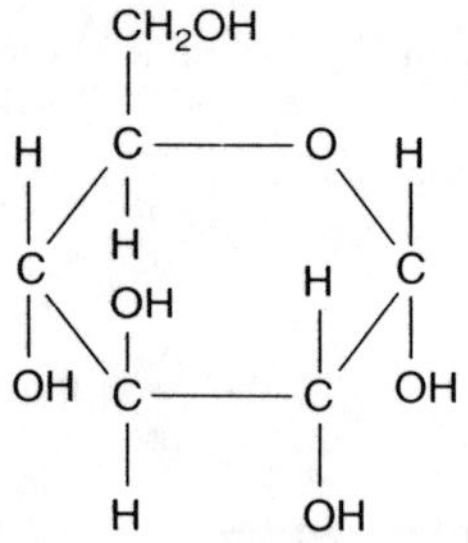

glucose
(a monosaccharide)

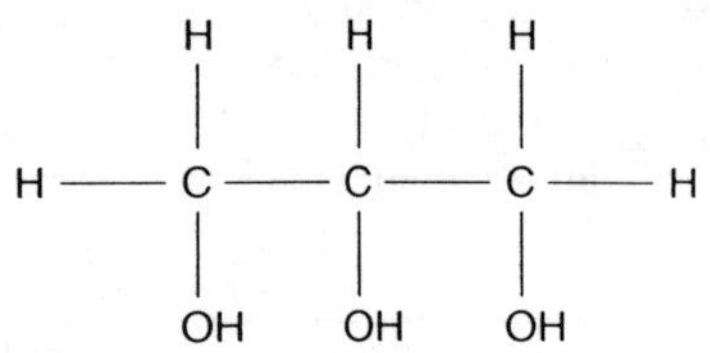

glycerol

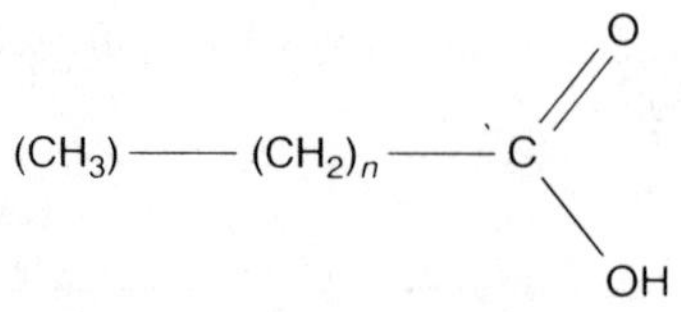

fatty acids
(general structure)

Building macromolecules

CONDENSATION REACTIONS

In a condensation reaction two molecules are joined together to form a larger molecule. Water is also formed in the reaction. For example, two amino acids can be joined together to form a dipeptide by a condensation reaction. The new bond formed is a **peptide linkage**.

Condensation of two amino acids to form a dipeptide and water

Further condensation reactions can link amino acids to either end of the dipeptide, eventually forming a chain of many amino acids This is called a **polypeptide**.

In a similar way, condensation reactions can be used to build up carbohydrates and lipids. The basic subunits of lipids are monosaccharides. Two monosaccharides can be linked to form a disaccharide and more monosaccharides can be linked to a disaccharide to form a large molecule called a **polysaccharide**. Fatty acids can be linked to glycerol by condensation reactions to produce lipids called **glycerides**. A maximum of three fatty acids can be linked to each glycerol, producing a triglyceride.

HYDROLYSIS REACTIONS

Large molecules such as polypeptides, polysaccharides and triglycerides can be broken down into smaller molecules by hydrolysis reactions. Water molecules are used up in hydrolysis reactions. Hydrolysis reactions are the reverse of condensation reactions.

Polypeptides + Water $\longrightarrow$ Dipeptides or Amino acids

Polysaccharides + Water $\longrightarrow$ Disaccharides or Monosaccharides

Glycerides + Water $\longrightarrow$ Fatty acids + Glycerol

EXAMPLES OF CARBOHYDRATES

Monosaccharides	Glucose, fructose and ribose
Disaccharides	Sucrose (glucose + fructose) Maltose (glucose + glucose)
Polysacharides	Starch (made of glucose subunits) Glycogen (made of glucose subunits, but linked differently from starch)

FUNCTIONS OF LIPIDS

- **Energy storage** – in the form of fat in humans and oil in plants
- **Heat insulation** – a layer of fat under the skin reduces heat loss
- **Buoyancy** – lipids are less dense than water so help animals to float

FUNCTIONS OF CARBOHYDRATES

- **Transport** – glucose is carried by the blood to transport energy to cells throughout the body
- **Energy storage** – energy is stored in the form of glycogen in liver cells

USING CARBOHYDRATES AND LIPIDS IN ENERGY STORAGE

Both lipids and carbohydrates can be used for energy storage in living organisms. Both types of storage compound have advantages. Carbohydrates are usually used for energy storage over short periods and lipids for long-term storage.

Advantages of lipids	Advantages of carbohydrates
1. Lipids contain more energy per gram than carbohydrates so stores of lipid are lighter than stores of carbohydrate that contain the same amount of energy	1. Carbohydrates are more easily digested than lipids so the energy stored by them can be released more rapidly
2. Lipids are insoluble in water, so they do not cause problems with osmosis in cells	2. Carbohydrates are soluble in water so are easier to transport to and from the store

Enzymes and substrates

INTRODUCING ENZYMES

Catalysts speed up chemical reactions without being changed themselves. Living organisms make biological catalysts called **enzymes**. *Enzymes are globular proteins which act as catalysts of chemical reactions.*

Without enzymes to catalyse them, many chemical processes happen at a very slow rate in living organisms. By making some enzymes and not others, cells can control what chemical reactions happen in their cytoplasm.

The structure of enzymes is quite delicate and can be damaged by various things. This is called **denaturation**. *Denaturation is changing the structure of an enzyme (or other protein) so that it can no longer carry out its function.* Denaturation is usually permanent.

In chemical reactions, one or more reactants are converted into one or more products. In reactions catalysed by enzymes, the reactants are called **substrates**.

ENZYME–SUBSTRATE SPECIFICITY

Most enzymes are specific – they catalyse very few different reactions. They therefore only have a very small number of possible substrates. This is called enzyme–substrate specificity. The substrates bind to a special region on the surface of the enzyme called the **active site**. *An active site is a region on the surface of an enzyme to which substrates bind and which catalyses a chemical reaction involving the substrates.*

The active site of an enzyme has a very intricate and precise shape. It also has distinctive chemical properties. Active sites match the shape and chemical properties of their substrates. Molecules of substrate fit the active site and are chemically attracted to it (right). Other molecules either do not fit or are not chemically attracted. They do not therefore bind to the active site. This is how enzymes are substrate-specific.

The way in which the enzyme and substrate fit together is similar to the way in which a key fits a lock. The enzyme is like the lock and the substrate is like the key that fits it.

EFFECT OF SUBSTRATE CONCENTRATION ON ENZYME ACTIVITY

At low substrate concentrations, enzyme activity is directly proportional to substrate concentration. This is because random collisions between substrate and active site happen more frequently with higher substrate concentrations.

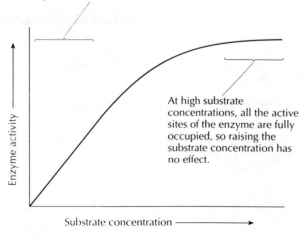

At high substrate concentrations, all the active sites of the enzyme are fully occupied, so raising the substrate concentration has no effect.

Stages in enzyme catalysis

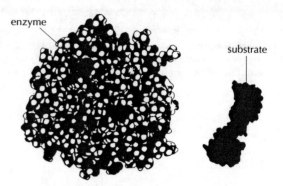

Substrate molecules are in continual random motion. If one collides with the active site it can bind to it.

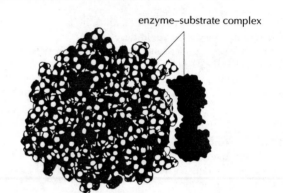

The substrate fits the active site. If other molecules collide with the active site they do not fit and fail to bind.

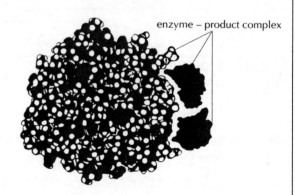

The active site catalyses a chemical reaction. The substrates are turned into products.

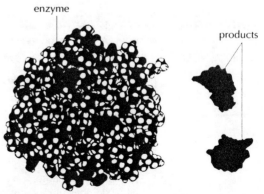

The products detach from the active site, leaving it free for more substrate to bind.

Enzymes in action

FACTORS AFFECTING ENZYME ACTIVITY

Wherever enzymes are used, it is important that they have the conditions that they need to work effectively. Temperature, pH and substrate concentration all affect the rate at which enzymes catalyse chemical reactions. The figures on page 14 and below show the relationships between enzyme activity and substrate concentration, temperature and pH.

EFFECT OF TEMPERATURE

Enzyme activity increases as temperature increases, often doubling with every 10 °C rise. This is because collisions between substrate and active site happen more frequently at higher temperatures due to faster molecular motion.

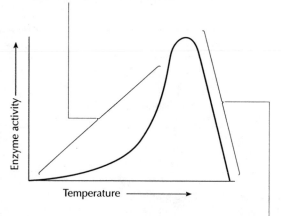

At high temperatures enzymes are denatured and stop working. This is because heat causes vibrations inside enzymes which break bonds needed to maintain the structure of the enzyme.

EFFECT OF pH

Optimum pH at which enzyme activity is fastest (pH 7 is optimum for most enzymes).

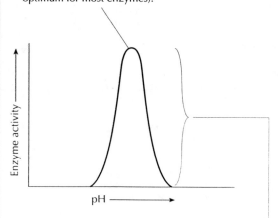

As pH increases or decreases from the optimum, enzyme activity is reduced. Both acids and alkalis can denature enzymes.

USING ENZYMES IN BIOTECHNOLOGY

Biotechnology is the use of organisms or parts of organisms to produce things or to carry out useful processes. There are many ways in which enzymes, obtained from living organisms, can be used in biotechnology. Two examples are described below.

The use of pectinase in fruit juice production

Pectin is a complex polysaccharide, found in the cell walls of plants. Pectinase is an enzyme that breaks down pectin by hydrolysis reactions.

Source of enzyme

Pectinase is obtained by artificially culturing a fungus (*Aspergillus niger*). The fungus grows naturally on fruits, where it uses pectinase to soften the cell walls of the fruit so that it can grow through it.

Use of pectinase in biotechnology

Fruit juices are produced by crushing ripe fruits to separate liquid juice from solid pulp. When ripe fruits are crushed, pectin forms links between the cell wall and the cytoplasm of the fruit cells, making the juice viscous and more difficult to separate from the pulp. Pectinase is added during crushing of fruit to break down the pectin.

Advantages

Pectinase makes juice more fluid and easy to separate from the pulp. It therefore increases the volume of juice that is obtained. It also makes the juice less cloudy by helping solids suspended in the juice to settle and be separated from the fluid.

The use of protease in biological washing powder

Protease enzymes break down proteins into soluble peptides and amino acids. Laundry washing powders that contain protease are called biological washing powders.

Source of enzyme

Protease is obtained by culturing a bacterium, *Bacillus licheniformis*, that is adapted to grow in alkaline conditions. This bacterium feeds on proteins in its habitat by secreting protease. The protease has a high pH optimum of between 9 and 10.

Use of protease in biotechnology

Detergents in laundry washing powders remove fats and oils during the washing of clothes, but much of the dirt on clothing is made of protein, not lipids. If protease is added to the washing powder, this protein is digested during the wash. The high pH optimum of the protease allows it to remain active, despite the high pH caused by alkalis in the washing powder.

Advantages

If protease is not used, protein stains on clothes can only removed by using a very high temperature wash. Protease allows much lower temperatures to be used, with lower energy use and less risk of shrinkage of garments or loss of coloured dyes.

Cell respiration and energy

ENERGY AND CELLS

All living cells need a continual supply of energy. This energy is used for a wide range of processes including active transport and protein synthesis. Most of these processes require energy in the form of ATP (adenosine triphosphate). ATP is a chemical substance that can diffuse to any part of the cell and release energy.

Every cell produces its own ATP, by a process called **cell respiration**. In cell respiration, organic compounds such as glucose or fat are carefully broken down. Energy from them is used to make ATP. Cell respiration is defined as *controlled release of energy, in the form of ATP, from organic compounds in cells*.

Cell respiration can be aerobic or anaerobic. Aerobic cell respiration involves the use of oxygen and anaerobic cell respiration does not.

THE USE OF GLUCOSE IN RESPIRATION

Glucose is often the organic compound that is used in cell respiration. Chemical reactions in the cytoplasm break down glucose into a simpler organic compound called pyruvate. In these reactions a small amount of ATP is made using energy released from glucose.

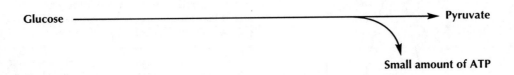

ANAEROBIC CELL RESPIRATION

If no oxygen is available, the pyruvate remains in the cytoplasm and is converted into a waste product that can be removed from the cell. No ATP is produced in these reactions. In humans the waste product is lactate (lactic acid). In yeast the products are ethanol and carbon dioxide.

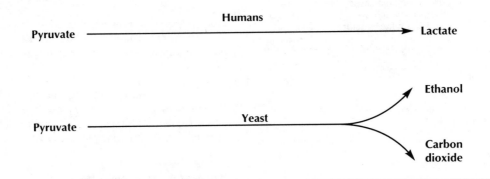

AEROBIC CELL RESPIRATION

If oxygen is available, the pyruvate is absorbed by the mitochondrion. Inside the mitochondrion the pyruvate is broken down into carbon dioxide and water. A large amount of ATP is produced as a result of these reactions. Aerobic cell respiration therefore has a much higher yield of ATP per gram of glucose than anaerobic cell respiration.

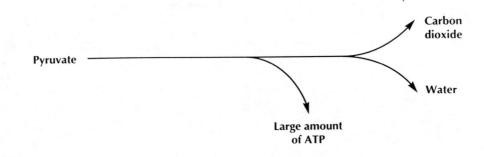

Photosynthesis

INTRODUCING PHOTOSYNTHESIS

Photosynthesis is the process used by plants and some other organisms to produce all their own organic substances (food), using only light energy and simple inorganic substances. It involves many stages and some complex chemical reactions, but it can be outlined in a series of statements.

- Photosynthesis involves an energy conversion. Light energy, usually sunlight, is converted into chemical energy.
- Sunlight is called white light, but it is actually made up of a wide range of wavelengths, including red, green and blue.
- Some substances called pigments can absorb light. The main pigment used to absorb light in photosynthesis is chlorophyll.
- The structure of chlorophyll allows it to absorb some colours or wavelengths of light better than others. Red and blue light are absorbed more than green.
- The green light that chlorophyll cannot absorb is reflected. This makes chlorophyll and therefore chloroplasts and plant leaves look green.
- Some of the energy absorbed by chlorophyll is used to produce ATP.
- Some of the energy absorbed by chlorophyll is used to split water molecules. This is called photolysis of water.
- Photolysis of water results in the formation of oxygen and hydrogen. The oxygen is released as a waste product.
- Carbon dioxide is absorbed for use in photosynthesis. The carbon from it is used to make a wide range of organic substances. The conversion of carbon in a gas to carbon in solid compounds is called carbon fixation.
- Carbon fixation involves the use of hydrogen from photolysis and energy from ATP.

MEASURING RATES OF PHOTOSYNTHESIS

Photosynthesis involves the production of oxygen, the uptake of carbon dioxide and an increase in biomass. Any of these can be used as a measure of the rate of photosynthesis.

Production of oxygen

Aquatic plants (e.g. *Myriophyllum*) release bubbles of oxygen when they carry out photosynthesis. If these bubbles are collected, their volume can be measured.

Uptake of carbon dioxide

Leaves take in CO_2 from the air or water around them, but this is difficult to measure directly. If CO_2 is absorbed from water, the pH of the water rises. This can be monitored with pH indicators or with pH meters.

Increases in biomass

If batches of plants are harvested at a series of times and the biomass of the batches is determined, the rate of increase in biomass gives an indirect measure of the rate of photosynthesis in the plants.

Effect of light intensity on photosynthesis

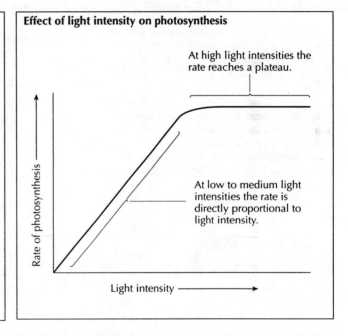

At high light intensities the rate reaches a plateau.

At low to medium light intensities the rate is directly proportional to light intensity.

Effect of CO_2 concentration on photosynthesis

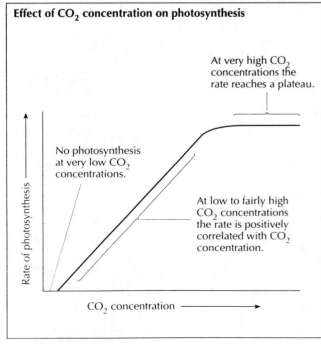

At very high CO_2 concentrations the rate reaches a plateau.

No photosynthesis at very low CO_2 concentrations.

At low to fairly high CO_2 concentrations the rate is positively correlated with CO_2 concentration.

Effect of temperature on photosynthesis

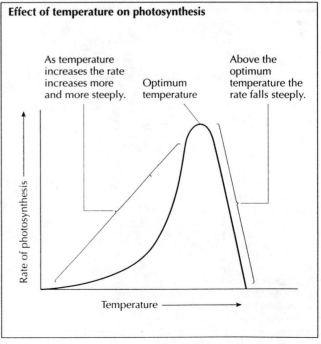

As temperature increases the rate increases more and more steeply.

Optimum temperature

Above the optimum temperature the rate falls steeply.

Introducing DNA

THE NUCLEOTIDE SUBUNITS OF DNA

Although DNA is the genetic material of living organisms and is therefore of immense importance, it is made of relatively simple subunits. These are called **nucleotides**. Each nucleotide consists of three parts – a sugar (called deoxyribose), a phosphate group and a base. In diagrams of DNA structure these are usually shown as pentagons, circles and rectangles, respectively. The figure (below) shows how the sugar, the phosphate and the base are linked up in a nucleotide.

phosphate

base

sugar

DNA nucleotides do not all have the same base. Four different bases are found – adenine, cytosine, guanine and thymine. These are usually simply referred to as A, C, G and T.

BUILDING DNA MOLECULES

Two DNA nucleotides can be linked together by a covalent bond between the sugar of one nucleotide and the phosphate group of the other. More nucleotides can be added in a similar way to form a strand of nucleotides.

DNA molecules consist of two strands of nucleotides wound together into a double helix. Hydrogen bonds link the two strands together. These form between the bases of the two strands. However, adenine only forms hydrogen bonds with thymine and cytosine only forms hydrogen bonds with guanine. This is called **complementary base pairing**.

DNA REPLICATION

DNA replication is a way of copying DNA to produce new molecules with the same base sequence. It is **semi-conservative** – each molecule formed by replication consists of one new strand and one old strand conserved from the parent DNA molecule.

Stage 1
The DNA double helix is unwound and separated into strands by breaking the hydrogen bonds. Helicase is the main enzyme involved.

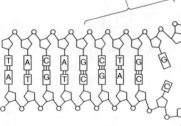

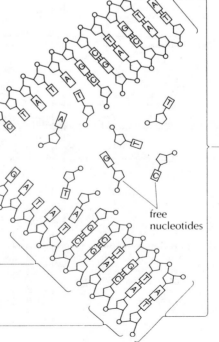

free nucleotides

The two daughter DNA molecules are identical in base sequence to each other and to the parent molecule, because of complementary base pairing (A pairs with T and C with G).
Each of the new strands is **complementary** to the template on which it was made and **identical** to the other template.

Stage 2
The single strands act as templates for new strands. Free nucleotides are present in large numbers around the replication fork. The bases of these nucleotides form hydrogen bonds with the bases on the parent strand. The nucleotides are linked up to form the new strand. DNA polymerase is the main enzyme involved.

Stage 3
The daughter DNA molecules each rewind into a double helix.

Genes to polypeptides

THE RELATIONSHIP BETWEEN GENES AND POLYPEPTIDES
Polypeptides are long chains of amino acids. There are 20 different amino acids that can form part of a polypeptide. To make one particular polypeptide, amino acids must be linked up in a precise sequence. Genes store the information needed for making polypeptides. The information is stored in a coded form. The sequence of bases in a gene codes for the sequence of amino acids in a polypeptide. The information in the gene is decoded during the making of the polypeptide. There are two stages in this process: **transcription** and **translation**.

TRANSCRIPTION
Genes made of DNA are too valuable to a cell to be used directly when polypeptides are being made. Instead a copy of the gene is made. The copy is RNA, not DNA. It carries the information needed for making a polypeptide out into the cytoplasm so is called messenger RNA (mRNA). The copying of the base sequence of a gene by making an RNA molecule is called **transcription.**

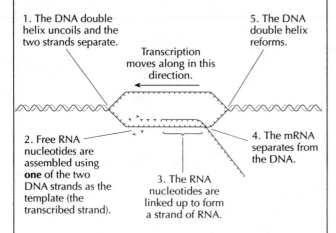

1. The DNA double helix uncoils and the two strands separate.

Transcription moves along in this direction.

5. The DNA double helix reforms.

2. Free RNA nucleotides are assembled using **one** of the two DNA strands as the template (the transcribed strand).

3. The RNA nucleotides are linked up to form a strand of RNA.

4. The mRNA separates from the DNA.

Stages 1, 2 and 3 are all carried out by the enzyme RNA polymerase.
In transcription, the same rules of complementary base pairing are followed as in replication, except that uracil pairs with adenine, as RNA does not contain thymine. The RNA molecule produced therefore has a base sequence that is complementary to the transcribed strand and identical to the other DNA strand except that U replaces T.

THE GENETIC CODE
- The genetic code is a **triplet code** – three bases code for one amino acid. A group of three bases is called a **codon**. There are 64 different codons.
- The genetic code is **degenerate**. This means that it is possible for two or more codons (triplets of bases) to code for the same amino acid.
- The genetic code is **universal**. This means that all living organisms use the same code. Viruses also use this code.

DIFFERENCES BETWEEN DNA AND RNA
DNA and RNA both consist of chains of nucleotides, each composed of a sugar, a base and a phosphate. There are three differences between them.

Feature	DNA	RNA
Number of strands in the molecule	Two strands forming a double helix	One strand only
Type of sugar in each nucleotide	Deoxyribose	Ribose
Types of bases contained	A, C, G and T	A, C, G and U Uracil replaces thymine

TRANSLATION
Translation is carried out by ribosomes, using mRNA and tRNA.

1. Messenger RNA binds to the small subunit of the ribosome. The mRNA contains a series of **codons**, each of which codes for one amino acid.

2. Transfer RNA molecules are present around the ribosome in large numbers Each tRNA has a special triplet of bases called an **anticodon** and carries the amino acid corresponding to this anticodon.

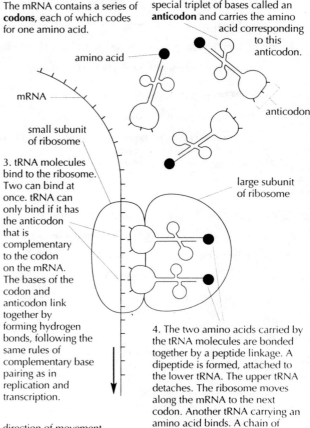

amino acid

mRNA

anticodon

small subunit of ribosome

large subunit of ribosome

3. tRNA molecules bind to the ribosome. Two can bind at once. tRNA can only bind if it has the anticodon that is complementary to the codon on the mRNA. The bases of the codon and anticodon link together by forming hydrogen bonds, following the same rules of complementary base pairing as in replication and transcription.

direction of movement of ribosome

4. The two amino acids carried by the tRNA molecules are bonded together by a peptide linkage. A dipeptide is formed, attached to the lower tRNA. The upper tRNA detaches. The ribosome moves along the mRNA to the next codon. Another tRNA carrying an amino acid binds. A chain of three amino acids is formed. These stages are repeated until a polypeptide is formed.

EXAM QUESTIONS ON TOPIC 2

1 The table below shows the base composition of genetic material from ten sources.

Source of genetic material	Base composition (%)				
	Adenine	Guanine	Thymine	Cytosine	Uracil
Cattle thymus gland	28.2	21.5	27.8	22.5	0.0
Cattle spleen	27.9	22.7	27.3	22.1	0.0
Cattle sperm	28.7	22.2	27.2	22.0	0.0
Pig thymus gland	30.0	20.4	28.9	20.7	0.0
Salmon	29.7	20.8	29.1	20.4	0.0
Wheat	27.3	22.7	27.1	22.8	0.0
Yeast	31.3	18.7	32.9	17.1	0.0
E. coli (bacteria)	26.0	24.9	23.9	25.2	0.0
human sperm	31.0	19.1	31.5	18.4	0.0
influenza virus	23.0	20.0	0.0	24.5	32.5

a) Deduce the type of genetic material used by

   (i) cattle [1]

   (ii) E. coli [1]

   (iii) influenza viruses. [1]

b) Suggest a reason for the difference between cattle thymus gland, spleen and sperm in the measurements of their base composition. [1]

c) (i) Explain the reasons for the total amount of adenine plus guanine being close to 50% in the genetic material of many of the species in the table. [3]

   (ii) Identify two other trends in the base composition of the species that have 50% adenine and guanine. [2]

d) (i) Identify a species shown in the table that does not follow the trends in base composition described in (c). [1]

   (ii) Explain the reasons for the base composition of this species being different. [2]

2 a) Draw the ring structure of glucose. [2]

   b) Explain briefly why glucose is so soluble in water. [2]

   c) State one element that is present in amino acids but not in glucose. [1]

3 The diagram shows the basic structure of amino acids

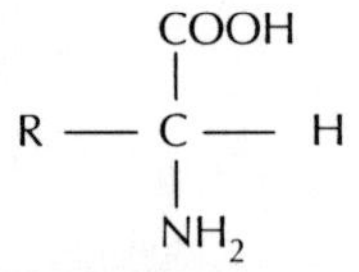

$$
\begin{array}{c}
COOH \\
| \\
R - C - H \\
| \\
NH_2
\end{array}
$$

a) State what is represented in the diagram by the letter R. [1]

b) Draw a simple diagram to show how two amino acids are linked together. [2]

c) Amino acids are linked together to form polypeptides at special sites in the cytoplasm of both prokaryotic and eukaryotic cells.

   Compare the sites where polypeptides are formed in prokaryotic cells with the sites in eukaryotic cells. [2]

Genes and chromosomes

GENES

Genetics is the study of variation and inheritance. The basic unit of inheritance is the **gene**. *A gene is a heritable factor that controls a specific characteristic.*

A typical animal or plant cell nucleus contains thousands of genes. The total number of genes in humans is not yet known but is probably between 30 000 and 40 000. All of the genes of an organism are known collectively as the **genome**. *A genome is the whole of the genetic information of an organism.*

CHROMOSOMES

Genes are made of DNA. They are part of much larger DNA molecules called **chromosomes**. In eukaryotes, proteins are always associated with the DNA in chromosomes.

A typical animal or plant chromosome contains about a 1000 genes, which are arranged in a linear sequence. In any particular type of chromosome the same genes are found arranged in the same sequence. The position of a gene on a chromosome is called the **gene locus**.

ALLELES

Although one particular chromosome type always has the same genes in the same sequence, the genes themselves can vary. Different forms of many genes can be found. These are called **alleles** of the gene. *An allele is a form of a gene, differing from other alleles of the gene by a few bases at most and occupying the same locus as the other alleles of that gene.*

REPLICATION OF CHROMOSOMES

If a nucleus is going to divide by mitosis or meiosis, all DNA in the nucleus is replicated. When mitosis or meiosis begins, each chromosome is visible as a double structure (see below). The two parts are called **chromatids** and are connected by a centromere. Some types of chromosome have a **centromere** in the centre and others have a centromere nearer to one end.

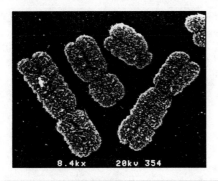

8.4kx   20kv 354

HAPLOID AND DIPLOID

In most cells the nucleus contains two of each type of chromosome (top right). The cell therefore has two full sets of chromosomes. This is called **diploid**. Some cells only contain one of each type of chromosome and therefore have just one set. This is called **haploid**.

In diploid cells each pair of chromosomes have the same genes, arranged in the same sequence. However, they do not usually have the same alleles of all of these genes. They are therefore not identical but instead are **homologous**.

Homologous chromosomes have the same genes as each other, in the same sequence, but not necessarily the same alleles of those genes.

The number of chromosomes in a cell can be reduced from diploid to haploid by the process of meiosis. Meiosis is described as a **reduction division**. Living organisms that reproduce sexually have to halve their chromosome number at some stage in the life cycle because the fusion of gametes during fertilization doubles it.

HUMAN FEMALE KARYOTYPE

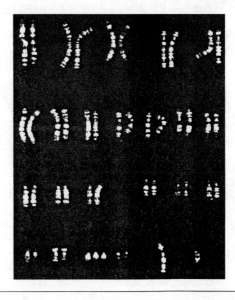

1 2 3 4 5
6 7 8 9 10 11 12
13 14 15 16 17 18
19 20 21 22

KARYOTYPES AND KARYOTYPING

The number and appearance of the chromosomes in an organism is called the **karyotype**. Living organisms that are members of the same species usually have the same karyotype. The karyotype of a human female is shown above. A small proportion of humans have a different karyotype. A procedure called karyotyping is used to test for this. One example is the testing of babies before birth (fetuses) to find out if they have Down's syndrome.

- A sample of amniotic fluid is removed from the mother. It contains cells from the fetus.
- The cells are incubated with chemicals that stimulate them to divide by mitosis.
- Another chemical is used which stops mitosis in metaphase of mitosis. Chromosomes are most easily visible in metaphase.
- A fluid is used to burst the cells and spread out the chromosomes.
- The burst cells are examined using a microscope and a photograph is taken of the chromosomes from one cell.
- The chromosomes in the photograph are cut out and arranged into pairs. This is called karyotyping.

The chromosomes of a boy with Down's syndrome are shown (below). There is an extra chromosome 21 – three are present instead of the usual two.

Karyotype of a person with Down's syndrome

Meiosis

STAGES OF MEIOSIS

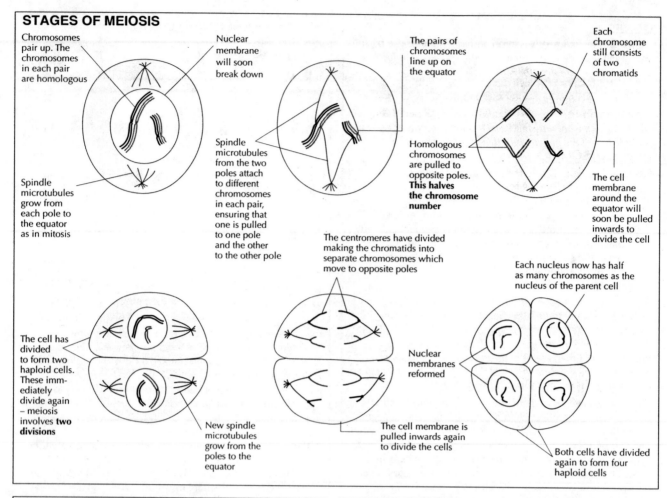

Chromosomes pair up. The chromosomes in each pair are homologous

Nuclear membrane will soon break down

Spindle microtubules grow from each pole to the equator as in mitosis

Spindle microtubules from the two poles attach to different chromosomes in each pair, ensuring that one is pulled to one pole and the other to the other pole

The pairs of chromosomes line up on the equator

Each chromosome still consists of two chromatids

Homologous chromosomes are pulled to opposite poles. **This halves the chromosome number**

The cell membrane around the equator will soon be pulled inwards to divide the cell

The cell has divided to form two haploid cells. These immediately divide again – meiosis involves **two divisions**

New spindle microtubules grow from the poles to the equator

The centromeres have divided making the chromatids into separate chromosomes which move to opposite poles

The cell membrane is pulled inwards again to divide the cells

Nuclear membranes reformed

Each nucleus now has half as many chromosomes as the nucleus of the parent cell

Both cells have divided again to form four haploid cells

CHROMOSOME MOVEMENTS IN MEIOSIS AND GENETIC VARIATION

During the first division of meiosis one chromosome of each pair moves to one pole and the other chromosome to the other pole of the cell. The position of each pair of chromosomes in the nucleus when the spindle microtubules become attached is random. This is called **random orientation**. Microtubules from each pole attach to whichever chromosome of a pair is closer.

Because of random orientation, each pole could receive either chromosome of a pair – there are two equally likely possibilities. All cells have at least two pairs of chromosomes. The second pair is also randomly orientated, giving two possibilities and therefore in total each pole could receive four (2×2) possible combinations of two chromosomes.

With three pairs of chromosomes there are eight possible combinations ($2 \times 2 \times 2$). If the number of pairs of chromosomes is n the number of possible combinations of chromosomes that can be formed because of random orientation during meiosis is 2^n. In humans for example, where n is 23, there are over 8 million possible combinations. Each of these is genetically different, so the movements of chromosomes in meiosis generate much genetic variety.

NON-DISJUNCTION AND DOWN'S SYNDROME

Sometimes chromosomes that should separate and move to opposite poles during meiosis do not separate and instead move to the same pole. This can happen in either the first (left) or the second division of meiosis (right).

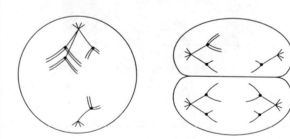

Non-separation of chromosomes is called **non-disjunction**. The result is that gametes are produced with either one chromosome too many or too few.

Gametes with one chromosome too few usually quickly die. Gametes with one chromosome too many sometimes survive. When they are fertilized, a zygote is produced with three chromosomes of one type instead of two. This is called **trisomy**. Down's syndrome is caused by trisomy of chromosome 21. It can be due either to non-disjunction during the formation of the sperm or egg. The chance of Down's syndrome increases with the age of the parents.

Mendel's Law of Segregation

MENDEL'S MONOHYBRID CROSSES

Gregor Mendel investigated inheritance by crossing varieties of pea plants that had different characteristics. For example, he crossed a variety that had round seeds with a variety that had wrinkled seeds. He found that all the offspring (called the F_1 generation) had the same characteristic as one of the parents. He allowed the F_1 generation to self-fertilize – each plant produced offspring by fertilizing its female gametes with its own male gametes. The offspring (called the F_2 generation) contained both of the original parental types. The characteristic that disappeared in the F_1 generation reappeared in a quarter of the F_2 generation.

From the results of monohybrid crosses, Mendel discovered the Law of Segregation. The figure below shows an example of Mendel's monohybrid crosses.

DEFINITIONS OF TERMS USED BY GENETICISTS

There are two pairs of terms that are often used by geneticists:

- **Homozygous** – *having two identical alleles of a gene.* All the gametes of a homozygote have the same allele.
- **Heterozygous** – *having two different alleles of a gene.* Half of the gametes of a heterozygote have one of the alleles and half have the other allele.
- **Dominant allele** – *an allele that has the same effect on the phenotype in a heterozygous individual* (where it is combined with a recessive allele) *as in a homozygous individual* (where there are two copies of the dominant allele).
- **Recessive allele** – *an allele that only has an effect on the phenotype in homozygous individuals* (where there are two copies of the recessive allele). In heterozygous individuals the recessive allele is hidden by the dominant allele.

Monohybrid cross between smooth and wrinkled seed pea plants

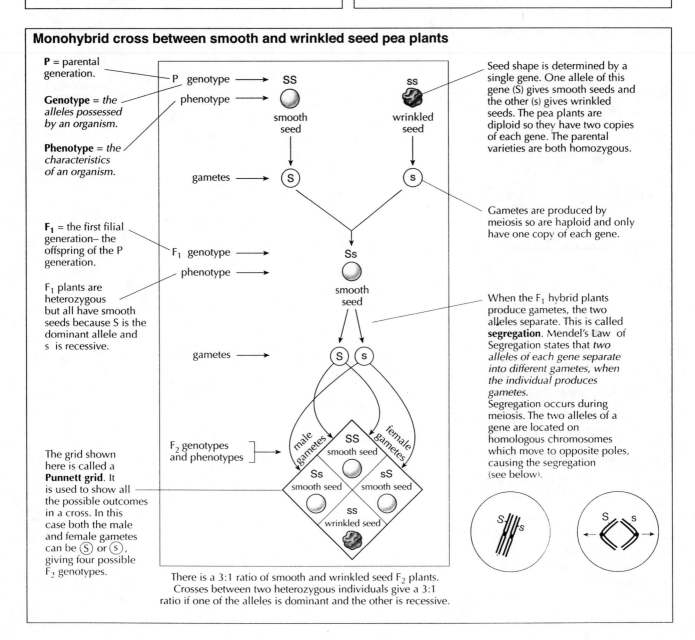

P = parental generation.

Genotype = *the alleles possessed by an organism.*

Phenotype = *the characteristics of an organism.*

F_1 = the first filial generation– the offspring of the P generation.

F_1 plants are heterozygous but all have smooth seeds because S is the dominant allele and s is recessive.

The grid shown here is called a **Punnett grid**. It is used to show all the possible outcomes in a cross. In this case both the male and female gametes can be (S) or (s), giving four possible F_2 genotypes.

Seed shape is determined by a single gene. One allele of this gene (S) gives smooth seeds and the other (s) gives wrinkled seeds. The pea plants are diploid so they have two copies of each gene. The parental varieties are both homozygous.

Gametes are produced by meiosis so are haploid and only have one copy of each gene.

When the F_1 hybrid plants produce gametes, the two alleles separate. This is called **segregation**. Mendel's Law of Segregation states that *two alleles of each gene separate into different gametes, when the individual produces gametes.*
Segregation occurs during meiosis. The two alleles of a gene are located on homologous chromosomes which move to opposite poles, causing the segregation (see below).

There is a 3:1 ratio of smooth and wrinkled seed F_2 plants. Crosses between two heterozygous individuals give a 3:1 ratio if one of the alleles is dominant and the other is recessive.

Inheritance of blood groups

The principles of inheritance discovered by Mendel in pea plants also operate in other plants and in animals. There are, however, sometimes differences and two of these are demonstrated by the inheritance of ABO blood groups in humans –codominance and multiple alleles.

CROSS INVOLVING CODOMINANT ALLELES

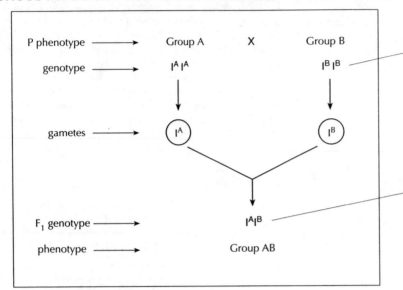

I^A is the allele for blood group A and I^B is the allele for blood group B. Neither allele is recessive, so both are given upper case letters as their symbol.

If I^A and I^B are present together, they both affect the phenotype because they are codominant. Codominant alleles are pairs of alleles that both affect the phenotype when present together in a heterozygote.

CROSS INVOLVING MULTIPLE ALLELES

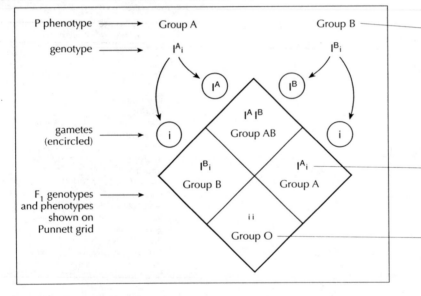

The gene that controls **ABO** blood groups has a third allele: i If there are more than two alleles of a gene, they are called *multiple alleles*.

i is recessive to both I^A and I^B so I^A i gives blood group A and I^B i gives blood group O.

Individuals who are homozygous for i are in blood group O.

DEDUCING GENOTYPES FROM PEDIGREE CHARTS

A pedigree chart shows the members of a family and how they are related to each other. Males are shown as squares and females as circles. If the phenotypes of the members of the family are known, the genotypes can often be deduced. The figure (right) is a pedigree chart that shows the blood group of each individual. All of the genotypes can be deduced. It is also possible to deduce the probability of the first child of the parents in the third generation being blood group A, B, AB and O.

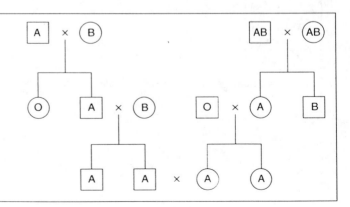

Genes and gender

SEX CHROMOSOMES AND GENDER
- Two chromosomes determine the gender of a child (whether it is male or female). These are called the sex chromosomes.
- The X chromosome is relatively large and carries many genes
- The Y chromosome is much smaller and carries only a few genes.
- If two X chromosomes are present in a human embryo, it develops into a girl.
- If one X chromosome and one Y chromosome are present, a human embryo develops into a boy.
- When women reproduce, they pass on one X chromosome in the egg.
- When men reproduce, they pass on either one X or one Y chromosome in the sperm, so the gender of a child depends on whether the sperm that fertilizes the egg is carrying an X or a Y chromosome (right).

SEX LINKAGE
If a gene is carried on the X chromosome, the pattern of inheritance is different for males and females – there is sex linkage. *Sex linkage is the association of a characteristic with gender, because the gene controlling the characteristic is located on a sex chromosome.* Sex-linked genes are almost always located on the X chromosome. Females have two X chromosomes and therefore have two copies of sex linked genes. Males only have one X chromosome and therefore only have one copy of sex linked genes. In humans, hemophilia (below) and red–green colour blindness are examples of sex-linked characteristics.

Inheritance of gender in humans

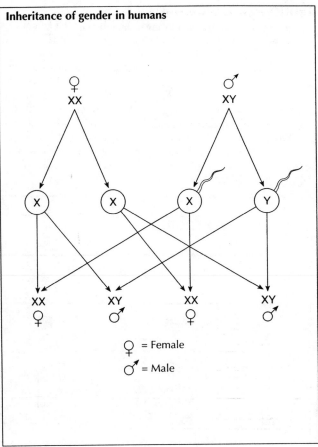

$\female$ = Female

$\male$ = Male

Example of a cross involving sex linkage

The diagram below shows how two parents who do not either have hemophilia could have a hemophiliac son.

The mother is heterozygous but is not a hemophiliac because H is dominant and h is recessive. She is a **carrier** of the allele for hemophilia.

A carrier has a recessive allele of a gene but it does not affect the phenotype because a dominant allele is also present.

KEY

X^H X chromosome carrying the allele for normal blood clotting

X^h X chromosome carrying the allele for hemophilia.

$X^H X^h$ $X^H Y$

X^h X^H X^H Y

$X^H X^H$
normal
$\female$

$X^H X^h$
carrier
$\female$

$X^H Y$
normal
$\male$

$X^h Y$
hemophiliac
$\male$

The Y chromosome does not carry either allele of the gene.

There is a 50% chance of a son being hemophiliac as half of the eggs produced by the mother carry X^h.
The chance of a daughter being hemophiliac is 0%, so the overall chance of offspring being hemophiliac is 25%.

None of the female offspring are hemophiliac because they all inherited the father's X chromosome which carries the allele for normal blood clotting (H), but there is a 50% chance of a daughter being a carrier.

Pedigree analysis

USING PEDIGREE CHARTS

Pedigree charts can be used to study the inheritance of a characteristic:
– whether it is caused by a dominant or recessive allele
– whether it is sex-linked or not.

The figures below are pedigree charts that each show a different pattern of inheritance. The most likely pattern of inheritance can be deduced in each case. Squares represent males and circles represent females. Black symbols represent individuals affected by the condition and white symbols represent unaffected individuals. In the bottom chart the grey symbols represent individuals who are partly affected.

The probability of the different phenotypes in the offspring of some of the couples in the pedigrees (marked with an asterisk *) can also be determined.

CHOOSING SYMBOLS FOR ALLELES

These rules are usually followed:

1. *Dominant and recessive alleles of a gene*
   One letter of the alphabet is chosen. The dominant allele is represented by the upper-case letter and the recessive allele by the lower-case letter (e.g. A and a).

2. *Codominant alleles*
   One letter of the alphabet is chosen. This letter and a superscript letter represent each allele (e.g. C^W and C^r).

3. *Sex-linked dominant and recessive alleles*
   The letter X is used to symbolize the X chromosome. Each allele is shown superscripted (e.g. X^H and X^h).

MUSCULAR DYSTROPHY

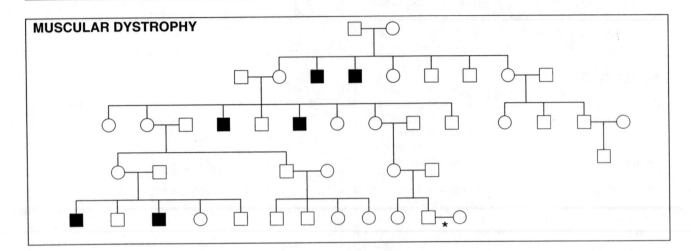

ALBINISM

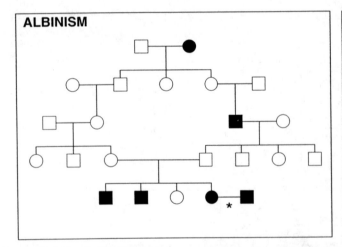

HUNTINGTON'S DISEASE

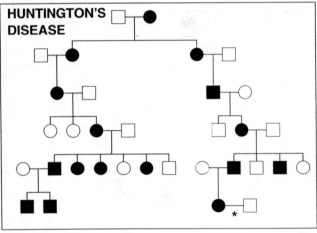

GLUCOSE PHOSPHATE DEHYDROGENASE DEFICIENCY

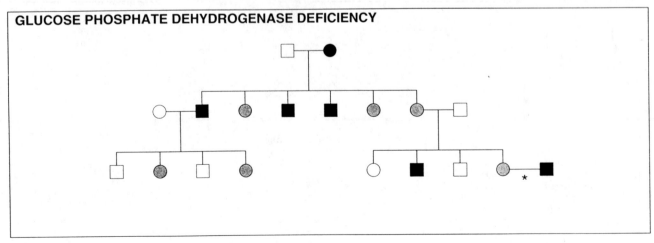

Genetic screening

Genetic screening is the testing of an individual for the presence or absence of a gene. Plant and animal breeders can test for the presence of a recessive allele using a test cross. Genetic screening for humans involves more modern analytical techniques but is a more controversial procedure.

TEST CROSSES

It is not always possible to discover whether an individual does or does not have a gene by looking at the individual's phenotype. If one allele of a gene is dominant and another allele is recessive, an individual with two copies of the dominant allele has the same phenotype as an individual with one dominant and one recessive allele. These two genotypes can be distinguished by carrying out a test cross. *In a test cross an individual that might be heterozygous is crossed with an individual that is homozygous recessive.*

AN EXAMPLE OF A TEST CROSS

A farmer is unsure whether his bull is a pure-bred Hereford or whether it is a Hereford x Aberdeen Angus hybrid. Hereford cattle have a white head caused by a dominant allele (H). Aberdeen Angus cattle have black heads caused by a recessive allele of the same gene (h). The farmer crosses his bull with 100 Aberdeen Angus cows. The figure (below left) shows the outcome if the bull is pure-bred Hereford and the figure (below right) shows the outcome if the bull is a Hereford x Aberdeen Angus hybrid.

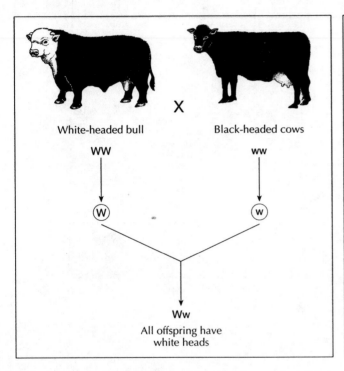

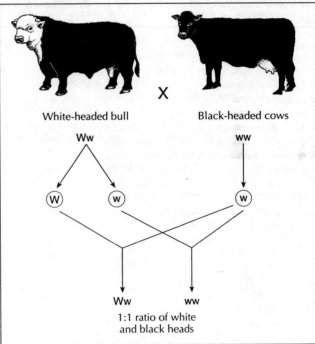

GENETIC SCREENING IN HUMANS

The question of whether genetic screening techniques should be used in human populations has been widely discussed. There are potential advantages but also possible disadvantages. Some of these are shown in the table below.

Advantages of genetic screening	Disadvantages of genetic screening
1. *Fewer children with genetic diseases are born* Men or women who are carriers of an allele that causes a genetic disease could avoid having children with the disease by choosing a partner who has been screened and found not to be a carrier of the same allele	1. *Frequency of abortion may increase* If a genetic disease is diagnosed in a child before birth, the parents may decide to have it aborted. Some people believe that this is unethical
2. *Frequency of alleles causing genetic disease can be reduced* Couples who know that they are both carriers of a recessive allele that causes a genetic disease could use IVF to produce embryos and have the embryos screened for the allele. Embryos that do not carry the allele could be selected for implantation	2. *Harmful psychological effects* If a person discovers by genetic screening that they have a genetic disease or will develop a disease when they are older, this knowledge might cause the person to become depressed
3. *Genetic diseases can be found and treated more effectively.* If some genetic diseases are diagnosed when a child is very young, treatments can be given which prevent some or all of the symptoms of the disease. PKU is an example of this	3. *Creation of a genetic underclass* People who are found to have a genetic disease may be refused jobs, life insurance and health insurance and be less likely to find a partner

Genetic disease and gene therapy

GENE MUTATION AND GENETIC DISEASE

Genes are almost always passed from parent to offspring without being changed. Occasionally genes do change and this is called **gene mutation**.

Gene mutation is a change to the base sequence of a gene. The smallest possible change is when one base in a gene is replaced by another base. This type of gene mutation is called a **base substitution**. Although only one base is changed, the consequences can be very significant. Many gene mutations cause a genetic disease. More than 4000 genetic diseases have been discovered. One example is sickle cell anemia.

GENE THERAPY

Gene therapy is the treatment of genetic disease by altering the genotype. It might be possible in the future to eliminate genetic disease by changing the base sequence of the allele causing the disease. An easier possible technique, if the allele causing the disease is recessive, is to insert the dominant allele that prevents the disease into affected cells. This could be done at various stages in the human life cycle – in sperm, eggs, early embryos or body cells. The best body cells to use are stem cells. Stem cells can divide again and again to replace body cells that have been lost.

Most attempts at gene therapy so far have not been successful. There has been some success in treating patients with severe combined immune deficiency (SCID) caused by lack of an enzyme called ADA. If this enzyme is not present, healthy lymphocytes cannot be produced by bone marrow and the immune system cannot fight diseases. The gene causing SCID is a recessive allele of a gene on chromosome 20. Most people have a different allele of this gene and can use it to make ADA. A famous case of gene therapy involved a baby called Andrew.

Gene therapy for SCID

Genetic screening before birth shows that Andrew has SCID

The allele that codes for ADA is obtained. This gene is inserted into a retrovirus

Blood removed from Andrew's placenta and umbilical cord immediately after birth contains stem cells. These are extracted from the blood

Retroviruses are mixed with the stem cells. They enter them and insert the gene into the stem cells' chromosomes

Stem cells containing the working ADA gene are injected into Andrew's blood system via a vein.

For four years T-cells (white blood cells), produced by the stem cells, made ADA enzymes, using the ADA gene. After four years more treatment was needed.

Sickle cell anemia – the consequences of a base substitution mutation

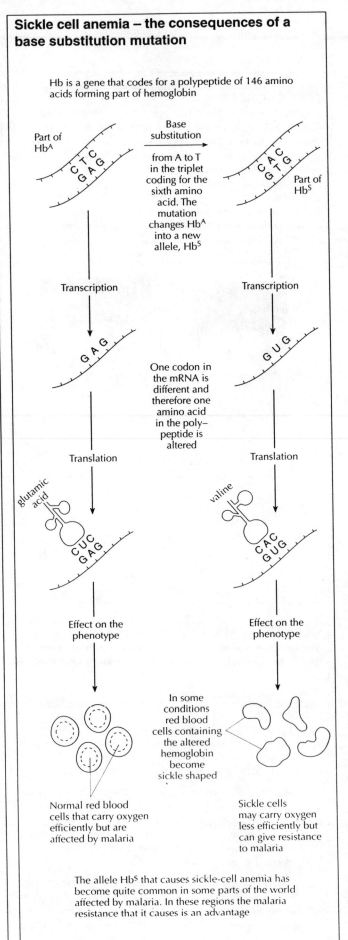

Hb is a gene that codes for a polypeptide of 146 amino acids forming part of hemoglobin

Part of Hb^A

C T C
G A G

Base substitution

from A to T in the triplet coding for the sixth amino acid. The mutation changes Hb^A into a new allele, Hb^S

Part of Hb^S

C A C
G T G

Transcription

G A G

Transcription

G U G

One codon in the mRNA is different and therefore one amino acid in the poly–peptide is altered

Translation

glutamic acid

C U C
G A G

Translation

valine

C A C
G U G

Effect on the phenotype

Effect on the phenotype

In some conditions red blood cells containing the altered hemoglobin become sickle shaped

Normal red blood cells that carry oxygen efficiently but are affected by malaria

Sickle cells may carry oxygen less efficiently but can give resistance to malaria

The allele Hb^S that causes sickle-cell anemia has become quite common in some parts of the world affected by malaria. In these regions the malaria resistance that it causes is an advantage

Genetic modification

GENETIC MODIFICATION AND ITS USES

The genetic code is universal, so genes can be transferred from one organism to another, even if they are members of different species. A gene codes for the same polypeptide whether it is in a human cell, a bacterium or any other cell. Organisms that have had genes transferred to them are called **genetically modified organisms** (GMO) or transgenic organisms. The process of transferring genes is called genetic modification. An example is the transfer from cattle to chickens of a gene for making growth hormone. Another example is the transfer of the gene for making human insulin to bacteria (see below).

Techniques used for gene transfer into bacteria

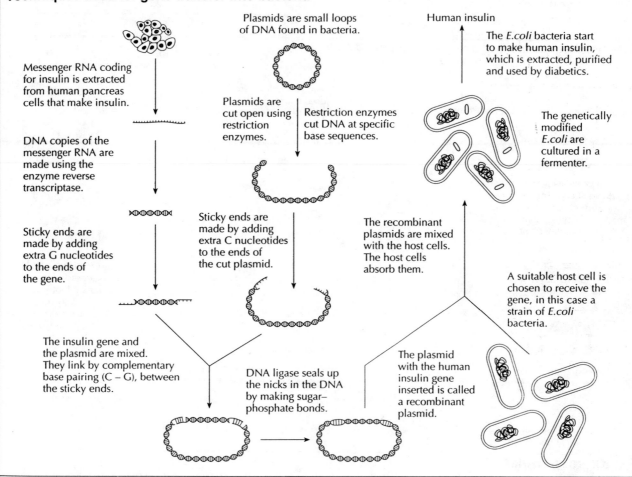

Messenger RNA coding for insulin is extracted from human pancreas cells that make insulin.

DNA copies of the messenger RNA are made using the enzyme reverse transcriptase.

Sticky ends are made by adding extra G nucleotides to the ends of the gene.

The insulin gene and the plasmid are mixed. They link by complementary base pairing (C – G), between the sticky ends.

Plasmids are small loops of DNA found in bacteria.

Plasmids are cut open using restriction enzymes.

Restriction enzymes cut DNA at specific base sequences.

Sticky ends are made by adding extra C nucleotides to the ends of the cut plasmid.

DNA ligase seals up the nicks in the DNA by making sugar–phosphate bonds.

Human insulin

The *E.coli* bacteria start to make human insulin, which is extracted, purified and used by diabetics.

The genetically modified *E.coli* are cultured in a fermenter.

The recombinant plasmids are mixed with the host cells. The host cells absorb them.

A suitable host cell is chosen to receive the gene, in this case a strain of *E.coli* bacteria.

The plasmid with the human insulin gene inserted is called a recombinant plasmid.

BENEFITS AND RISKS OF GENETIC MODIFICATION

The production of human insulin using bacteria has enormous benefits and no obvious harmful effects. There are other examples of genetic modification that are more controversial. Maize crops are often seriously damaged by corn borer insects. A gene from a bacterium (*Bacillus thuringiensis*) has been transferred to maize. The gene codes for a bacterial protein called Bt toxin that kills corn borers feeding on the maize.

Potential benefits of Bt maize	Possible harmful effects of Bt maize
1. Less pest damage and therefore higher crop yields to help to reduce food shortages	1. Humans or farm animals that eat the genetically modified maize might be harmed by the bacterial DNA in it, or by the Bt toxin
2. Less land needed for crop production, so some could become areas for wildlife conservation	2. Insects that are not pests could be killed. Maize pollen containing the toxin is blown onto wild plants growing near the maize. Insects feeding on the wild plants, including Monarch butterfly caterpillars are therefore affected even if they do not feed on the maize
3. Less use of insecticide sprays, which are expensive and can be harmful to farm workers and to wildlife	3. Populations of wild plants might be changed. Cross-pollination will spread the Bt gene into some wild plants but not others. These plants would then produce the Bt toxin and have an advantage over other wild plants in the struggle for survival

Cloning

Cloning is producing identical copies of genes, cells or organisms. The products of cloning are called a **clone**. *A clone is a group of genetically identical organisms or a group of genetically identical cells derived from a single parent.* Cloning is very useful if an organism has a desirable combination of characteristics, and more organisms with the same characteristics are wanted. Most plants can be cloned quite easily from pieces of root, stem or leaf. Animals cannot be cloned in the same way from parts of their bodies. If animal embryos are divided up at an early stage into several pieces, each piece can develop into a separate animal. (This happens naturally when identical twins are formed.) However, it is hard to predict which embryos will develop into animals with desirable characteristics and should therefore be cloned. The first successful cloning of an adult with known characteristics produced Dolly the sheep (see below).

Techniques for cloning using differentiated cells

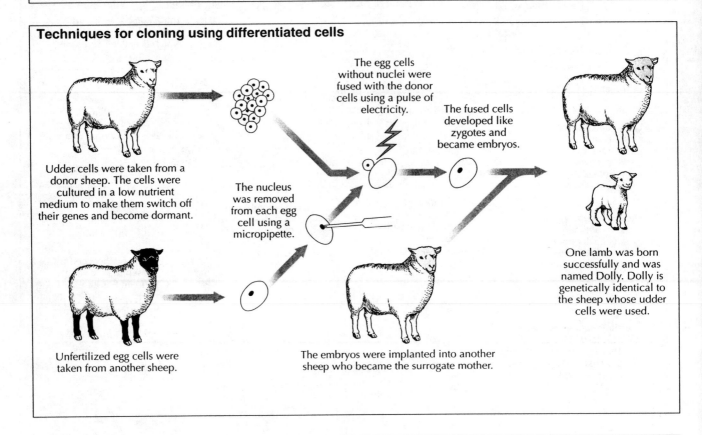

Udder cells were taken from a donor sheep. The cells were cultured in a low nutrient medium to make them switch off their genes and become dormant.

Unfertilized egg cells were taken from another sheep.

The nucleus was removed from each egg cell using a micropipette.

The egg cells without nuclei were fused with the donor cells using a pulse of electricity.

The fused cells developed like zygotes and became embryos.

The embryos were implanted into another sheep who became the surrogate mother.

One lamb was born successfully and was named Dolly. Dolly is genetically identical to the sheep whose udder cells were used.

CLONING IN HUMANS

Experiments have shown that it is possible to clone humans, but there are many ethical issues and human cloning has been banned in many countries.

Arguments for cloning in humans

1. Happens naturally when identical twins are formed, so it is not a new phenomenon

2. Cloning of embryos would make screening of embryos for genetic disease easier

3. Infertile couples might have more chance of success with IVF if their embryos were cloned

Arguments against cloning in humans

1. Groups of genetically identical people might suffer psychological problems of identity or individuality

2. Cloning using differentiated cells would often cause suffering because it carries a high risk of fetal abnormalities and a high rate of miscarriage

3. DNA taken from differentiated cells has already begun ageing and humans cloned from it might therefore grow old more quickly than is usual

THE HUMAN GENOME PROJECT

The human genome has been estimated to consist of between 30 000 and 40 000 genes. The Human Genome Project aims to find the location of all of these genes on the human chromosomes and the base sequence of all of the DNA that makes them up. The project is an international cooperative one, with laboratories in many countries involved.

The sequencing of the entire human genome will make it easier to study how genes control human development. It will allow easier identification of genetic diseases. It will allow the production of new drugs based on DNA base sequences of genes or the structure of proteins coded for by these genes.

DNA profiling

PCR –THE POLYMERASE CHAIN REACTION

In the polymerase chain reaction, DNA is copied again and again to produce many copies of the original molecules. Millions of copies of the DNA can be produced in a few hours. This is very useful when very small quantities of DNA are found in a sample and larger amounts are needed for analysis. DNA from very small samples of semen, blood or other tissue or even from long-dead specimens can be amplified using PCR.

GEL ELECTROPHORESIS

Gel electrophoresis is a method of separating mixtures of proteins, DNA or other molecules that are charged. The mixture is placed on a thin sheet of gel, which acts like a molecular sieve. An electric field is applied to the gel by attaching electrodes to both ends. Depending on whether the particles are positively or negatively charged, they move towards one of the electrodes or the other. The rate of movement depends on the size and charge of the molecules – small and highly charged molecules move faster than larger or less charged ones.

DNA PROFILING

Humans and other organisms have short sequences of bases that are repeated many times, called satellite DNA. This satellite DNA varies greatly between different individuals in the number of repeats. If it is copied using PCR and then cut up into short fragments using restriction enzymes, the lengths of the fragments vary greatly between individuals. Gel electrophoresis can be used to separate fragmented pieces of DNA according to their charge and size. The pattern of bands on the gel is very unlikely to be the same for any two individuals. This technique, called DNA profiling or DNA fingerprinting has many applications. These include criminal investigations, research into variation in populations and tracking individuals in populations such as migrating whales. Figures below show two examples of DNA profiling.

Testing whether samples of DNA show differences using DNA profiling

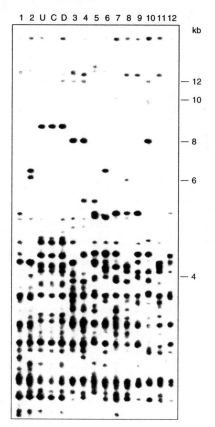

The results of DNA profiling of Dolly the sheep are shown above.
U = the udder cells from the donor sheep
C = cells in the culture derived from the udder cells
D = blood cells from Dolly the sheep
1 – 12 = results from 12 other sheep for comparison.
The results confirm that Dolly is a clone of the donor udder cells.

Testing parentage using DNA profiling

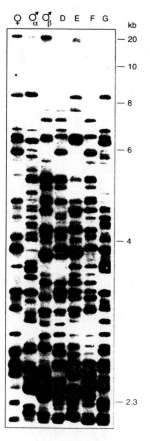

The DNA profiles of a family of dunnocks (*Prunella modularis*) are shown above. Dunnocks are small birds found in Europe, North Africa and Asia. The tracks from left to right are: the female, two resident males that might have been the father of the offspring and four offspring. The results show that the β male fathered three of the four offspring (D, E and F), despite being less dominant than the α male.

Genetics 31

EXAM QUESTIONS ON TOPIC 3

1 In humans the blood groups A, B, AB and O are determined by three alleles of an autosomal gene: I^A, I^B and i. Alleles I^A and I^B are codominant and allele i is recessive. The phenotypes of some individuals in the pedigree below are shown.

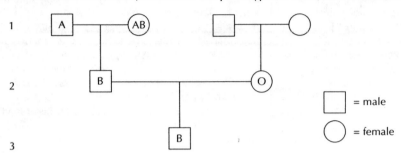

a) Explain the conclusions that can be drawn about the genotypes of the individuals in the pedigree in generations 2 and 3.   [3]

b) Explain to which blood groups the parents of the blood group O female in the pedigree could have belonged.   [3]

c) Suggest **one** reason for testing the blood groups in humans.   [1]

2 When red and white flowered *Mirabilis jalapa* plants are crossed together, all the offspring have pink flowers. The symbols for the two alleles involved are C^r (red) and C^w (white).

a) State the genotypes of the red- and white-flowered parents and the pink-flowered offspring.   [1]

b) When Mendel crossed red- and white-flowered pea plants together, all of the offspring had red flowers. Suggest a reason for the difference in results between pea plants and *Mirabilis jalapa* plants.   [1]

c) Predict the outcome of a cross between two pink-flowered *Mirabilis jalapa* plants, using a Punnett grid.   [3]

Gametes ↓ and →		

3 a) Define clone.   [1]

   b) Outline one technique used to clone farm animals.   [2]

   c) Some people believe that the cloning of human embryos is unethical. Suggest **two** reasons for this belief.   [2]

Classification

WHY CLASSIFY ORGANISMS?

Classification in biology is arranging living organisms into groups. There are many advantages gained by classifying organisms.

- **Species identification** – it is easier to find out to which species an organism belongs with organisms classified rather than in a disorganized catalogue.
- **Predictive value** – if several members of a group have a characteristic, another species in this group will probably also have this characteristic.
- **Evolutionary links** – species that are in the same group probably share characteristics because they have evolved from a common ancestor, so the classification of groups can be used to predict how they evolved.

SPECIES AND GENUS

In the classification of living organisms the basic group is the **species**. *A species is a group of organisms with similar characteristics, which can interbreed and produce fertile offspring.* Every species is classified into a **genus**. A genus is a group of similar species.

Each species needs an international name, so that biologists throughout the world can refer to it. The naming of species is called nomenclature. The **nomenclature** that biologists use is called the **binomial system** because two names are used to refer to each species. The key features of the binomial system of nomenclature are:

- the first name is the genus name
- the genus name is given an upper-case first letter
- the second name is the species name
- the species name is given a lower-case first letter
- italics are used when the name is printed
- the name is underlined if it is handwritten.

FIVE KINGDOMS

Taxonomists do not always agree about how living organisms should be classified into kingdoms. One system that is widely used has five kingdoms.

1. **Prokaryotae** – including all types of bacteria.
2. **Protoctista** – including unicellular organisms like Amoeba and algae.
3. **Fungi** – the moulds and the yeasts.
4. **Plantae** – including mosses, ferns, conifers and flowering plants.
5. **Animalia** – including sponges, corals, insects, birds and mammals.

Evidence from DNA sequences used to help in classification

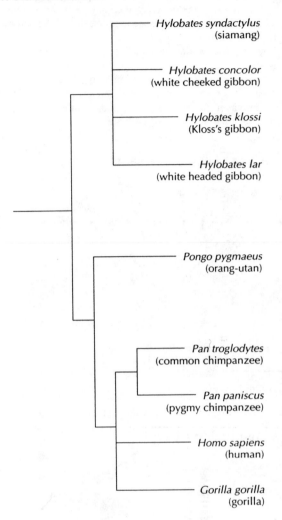

Hylobates syndactylus (siamang)

Hylobates concolor (white cheeked gibbon)

Hylobates klossi (Kloss's gibbon)

Hylobates lar (white headed gibbon)

Pongo pygmaeus (orang-utan)

Pan troglodytes (common chimpanzee)

Pan paniscus (pygmy chimpanzee)

Homo sapiens (human)

Gorilla gorilla (gorilla)

Tree diagram showing relationships between humans and the species most similar to humans, based on comparisons between DNA of each species. Chimpanzees and gorillas were placed in a family with orang-utans, but should probably be placed in the same family as humans according to the DNA evidence.

CLASSIFICATION FROM SPECIES TO KINGDOM

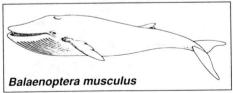

Balaenoptera musculus

A group of organisms, such as a species or a genus is called a **taxon**. Species are classified into a series of taxa, each of which includes a wider range of species than the previous one. This is called the **hierarchy of taxa**.

Sequoia sempervirens

	Animal example Balaenoptera musculus – the blue whale (left)	Plant example Sequoia sempervirens – the coast redwood (right)
Species that are similar are grouped into a genus	Genus Balaenoptera	Genus Sequoia
Genera that are similar are grouped into a family	Family Balaenopteridae	Family Taxodiaceae
Families that are similar are grouped into an order	Order Cetacea	Order Pinales
Orders that are similar are grouped into a class	Class Mammalia	Class Pinopsida
Classes that are similar are grouped into a phylum	Phylum Chordata	Phylum Coniferophyta
Phyla that are similar are grouped into a kingdom	Kingdom Animalia	Kingdom Plantae

Identifying living organisms

USING KEYS TO IDENTIFY ORGANISMS

The first stage in many ecological investigations is to find out what species of organism there are in the area being studied. This is called **species identification**. This can be done using **keys**.

Keys for species identification are usually constructed in this way:
• the key consists of a series of numbered stages
• each stage consists of a pair of alternative characteristics
• some alternatives give the next stage of the key to go to
• some alternatives give the identification.

Identifying aquarium plants using a key

Many aquatic plants in aquariums in biology laboratories belong to one of these four genera:
• *Cabomba*
• *Ceratophyllum*
• *Elodea*
• *Myriophyllum*

All of these plants have cylindrical stems with whorls of leaves. The shape of four leaves is shown in the figure (left). A key can be used to identify which of the four genera a plant belongs to, if it is known to be in one of them.

1. Simple undivided leaves . . . .*Elodea*

   Leaves forked or divided into segments 2

2. Leaves forked once or twice to form
   two or four segments . . . .*Ceratophyllum*

   Leaves divided into more than four segments 3

3. Leaves divided into many flattened segments . . . .*Cabomba*

   Leaves divided into many
   filamentous segments . . . .*Myriophyllum*

Some species of Elodea have recently been moved by taxonomists to other genera:
Elodea densa is now *Egeria densa*.
Elodea crispa is now *Lagarosiphon major*.

Leaves of aquarium plants

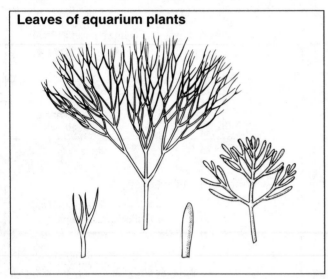

Constructing a key

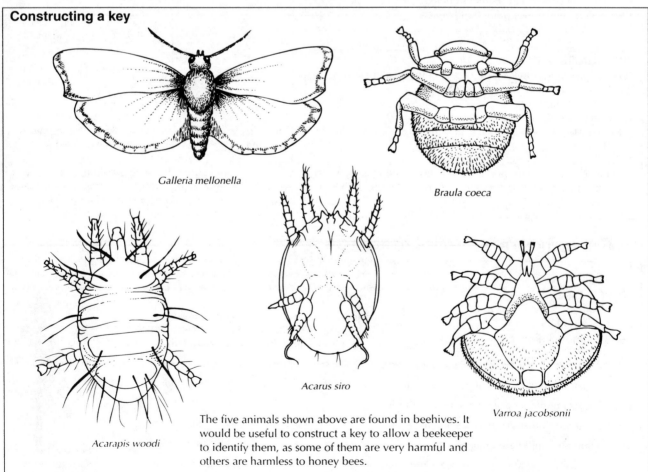

Galleria mellonella

Braula coeca

Acarus siro

Varroa jacobsonii

Acarapis woodi

The five animals shown above are found in beehives. It would be useful to construct a key to allow a beekeeper to identify them, as some of them are very harmful and others are harmless to honey bees.

Measuring populations

POPULATIONS

A human population is the people who live in a town, a country or some other defined area. In biology, populations can be of humans, animals, plants or any living organism. *A population is a group of organisms of the same species living in the same area at the same time.* It is usually impossible to count every individual in a population. Instead, an accurate estimate is made. Ecologists often need to measure the size of a population. There are many methods for making estimates of population size. The **capture–mark –release–recapture method** is suitable for animals that move around and are difficult to find (see below left).

Random sampling is suitable for plants that do not move around and are easy to find (below right).

RANDOM SAMPLES

A sample is a part of a population, part of an area or part of some other whole thing, chosen to illustrate what the whole population, area or other thing is like. For example, a sample of a population is some individuals in the population but not all of them. *In a random sample, every individual in a population has an equal chance of being selected.* Random sampling of plant populations involves counting numbers in small, randomly located parts of the total area. The sample areas are usually square and are marked out using frames called **quadrats**.

CALCULATING THE MEAN OF A SET OF VALUES

To calculate a mean of a set of values, all of the values must be added together to give a total (Σx). Divide this total by the number of values (n):

$$\text{Mean} = \frac{(\Sigma x)}{n}$$

CAPTURE–MARK–RELEASE–RECAPTURE METHOD

1. Capture as many individuals as possible in the area occupied by the animal population, using netting, trapping or careful searching

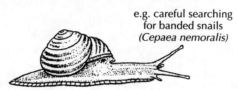

e.g. careful searching for banded snails (*Cepaea nemoralis*)

2. Mark each individual, without making them more visible to predators.

e.g. marking the inside of the snail shell with a dot of non-toxic paint.

3. Release all the marked individuals and allow them to settle back into their habitat.

4. Recapture as many individuals as possible and count how many are marked and how many unmarked.

24 marked

16 unmarked

5. Calculate the estimated population size by using the Lincoln index:

$$\text{population size} = \frac{n_1 \times n_2}{n_3}$$

n_1 = number caught and marked initially
n_2 = total number caught on the second occasion
n_3 = number of marked individuals recaptured

RANDOM SAMPLING USING QUADRANTS

1. Mark out gridlines along two edges of the area.

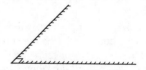

2. Use a calculator or tables to generate two random numbers, to use as co-ordinates and place a quadrat on the ground with its corner at these co-ordinates.

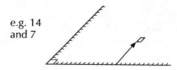

e.g. 14 and 7

3. Count how many individuals there are inside the quadrat of the plant population being studied. Repeat stages 2 and 3 as many times as possible.

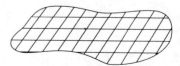

= 5 individuals

4. Measure the total size of the area occupied by the population, in square metres.

5. Calculate the mean number of plants per quadrat. Then calculate the estimated population size using this equation:

$$\text{population size} = \frac{\text{mean number per quadrat} \times \text{total area}}{\text{area of each quadrat}}$$

Variation in populations

Although members of a population show similarities because they are members of the same species, they also show differences – **variation**. For example, humans vary in height and in skin colour.

The range of variation can be shown using a graph called a frequency distribution. Most variation gives a bell-shaped frequency distribution called the **normal distribution**.

The mean value is in the middle of the distribution. Another statistic called the **standard deviation** is used to assess how far the values are spread above and below the mean. A high standard deviation shows that the data is widely spread and a low standard deviation shows that the data are clustered closely around the mean. A useful rule is that 68% of the values lie between one standard deviation above and below the mean in a normal distribution (right).

The standard deviation can be used to help decide whether the difference between two means is likely to be significant. Two examples are described below.

The normal distribution

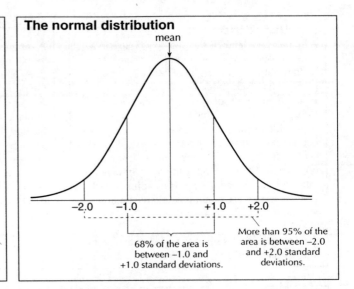

68% of the area is between –1.0 and +1.0 standard deviations.

More than 95% of the area is between –2.0 and +2.0 standard deviations.

Variation in Holly berries

A group of students collected holly berries from two *Ilex aquifolium* trees. One tree had green leaves and the other variegated leaves (leaves that were partly green and partly yellow).

Hypothesis: the berries from the tree with variegated leaves will be smaller than the berries from the tree with all-green leaves. The mass of some berries from each tree was found. The mean mass and the standard deviation were calculated for each tree.

Tree	Mean berry mass	Standard deviation
Green leaves	427 mg	73 mg
Variegated leaves	399 mg	80 mg

The berries from the tree with green leaves had a 28 mg larger mean mass than those from the tree with variegated leaves. However the standard deviations (73 mg and 80 mg) are much larger than the difference between the means. The difference in the mean mass of the berries is therefore unlikely to be significant.

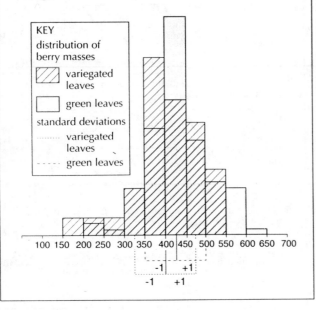

Variation in Bank voles

Ecologists noticed that bank voles (*Clethrionomys glareolus*) seemed to grow to a larger size on Raasay, a small Scottish island than on the mainland.

Hypothesis: adult bank voles are larger on Raasay than on mainland Britain.

Adult voles were caught using small mammal traps on Raasay and on the mainland. The length of each vole was measured and the mean lengths and standard deviations were calculated.

Vole population	Mean length	Standard deviation
Mainland Britain	92 mm	5.2 mm
Raasay	110 mm	7.1 mm

The mean length of the bank voles on Raasay is 18 mm greater than mean length of those on the mainland. The standard deviations (5.2 and 7.1) are much smaller than the difference in the means. The difference in the length of the bank voles was therefore almost certainly significant – the population on Raasay grew to a larger size.

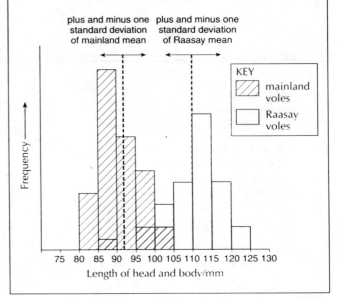

Population dynamics

CHANGES TO THE SIZE OF A POPULATION

There are four ways in which the size of a population can change:
- Offspring are produced and are added to the population – **natality**.
- Individuals die and are lost from the population – **mortality**.
- Individuals move into the area from elsewhere and are added to the population – **immigration**.
- Individuals move out of the area to live elsewhere – **emigration**.

Populations are often affected by all four of these things and the overall change can be calculated using an equation:

$$\text{Population change} = (\text{natality} + \text{immigration}) - (\text{mortality} + \text{emigration})$$

POPULATION GROWTH CURVES

If the size of a population is measured regularly, a curve can be plotted. When a species spreads into a new area, the population growth curve is often sigmoid (S-shaped). The three phases of this curve are explained by changes in natality and mortality.

1. Exponential phase

The population increases exponentially because the natality rate is higher than the mortality rate. The resources needed by the population such as food are abundant, and diseases and predators are rare.

2. Transitional phase

The natality rate starts to fall and/or the mortality rate starts to rise. Natality is still higher than mortality so the population still rises, but less and less rapidly.

3. Plateau phase

Natality and mortality are equal so the population size is constant. Something has limited the population such as:

- shortage of resources, e.g. food,
- more predators,
- more disease or parasites.

All of these factors limit population increase because they become more intense as the population rises and becomes more crowded. They either reduce the natality rate or increase the mortality rate.

If the population is limited by a shortage of resources, it has reached the **carrying capacity** of the environment. The carrying capacity is the maximum population size that can be supported by the environment.

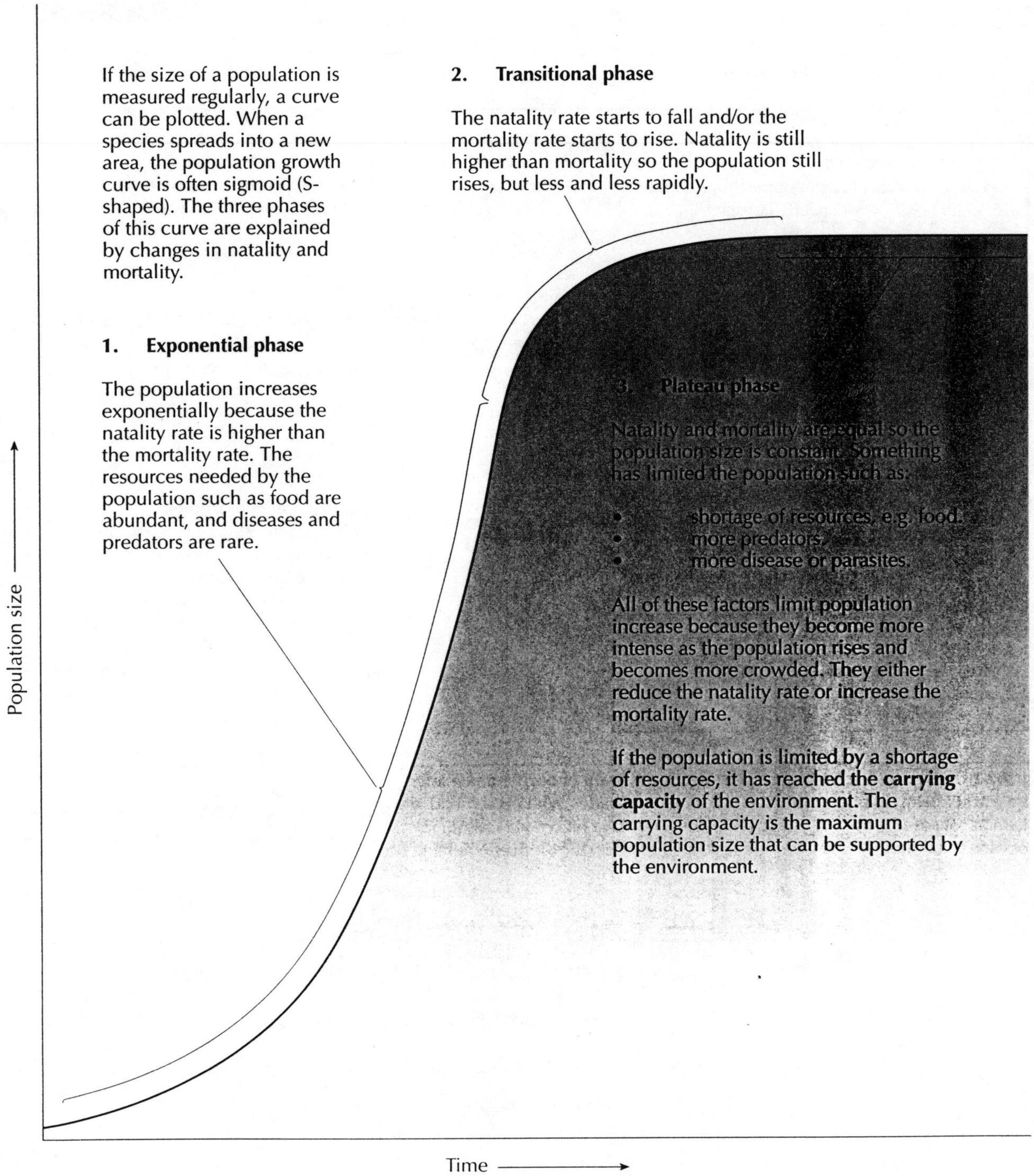

Population size

Time

Natural selection

DARWIN, WALLACE AND EVOLUTION BY NATURAL SELECTION

Evolution is the accumulation of changes in the heritable characteristics of a population.

Charles Darwin developed the theory that evolution occurs as a result of natural selection. He explained his theory in *The Origin of Species*, published in 1859. He had done many years of research and had collected much evidence for the theory before then.

Darwin delayed publication of his ideas for many years, fearing a hostile reaction. He might never have published them if another biologist, Alfred Wallace, had not written a letter to him in 1858 suggesting very similar ideas.

The theory of evolution by natural selection can be explained in a series of observations and deductions.

The photograph on the right shows a statue of Charles Darwin at Shrewsbury School, where he was a pupil from 1818–1825.

Observations	Deductions
* Populations of living organisms tend to increase exponentially * Yet, on the whole, the number of individuals in populations remains nearly constant	* More offspring are produced than the environment can support * There is a struggle for existence in which some individuals survive and some die
* Living organisms vary. The members of a species are different from each other in many ways * Some individuals have characteristics that make them well adapted to their environment and other individuals have characteristics that make them less well adapted to their environment	* The better adapted individuals tend to survive and reproduce more than the less well-adapted individuals **This is natural selection**
* Much variation is heritable – it can be passed on to offspring	* The better-adapted individuals pass on their characteristics to more offspring than the less well adapted individuals. The results of natural selection therefore accumulate * As one generation follows another, the characteristics of the species gradually change – the species evolves

In 1828 Darwin, as a young man was struggling to learn enough mathematics to pass a university exam.
The extract below is from a letter that he wrote to Charles Whitley, a friend and eminent mathematician.
' I am as idle as idle can be: one of the causes you have hit on, viz irresolution the other being made fully aware that my noddle is not capacious enough to retain or comprehend Mathematics. – Beetle hunting & such things I grieve to say is my proper sphere…'

Evolution in action

SEXUAL REPRODUCTION AND EVOLUTION

Variation is essential for natural selection and therefore for evolution. Although mutation is the original source of new genes or alleles, sexual reproduction promotes variation by allowing the formation of new combinations of alleles. Two stages in sexual reproduction promote variation.

1. Meiosis allows a huge variety of genetically different gametes to be produced by each individual.
2. Fertilization allows alleles from two different individuals to be brought together in one new individual.

Prokaryotes do not reproduce sexually but have other ways to promote variation by exchanging genes.

Some species of organisms only reproduce asexually. Mutations still produce some variation in these species, but without sexual reproduction the variation and the capacity for evolution is less.

MULTIPLE ANTIBIOTIC RESISTANCE IN BACTERIA

Antibiotics are used to control diseases caused by bacteria in humans. There have been increasing problems with disease-causing bacteria being resistant to antibiotics. The figure below shows the percentage of cases of gonorrhea (a sexually transmitted disease) in the United States that were caused by antibiotic-resistant strains of *Neisseria gonorrhoeae* between 1980 and 1990. The trend with many other diseases has been similar.

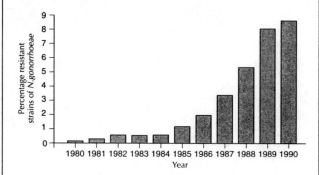

Genes that give resistance to an antibiotic can be found in the micro-organisms that naturally make that antibiotic. The evolution of multiple antibiotic resistance involves the following steps.

- A gene that gives resistance to an antibiotic is transferred to a bacterium by means of a plasmid or in some other way. There is then variation in this type of bacterium – some of the bacteria are resistant to the antibiotic and some are not.
- Doctors or vets use the antibiotic to control bacteria. Natural selection favours the bacteria that are resistant to it and kills the non-resistant ones.
- The antibiotic-resistant bacteria reproduce and spread, replacing the non-resistant ones. Eventually, most of the bacteria are resistant.
- Doctors or vets change to a different antibiotic to control bacteria. Resistance to this soon develops, so another antibiotic is used, and so on until multiply resistant bacteria have evolved.

The more an antibiotic is used, the more bacteria resistant to it there will be and the fewer non-resistant.

UNCERTAINTIES ABOUT EVOLUTION BY NATURAL SELECTION

There is much evidence for the theory that species evolve by natural selection. For example, there are some well-known cases where species have been observed to change their characteristics in response to changes in their environment. Two examples – the development of antibiotic resistance in bacteria and of metal tolerance in plants are explained below. (Other examples are given in Option D.)

However, recent cases of observed evolution all involve relatively small changes. Despite the strength of the evidence, it is not possible to *prove* that modern species have evolved by natural selection and so evolution remains a theory.

METAL TOLERANCE IN PLANTS

Waste material from the mining of metal ores and smelting often contains high levels of metals such as lead, nickel or copper. These wastes are often dumped and because of the metal pollution, few plants grow on it. Some plants do colonise the waste heaps and when they are tested they are found to have higher tolerance to the metals in the waste than usual for their species.

Evidence for the evolution of metal tolerance in a grass (*Agrostis tenuis*) was obtained in the following way.

- An area of copper pollution around an old copper mine in North Wales was mapped.
- A transect line was marked out, which ran from an unpolluted area to a heavily polluted area.
- Samples of *Agrostis tenuis* plants were collected along the transect line and were tested for copper tolerance.
- Seeds were collected from the same plants. The seeds were germinated and the plants that grew from them were also tested for copper tolerance.

The figure below shows the results.

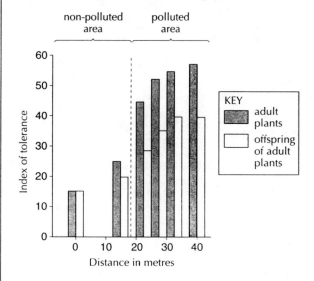

- The plants growing in the polluted area were more copper tolerant than the plants in the unpolluted area.
- The offspring of these plants inherited at least some of the copper tolerance, showing that genes are involved.
- Other experiments showed that, if plants were raised using seed collected from adult plants growing down-wind of the area of copper pollution, these plants also showed copper tolerance. Pollen carrying copper tolerance genes is blown to plants down-wind.

Trophic levels

Populations do not live in isolation – they live together with other populations in **communities**. *A community is a group of populations living together and interacting with each other in an area.*

There are many types of interaction between populations in a community. Trophic relationships are very important – where one population of organisms feeds on another population. Sequences of trophic relationships, where each member in the sequence feeds on the previous one, are called **food chains**.

An example from rainforest at Iguazu in north-east Argentina.

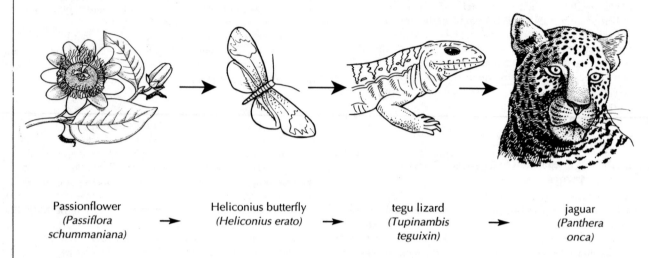

Passionflower (*Passiflora schummaniana*) →	Heliconius butterfly (*Heliconius erato*) →	tegu lizard (*Tupinambis teguixin*) →	jaguar (*Panthera onca*)

An example from chalk grassland and the air above it in Europe.

carrot plant (*Daucus carota*) →	carrot fly (*Psila rosea*) →	flycatcher (*Muscicapa striata*) →	sparrowhawk (*Accipiter nisus*) →	goshawk (*Accipiter gentilis*)

The first organism in a food chain does not feed on other organisms so must be a producer – an organism that makes its own food. The other organisms are all consumers and are called primary, secondary, tertiary and so on, depending on their position in the chain.

Producer, primary consumer, secondary consumer and tertiary consumer are examples of **trophic levels**. *The trophic level of an organism is its position in the food chain.*

Example:

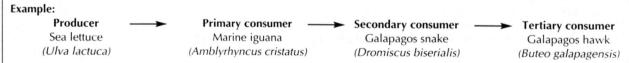

Producer →	**Primary consumer** →	**Secondary consumer** →	**Tertiary consumer**
Sea lettuce (*Ulva lactuca*)	Marine iguana (*Amblyrhyncus cristatus*)	Galapagos snake (*Dromiscus biserialis*)	Galapagos hawk (*Buteo galapagensis*)

A food chain shows only some of the trophic relationships in a community. Organisms rarely feed on only one other organism and are usually fed on by more than one organism. The complex network of trophic relationships in a community is shown in full in a complex diagram called a food web. An example of a food web is shown on page 42.

Energy flow

OBTAINING ENERGY

The organisms in a community all need a supply of energy. Organisms are divided into two groups according to their source – **autotrophs** and **heterotrophs**.

Autotrophs

Autotrophs are organisms that use an external energy source to produce organic matter from inorganic raw materials

Autotrophs make their own food, so are also called producers

Examples of producers – oak trees, maize plants, algae, blue–green bacteria

All food chains start with a producer. In almost all communities the producers make organic matter by photosynthesis

Light is therefore the **initial energy source** for the whole community

Heterotrophs

Heterotrophs are organisms that use the energy in organic matter, obtained from other organisms

Heterotrophs obtain their food from other organisms. There are three types of heterotroph – **consumers, detritivores** and **saprotrophs**

Consumers feed on other living organisms

Examples of consumers – locusts, sheep, lions,

Detritivores feed on dead organic matter by ingesting it

Examples of detritivores – dung beetles, earthworms

Saprotrophs feed on dead organic matter by secreting digestive enzymes into it and absorbing the products of digestion

Examples of saprotrophs – bread mould, mushrooms

ENERGY FLOW THROUGH COMMUNITIES

Energy flow through producers

Producers convert light energy into the chemical energy of sugars and other organic compounds. This energy trapped by the producers eventually leaves them in one of three ways.

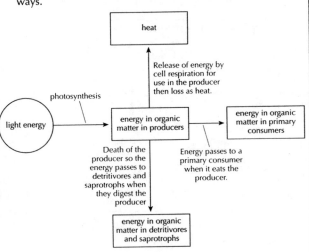

Energy flow through consumers

Primary consumers eat producers and so obtain energy from them. They do not absorb all of the energy in the food that they eat. The energy that they do take into their tissues leaves them in one of three ways.

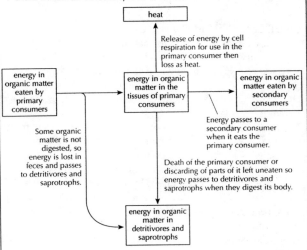

Energy flow through a food chain

The energy that passes to detritivores and saprotrophs is eventually released by cell respiration and lost as heat. In most communities all the light energy that was trapped by producers is ultimately lost as heat after flowing through the food chain. A summary of energy flow for a three-stage food chain is shown here.

Food webs and energy pyramids

FOOD WEBS

A food web is a diagram that shows all the feeding relationships in a community. The arrows indicate the direction of energy flow. Complete food web diagrams are very complex. The figure (below) shows a simplified food web for a community that lives in an area of Arctic tundra in Ogotoruk Valley.

Food web for Arctic tundra

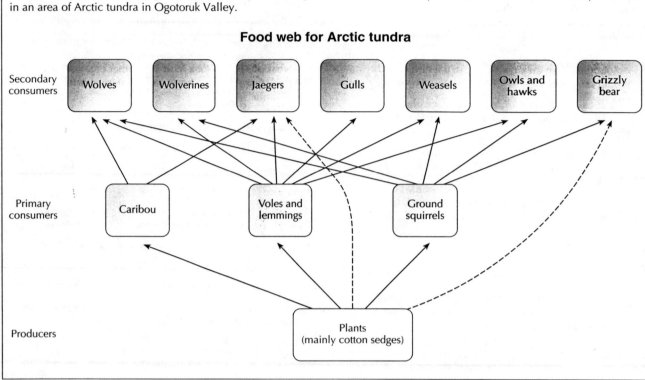

ENERGY PYRAMIDS

Energy pyramids are diagrams that show how much energy flows through each trophic level in a community. The amounts of energy are shown per square metre of area occupied by the community and per year ($kJ\,m^{-2}\,year^{-1}$). The figure (right) is a pyramid of energy for Silver Springs, a stream in Florida.

The figure (below right) is a pyramid of energy for a salt marsh in Georgia. Pyramids of energy are always pyramid shaped – each level is smaller than the one below it. This is because less energy flows through each successive trophic level. Energy is lost at each trophic level, so less remains for the next level. Note that mass is lost as well as energy, so the energy content per gram of the tissues of each successive trophic level is *not* lower.

Energy is lost in various ways. In each of the first three ways the energy is not completely lost from the community as it passes to detritivores and saprotrophs.

- Some organisms die before an organism in the next trophic level eats them.
- Some parts of organisms such as bones or hair are not eaten.
- Some parts of organisms are indigestible and pass out as feces.
- Much of the energy absorbed by an organism is released in cell respiration. The energy, in the form of ATP, is used in processes such as muscle contraction or active transport that require energy. These processes involve energy transformations, which are never 100% efficient. Some of the energy is converted to heat. 10–20% is a typical efficiency level. Most of the energy released by cell respiration is lost from the organism as heat.

Energy absorbed by living organisms is only available to the next trophic level if it remains as chemical energy in the growth of the organism. This is only a small proportion of the energy absorbed.

Energy pyramid for a stream

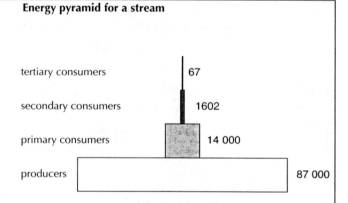

Energy pyramid for a salt marsh

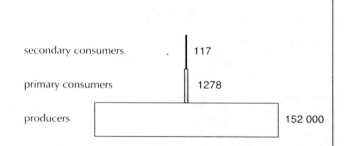

Nutrient recycling

ECOSYSTEMS, ECOLOGISTS AND ECOLOGY

Communities of living organisms interact in many ways with the soil, water and air that surround them. The non-living surroundings of a community are its abiotic environment. A community and its **abiotic environment** function together as a system called an **ecosystem**. *An ecosystem is a community and its abiotic environment.*

Ecologists study the complex relationships within ecosystems. This area of study is called **ecology**. *Ecology is the study of relationships in ecosystems – both relationships between organisms and between organisms and their environment.*

NUTRIENT RECYCLING IN ECOSYSTEMS

The recycling of nutrients is one example of the interactions between living organisms and the abiotic environment in an ecosystem. Energy is not recycled. It is supplied to ecosystems in the form of light, flows through food chains and is lost as heat. Nutrients are not usually resupplied to ecosystems – they must be used again and again by recycling. Carbon, nitrogen, phosphorus and all the other essential elements must be recycled. They are absorbed from the environment, used by living organisms and then returned to the environment.

The processes involved in the carbon cycle are shown below.

The carbon cycle

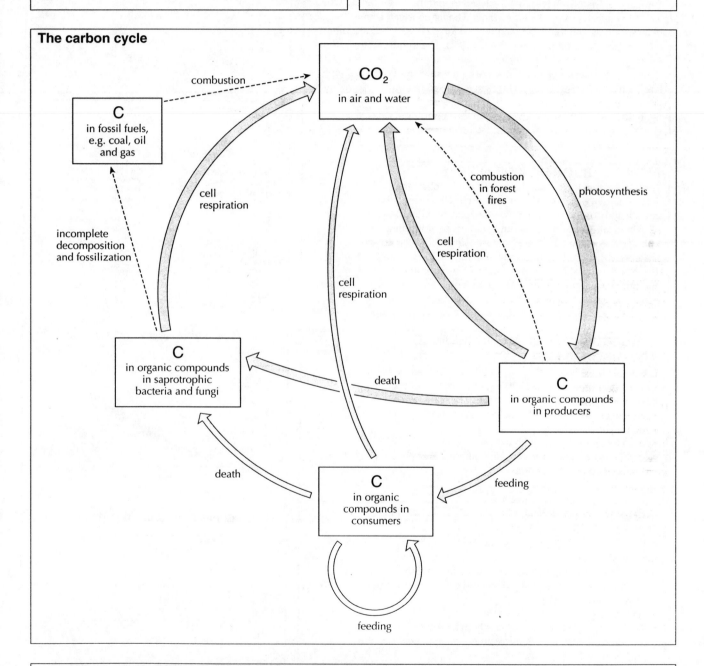

THE ROLE OF SAPROTROPHS IN RECYCLING OF NUTRIENTS

Saprotrophic bacteria and fungi have an essential role in nutrient cycles. They feed by secreting digestive enzymes into dead organic matter, including dead plants and animals and feces. The enzymes gradually break down the organic matter and the nutrients that were locked up in complex organic compounds are released. The saprotrophs absorb the substances that they need from the digested organic matter.

Without saprotrophs, nutrients would remain locked up permanently in dead organic matter and organisms that need the nutrients would soon become deficient.

Global impacts

HUMANS AND THE BIOSPHERE
The ecosystems of the world are not isolated. They have effects on each other and sometimes even depend on each other. For example, carbon dioxide produced by one ecosystem can be carried in winds to another ecosystem and there be used in photosynthesis. Ecosystems function together as a system called the **biosphere**. The biosphere is the thin layer of interdependent and interrelated ecosystems that cover the Earth.

Many human activities have affected the biosphere. Even the ice of the Antarctic has been affected – it contains lead from vehicle exhausts and large amounts of it have melted as a result of the increased greenhouse effect. The increased greenhouse effect is an example of human activities having a global impact.

RISING GLOBAL TEMPERATURES
Temperature records have been analysed to find the mean for the whole world in each year from 1856 onwards. The figure (right) shows the difference between the mean temperature for each year and an overall mean temperature for the years 1961 – 1990. The trends are that, from 1856 until about 1910, temperatures were relatively stable, from 1910 until 1940 temperatures rose and were then stable and from 1970 there has been a rapid rise. These changes in temperature could have various causes, but the most likely cause is an increased greenhouse effect. The figure (below right) shows how gases in the atmosphere cause the greenhouse effect on Earth. Carbon dioxide has the greatest overall effect.

RISING CARBON DIOXIDE LEVELS
The carbon dioxide concentration of bubbles of air trapped in Antarctic ice at different dates have been measured. These show that for 2000 years before 1880 the carbon dioxide concentration of the atmosphere remained fairly constant at about 270 parts per million (ppm). From 1880 onwards, the concentration rose. Since 1958 the concentration has been monitored continuously at Mauna Loa, Hawaii (below right). There is an annual fluctuation, but the overall trend has been upwards and the concentration is now 100 ppm higher than in 1880. This rise is enough to cause a significant increase in the greenhouse effect.

CONSEQUENCES OF THE INCREASED GREENHOUSE EFFECT
The whole biosphere is likely to be affected in many ways:
• Global warming by up to 3 °C over the next 50 years.
• Rising sea levels due mainly to thermal expansion of water.
• Flooding of low-lying land including coral atolls.
• Melting of glaciers and polar ice.
• More frequent storms and hurricanes.
• Changes to weather patterns, with different areas becoming warmer or colder and wetter or drier.

MEASURES NEEDED TO REDUCE THE GREENHOUSE EFFECT
Rising carbon dioxide levels are due to changes in the carbon cycle (page 43) including less photosynthesis and more burning of fossil fuels (right). To reduce the greenhouse effect, carbon dioxide absorption by photosynthesis must be encouraged and emissions from burning of fossil fuels must be reduced.
The following measures would help.
• Restoration of ecosystems where there has been deforestation, desertification or other damage, to encourage the growth of photosynthesizing plants.
• Spreading nutrients such as iron in nutrient-deficient oceans to encourage growth of photosynthesizing algae.
• Reducing energy consumption, for example by thermal insulation of homes, driving smaller vehicles or eating food grown locally rather than food transported great distances.
• Changing from fossil fuels as an energy source to solar, wind or nuclear power.

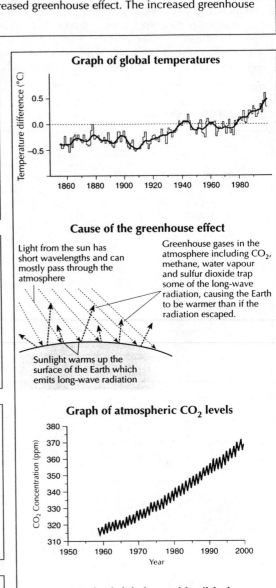

Graph of global temperatures

Cause of the greenhouse effect

Light from the sun has short wavelengths and can mostly pass through the atmosphere

Greenhouse gases in the atmosphere including CO_2, methane, water vapour and sulfur dioxide trap some of the long-wave radiation, causing the Earth to be warmer than if the radiation escaped.

Sunlight warms up the surface of the Earth which emits long-wave radiation

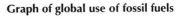

Graph of atmospheric CO_2 levels

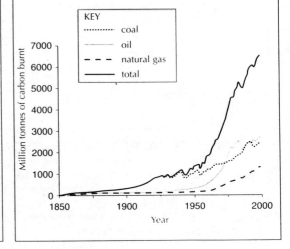

Graph of global use of fossil fuels

KEY
.... coal
‒‒‒‒ oil
– – – natural gas
——— total

Local impacts

HUMANS AND HABITATS

Every species has its **habitat**. *The habitat of a species is the environment or location where it normally lives.* For example, the habitat of *Pinus aristata* (bristlecone pine) is exposed, dry, rocky slopes and ridges in the sub-alpine zone of mountains in Colorado, New Mexico and California. The habitat of *Hippocampus ramulosus* (seahorse) is among seaweeds and sea-grasses on the seabed in shallow parts of the Mediterranean and the Atlantic as far north as the English Channel.

Many human activities have an impact on a specific habitat – a local impact. Introduction of alien species can have devastating effects on a habitat – for example, the introduction of rats to New Zealand.

THE INTRODUCTION OF RATS TO BIG SOUTH CAPE ISLAND

Three species of rat that were introduced to the mainland of New Zealand during the ninententh century eliminated many species of bird from the mainland. On islands that remained free of rats, some of these birds were able to survive. Until the 1950s Big South Cape Island (right) in the far south of New Zealand remained rat free and was a haven for many rare birds. Three types were, by then, found nowhere else: South Island saddleback (right), Stewart Island snipe and Stead's bush wren (bottom right).

In the mid-1950s black rats (*Rattus rattus*) reached Big South Cape Island. Their numbers rose exponentially and by 1964 there were huge numbers on the island. They attacked eggs, young birds in nests and even adult birds, which were not behaviourally adapted to resist them. It became obvious that human intervention was needed to save the three rarest species of bird. Ecologists from the New Zealand Wildlife Service trapped as many of the remaining individuals as they could. Only two Stewart Island snipe were trapped and they died soon after, so this species became extinct. Nine Stead's bush wrens were trapped and transferred to another island that was still rat free. Unfortunately they failed to breed and gradually died out, so this species also became extinct. Forty-one South Island saddlebacks were caught and transferred to two other rat-free islands. They survived and bred and were eventually distributed to other islands. In the 1980s they were re-introduced to Little Barrier Island after another alien species had been eliminated – wild cats.

The South Island saddleback was the first species of bird to be saved from extinction by human intervention. Its future for the moment seems relatively secure.

REDUCING THE IMPACT OF ALIEN SPECIES

Various lessons can be learned from Big South Cape Island.
- Alien species should never be introduced to habitats containing rare or endangered species.
- Alien species *can* sometimes be eliminated by trapping, poisoning or other methods.
- Human intervention is sometimes essential to save a species, for example moving a population to a safer area.
- Island nature reserves can play a vital role in ensuring the survival of rare and endangered species.

Some methods for controlling alien species that have been tried elsewhere have been found to have serious risks. For example predators have been introduced to try to control alien species, but they have sometimes attacked native species rather than the alien species. Diseases have also been introduced, but again this is a risky policy, as the spread of the disease cannot be predicted with certainty.

South Island Saddleback

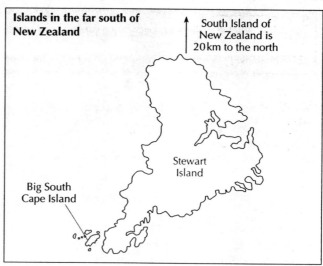

Islands in the far south of New Zealand

South Island of New Zealand is 20 km to the north

Stewart Island

Big South Cape Island

Stead's bush wren

EXAM QUESTIONS ON TOPIC 4

1 The graph below shows the growth of a population of ring-necked pheasants (*Phasianus colchicus*) on Protection Island off the north west coast of the United States. The original population released by the scientists consisted of two male and eight female birds. Two of the females died immediately after release.

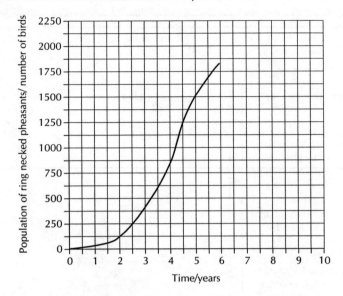

[Source of data: Elinarson A. S., Murrelet, (1945) 26: pages 39–44]

a) State the term used to describe the shape of a growth curve of this type. [1]

b) (i) The scientists predicted that the population would reach its carrying capacity of 2000 by year 8.
Draw a line on the graph to show the population growth between years 6 and 10. [2]

c) (i) Predict how the population growth would change if all the female birds in the original sample had survived. [1]

(ii) Predict the effect on the carrying capacity if all the female birds in the original sample had survived. [1]

2 The diagram below shows in simplified form the transfers of energy in a generalized ecosystem.
Each box represents a category of organisms, grouped together by their trophic position in the ecosystem.

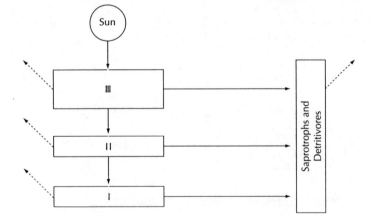

a) Deduce the trophic levels of the organisms in boxes I, II and III. [3 marks]

b) State the form in which energy enters organisms in box 1. [1 mark]

c) Identify which arrow represents the greatest transfer of energy per unit of time. (Add a large X to the arrow). [1 mark]

d) Explain what is represented by the dotted arrows leaving each box. [3 marks]

3 Methane acts as a greenhouse gas in the atmosphere. The main sources of methane are the digestive systems of cattle and sheep, bacterial action in rice paddies, burning of biomass (e.g. forest fires), bacterial action in swamps and marshes, burning of coal and release of natural gas.

a) Discuss whether methane emissions from these sources will cause a change in the Earth's temperature. [3]

b) Discuss whether release of methane is a natural process or an example of a human impact on the environment. [3]

c) Suggest measures that could be taken to reduce the emission of methane. [3]

Digestion

TAKING IN FOOD

Humans take food into their digestive system through the mouth and the esophagus. However, this food is not truly inside the body until it has passed through a layer of cells into the body's tissues. This happens in the small intestines and is called **absorption**. Small finger-like projections from the wall of the small intestine called villi are specially adapted to absorb food molecules. The structure of a villus is shown below. After food has been has been absorbed it is **assimilated** – it becomes part of the tissues of the body.

Structure of a villus

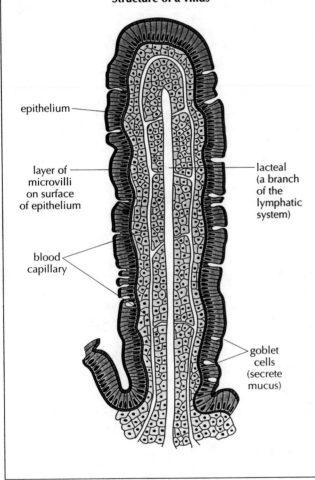

- epithelium
- layer of microvilli on surface of epithelium
- lacteal (a branch of the lymphatic system)
- blood capillary
- goblet cells (secrete mucus)

RELATIONSHIP BETWEEN STRUCTURE OF A VILLUS AND ITS FUNCTION

- Villi increase the surface area over which food is absorbed.
- An epithelium, consisting of only one thin layer of cells, is all that foods have to pass through to be absorbed.
- Protrusions of the exposed part of the plasma membranes of the epithelium cells increase the surface area for absorption. These projections are called microvilli.
- Protein channels in the microvilli membranes allow rapid absorption of foods by facilitated diffusion and pumps allow rapid absorption by active transport.
- Mitochondria in epithelium cells provide the ATP needed for active transport.
- Blood capillaries inside the villus are very close to the epithelium so the distance for diffusion of foods is very small.
- A lacteal (a branch of the lymphatic system) in the centre of the villus carries away fats after absorption.

THE NEED FOR DIGESTION

The food that humans eat contains substances made by other organisms, many of which are not suitable for human tissues. They must therefore be broken down and reassembled in a form that is suitable.

A second reason for digestion is that many of the molecules in foods are too large to be absorbed by the villi in the small intestine. These large molecules have to be broken down into small molecules that can then be absorbed by diffusion, facilitated diffusion or active transport. The three main types of food molecule that need to be digested are starch, protein and triglycerides (fats and oils).

Digestion of these large molecules happens naturally at body temperature, but only at a very slow rate. Enzymes are essential to speed up the process.

Enzymes of digestion

	Amylase	**Protease**	**Lipase**
Example of this enzyme	Salivary amylase	Pepsin	Pancreatic lipase
Source	Salivary glands	Wall of stomach	Pancreas
Substrate	Starch	Proteins	Triglycerides (fats or oils)
Products	Maltose	Small polypeptides	Fatty Acids and Glycerol
Optimum pH	pH 7	pH 1.5	pH 7

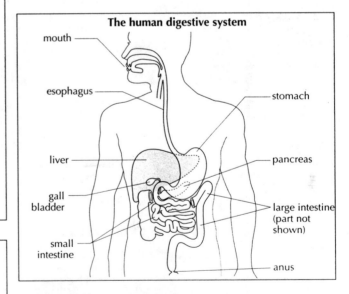

The human digestive system

- mouth
- esophagus
- liver
- gall bladder
- small intestine
- stomach
- pancreas
- large intestine (part not shown)
- anus

FUNCTIONS OF THE STOMACH AND INTESTINES

Digestion of proteins begins in the stomach, catalysed by pepsin. Bacteria, which could cause food poisoning, are mostly killed by the acid conditions of the stomach. The acidity also provides optimum conditions for pepsin to work. Enzymes secreted by the wall of the small intestine complete the process of digestion. The end products of digestion are absorbed by the villi protruding from the wall of the small intestine.

The indigestible parts of the food, together with a large volume of water, pass on into the large intestine. Water is absorbed here leaving solid feces, which are eventually egested through the anus.

The blood system

THE COMPOSITION OF BLOOD

Blood is composed of plasma, erythrocytes (red blood cells), leukocytes and platelets. The figure below shows the appearance of blood as seen using a light microscope. Two types of leukocyte are shown.

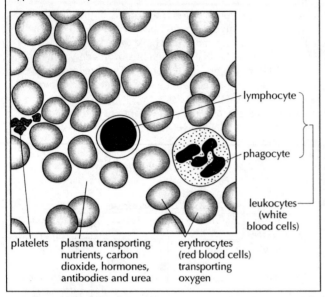

lymphocyte

phagocyte

leukocytes (white blood cells)

platelets

plasma transporting nutrients, carbon dioxide, hormones, antibodies and urea

erythrocytes (red blood cells) transporting oxygen

THE ACTION OF THE HEART

The atria are the collecting chambers – they collect blood from the veins. The ventricles are the pumping chambers – they pump blood out into the arteries at high pressure. The valves ensure that the blood always flows in the correct direction. Every heartbeat consists of a sequence of actions.

1. The walls of the atria contract, pushing blood from the atria into the ventricles through the atrio-ventricular valves, which are open. The semi-lunar valves are closed, so the ventricles fill with blood.
2. The walls of the ventricles contract powerfully and the blood pressure rapidly rises inside them. This rise in pressure first causes the atrio-ventricular valves to close, preventing back-flow of blood to the atria and then causes the semi-lunar valves to open, allowing blood to be pumped out into the arteries. At the same time the atria start to refill as they collect blood from the veins.
3. The ventricles stop contracting and as pressure falls inside them the semi-lunar valves close, preventing back-flow of blood from the arteries to the ventricles. When the ventricular pressure drops below the atrial pressure, the atrioventricular valves open. Blood entering the atrium from the veins then flows on to start filling the ventricles.

The next heartbeat begins when the walls of the atria contract again.

BLOOD VESSELS

Arteries

Thick outer layer of longitudinal collagen and elastic fibres to avoid bulges and leaks

Thick wall to withstand the high pressures

Thick layers of circular elastic and muscle fibres to help pump the blood on after each heart beat

Narrow lumen to help maintain the high pressures

Veins

Thin layers with a few circular elastic and muscle fibres because blood does not flow in pulses so the veins wall cannot help pump it.

Wide lumen is needed to accomodate the slow-flowing blood

Thin wall allows the vein to be pressed flat by adjacent muscles, helping to move the blood

Thin outer layer of longitudinal collagen and elastic fibres because there is little danger of bursting

N.B veins have valves to prevent back-flow

Capillaries

Wall consists of a single layer of thin cells so the distance for diffusion in or out is small.

Pores between cells in the wall allow some of the plasma to leak out and form tissue fluid. Phagocytes can also squeeze out.

Very narrow lumen – only about 10μm across so that capillaries fit into small spaces. Many small capillaries have a larger surface area than fewer wider ones

STRUCTURE OF THE HEART

aorta

pulmonary arteries

vena cava (superior)

left atrium

pulmonary veins

right atrium

semilunar valves

atrio-ventricular valve

vena cava (inferior)

left ventricle

atrio-ventricular valve

right ventricle

THE CONTROL OF THE HEART BEAT

Heart muscle tissue has a special property – it can contract on its own without being stimulated by a nerve. One region is responsible for initiating each contraction. This region is called the pacemaker and is located in the wall of the right atrium. Each time the **pacemaker** sends out a signal the heart carries out a contraction or beat. Nerves and hormones can transmit messages to the pacemaker.

- One nerve carries messages from the brain to the pacemaker that tell the pacemaker to speed up the beating of the heart.
- Another nerve carries messages from the brain to the pacemaker that tell the pacemaker to slow down the beating.
- Adrenalin, carried to the pacemaker by the bloodstream tells the pacemaker to speed up the beating of the heart.

Pathogens and disease

PATHOGENS

Humans sometimes develop diseases because a living organism or a virus gains entry and reproduces inside the body, causing harm. The human is called the **host** and organism that enters the body and causes harm is called the **pathogen**. *A pathogen is an organism or a virus that causes a disease.* Many types of organism can act as pathogens in humans. Six examples are shown in the table below.

Disease	Type of pathogen
Influenza	Viruses
Tuberculosis	Bacteria
Thrush (oral or vaginal)	Fungi
Malaria	Protozoa
Schistosomiasis (bilharzia)	Flatworms
Hookworm	Roundworms

Robert Koch's drawing of a culture of *Mycobacterium tuberculosis*

TRANSMISSION OF PATHOGENS

One of the main problems in the life of a pathogen is how to reach a new host and gain entry to the body. There are various possible methods.

- **Contact** – contagious diseases are transmitted when an uninfected person touches an infected person as the pathogen can enter the body through the skin.
- **Cuts** – pathogens enter the body when the skin is cut or punctured by any object that is contaminated with pathogens
- **Droplets** – diseases of the human ventilation system can be transmitted when an infected person coughs or sneezes out droplets containing pathogens, which are breathed in by an uninfected person.
- **Food or water** – pathogens in contaminated food or water enter the body through the soft gut wall.
- **Sexual intercourse** – sexually transmitted diseases gain entry through the soft mucous membranes of the penis and vagina during sexual intercourse.
- **Insects** – blood-sucking insects inject their mouthparts though the skin and can transmit pathogens that they sucked out of an infected person.

TUBERCULOSIS –AN EXAMPLE OF A BACTERIAL DISEASE

Cause
Tuberculosis is caused by *Mycobacterium tuberculosis*, a rod-shaped bacterium (left). Malnutrition, overcrowding and stress all increase the chance of infection.

Transmission
The most common form of tuberculosis is spread by droplet infection. People who have advanced tuberculosis tend to cough frequently and spread droplets containing the bacteria. Another person can become infected if they breathe in the droplets. The bacteria enter the lungs and start to grow and divide there.
A rarer form of tuberculosis is transmitted from cattle to humans in infected milk. Most milk is now pasteurized or sterilized to kill the bacteria and prevent this method of transmission.

Effects
Phagocytes move to the areas of infection in the lungs and take in the bacteria by endocytosis. The bacteria are usually able to survive and breed inside the phagocytes. Small rounded swellings containing these infected phagocytes form in the lungs. These are called tubercles and give the disease its name. In X-ray photographs the tubercles are clearly visible as white patches (below). The infection usually remains confined to the lungs and gradually becomes less severe. However if the person becomes re-infected, a chronic form of the disease usually develops. Lung tissue is gradually destroyed. The person develops a fever and loses their appetite, so becomes very thin. They develop a persistent cough and often cough up blood. The infection can spread from the lungs to the lymph nodes, bones and gut. Over three million deaths each year in the world are caused by tuberculosis.

X-ray of lungs affected by tuberculosis

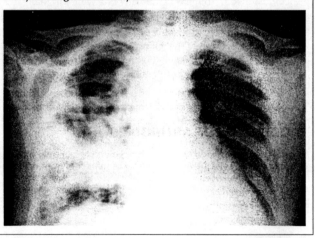

TREATMENT OF DISEASES WITH ANTIBIOTICS

Most bacterial diseases can be treated successfully with antibiotics. For example tuberculosis has been treated with streptomycin. There are many differences between human cells and bacterial cells and so there are many antibiotics that block a process in bacterial cells without causing any harm to human cells.
Viruses carry out very few processes themselves. They rely instead on a host cell such as a human cell to carry out the processes for them. It is not possible to block these processes with an antibiotic without also harming the human cells. For this reason, virus diseases cannot be treated with antibiotics.

Defence against infectious disease

BARRIERS TO INFECTION

The skin and mucous membranes form a barrier that prevents most pathogens from entering the body. The outer layers of the skin are tough and form a physical barrier. Sebaceous glands in the skin secrete lactic acid and fatty acids, which make the surface of the skin acidic. This prevents the growth of most pathogenic bacteria.

Mucous membranes are soft areas of skin that are kept moist with mucus. Mucous membranes are found in the nose, trachea, vagina and urethra. Although they do not form a strong physical barrier, many bacteria are killed by lysozyme, an enzyme in the mucus. In the trachea pathogens tend to get caught in the sticky mucus and cilia then push the mucus and bacteria up and out of the trachea.

Despite these barriers to infection, pathogens do sometimes enter the body so another defence is needed.

PHAGOCYTES

Some of the leukocytes in blood are phagocytes. These cells can identify pathogens and ingest them by endocytosis. The pathogens are then killed and digested inside the cell by enzymes from lysosomes. Phagocytes can ingest pathogens in the blood. They can also squeeze out through the walls of blood capillaries and move through tissues to sites of infection. They then ingest the pathogens causing the infection. Large numbers of phagocytes at a site of infection form pus.

Some pathogens are able to avoid being killed by phagocytes, so another defence is needed.

ANTIBODIES

Antibodies are proteins that recognize and bind to specific antigens. Antigens are foreign substances that stimulate the production of antibodies. Antibodies usually only bind to one specific antigen. Antigens can be any of a wide range of substances including cell walls of pathogenic bacteria or fungi and protein coats of pathogenic viruses.

Antibodies defend the body against pathogens by binding to antigens on surface of a pathogen and stimulating its destruction. The figure (below) shows how antibodies are produced.

AIDS – A SYNDROME CAUSED BY A VIRUS

AIDS shows how vital the body's defences against disease are.

Destruction of the immune system leads inevitably to death. AIDS is an example of a syndrome. A syndrome is a group of symptoms that are found together. Individuals with acquired immunodeficiency syndrome (AIDS) have low numbers of one type of lymphocyte together with weight loss and a variety of diseases caused by viruses, bacteria, fungi and protozoa. These diseases weaken the body and eventually cause death.

Cause

HIV (human immunodeficiency virus) causes AIDS. The virus infects a type of lymphocyte that plays a vital role in antibody production. Over a period of years these lymphocytes are destroyed and antibodies cannot then be produced. Without a functioning immune system, the body is vulnerable to pathogens that would normally be controlled easily.

Transmission

HIV does not survive for long outside the body and cannot easily pass through the skin. Transmission involves the transfer of body fluids from an infected person to an uninfected one.
- Through small cuts or tears in the vagina, penis, mouth or intestine during vaginal, anal or oral sex.
- In traces of blood on a hypodermic needle that is shared by intravenous drug abusers.
- Across the placenta from a mother to a baby, or through cuts during childbirth or in milk during breast-feeding.
- In transfused blood or with blood products such as Factor VIII used to treat hemophiliacs.

Social implications
- Families and friends suffer grief.
- Families become poorer if the individual with AIDS was the wage earner and is refused life insurance.
- Individuals infected with HIV may become stigmatized and not find partners, housing or employment.
- Sexual activity in a population may be reduced because of the fear of AIDS.

PRODUCTION OF ANTIBODIES

① Antibodies are made by lymphocytes, one of the two main types of leukocyte.

② A lymphocyte can only make one type of antibody so a huge number of different lymphocyte types is needed. Each lymphocyte puts some of the antibody that it can make into its cell surface membrane with the antigen-combining site projecting outwards.

③ When a pathogen enters the body, its antigens bind to the antibodies in the cell surface membrane of one type of lymphocyte.

lymphocyte

phagocyte

Variety of antibodies on lymphocyte surfaces.

inactive lymphocyte

④ When antigens bind to the antibodies on the surface of a lymphocyte, this lymphocyte becomes active and divides by mitosis to produce a clone of many identical cells.

mitosis

active lymphocyte

⑤ The clone of cells starts to produce large quantities of the same antibody – the antibody needed to defend the body against the pathogen.

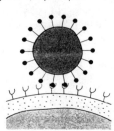

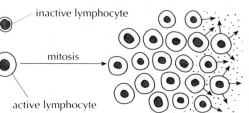

Gas exchange

THE NEED FOR GAS EXCHANGE AND VENTILATION IN HUMANS

Cell respiration happens in the cytoplasm and mitochondria of cells and releases energy in the form of ATP for use inside the cell. In humans oxygen is used in cell respiration and carbon dioxide is produced. Humans therefore must take in oxygen from their surroundings and release carbon dioxide. This process of swapping one gas for another is called **gas exchange**.

Gas exchange happens in the alveoli of human lungs. Oxygen diffuses from the air in the alveoli to the blood in capillaries. Carbon dioxide diffuses in the opposite direction. The figure (below) shows the adaptations of the alveolus for gas exchange. Diffusion of oxygen and carbon dioxide happens because there are concentration gradients of oxygen and carbon dioxide between the air and the blood. To maintain these concentration gradients, the air in the alveoli must be refreshed frequently. The process of bringing fresh air to the alveoli and removing stale air is called **ventilation**.

THE VENTILATION SYSTEM

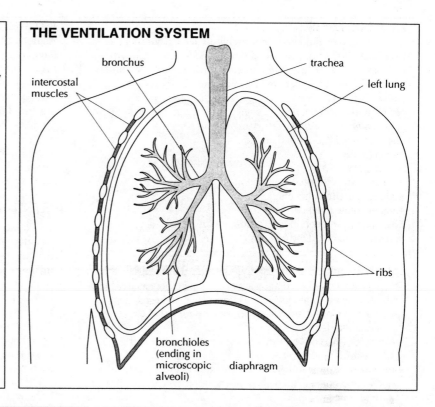

bronchus · trachea · intercostal muscles · left lung · ribs · bronchioles (ending in microscopic alveoli) · diaphragm

ADAPTATIONS OF THE ALVEOLUS TO GAS EXCHANGE

Although each alveolus is very small, the lungs contain hundreds of millions of alveoli in total, giving a huge overall surface area for gas exchange.

The wall of the alveolus consists of a single layer of very thin cells. The capillary wall also is a single layer of very thin cells, so the gases only have to diffuse a very short distance.

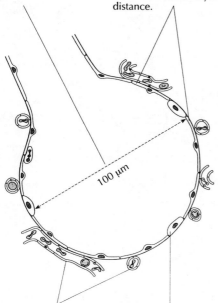

100 μm

The alveolus is covered by a dense network of blood capillaries with low oxygen and high carbon dioxide concentrations. Oxygen therefore diffuses into the blood and carbon dioxide diffuses out.

Cells in the alveolus wall secrete a fluid which keeps the inner surface of the alveolus moist, allowing the gases to dissolve. The fluid also contains a natural detergent, which prevents the sides of the alveoli from sticking together.

VENTILATION OF THE LUNGS

Air is inhaled into the lungs through the trachea, bronchi and bronchioles.

It is exhaled via the same route. Muscles are used to lower and raise the pressure inside the lungs to cause the movements of air.

Inhaling

- The external intercostal muscles contract, moving the ribcage up and out

- The diaphragm contracts, becoming flatter and moving down

- These muscle movements increase the volume of the thorax

- The pressure inside the thorax therefore drops below atmospheric pressure

- Air flows into the lungs from outside the body until the pressure inside the lungs rises to atmospheric pressure

Exhaling

- The internal intercostal muscles contract, moving the ribcage down and in

- The abdominal muscles contract, pushing the diaphragm up into a dome shape

- These muscle movements decrease the volume of the thorax

- The pressure inside the thorax therefore rises above atmospheric pressure

- Air flows out from the lungs to outside the body until the pressure inside the lungs falls to atmospheric pressure

Maintaining the internal environment

HOMEOSTASIS
Blood and tissue fluid derived from blood, flow around or close to all cells in the body. Blood and tissue fluid form the internal environment of the body. This internal environment is controlled and varies very little, despite large variations in the external environment. The control process is called **homeostasis**. *Homeostasis is maintaining the internal environment of the body at constant levels or between narrow limits.*
The parameters controlled include:
• temperature
• blood pH
• oxygen and carbon dioxide concentrations
• blood glucose concentration
• water/solute balance.
Two systems in the body play a major part in homeostasis – the nervous system and the endocrine system.

THE NERVOUS AND ENDOCRINE SYSTEMS
The nervous system is composed of cells called neurones. These cells are often very elongated and can carry messages at high speed in the form of electrical impulses. There are two parts of the nervous system:
• the central nervous system, consisting of the brain and spinal cord
• the peripheral nervous system, consisting of the peripheral nerves that connect all parts of the body to the central nervous system.
The endocrine system is composed of glands that secrete hormones. These hormones are secreted directly into the blood and are carried by the blood throughout the body.

EXCRETION
All body cells produce waste substances that have to be removed, because they can damage the body – they are toxic. These waste substances are the products of **metabolism**. Metabolism consists of all of the chemical reactions that occur inside the cells of the body. There are useful products of metabolism, but also waste products. For example, carbon dioxide is a waste product of aerobic cell respiration. The process of removing waste products is **excretion**. *Excretion is the removal from the body of the toxic waste products of metabolism.*

THE ROLE OF THE KIDNEYS
The kidneys have a major role in excretion. Waste products of metabolism are carried from body cells to the kidneys by the blood system. Urea is one of the main waste products. The kidneys remove the waste products from the blood and produce urine containing them. This is stored in the bladder and then passes out through the urethra.
The kidneys also have a role in homeostasis. By varying the composition and volume of urine, the kidneys help to keep the water and salt content of the blood and tissue fluid constant. If there is too much water or too little salt inside the body, the kidneys produce a large volume of urine with a low salt concentration. If there is too little water or too much salt inside the body, the kidneys produce a small volume of urine with a high salt content.

ORGAN SYSTEMS THAT MAINTAIN THE INTERNAL ENVIRONMENT
Each of the diagrams below shows organs that form part of an organ system.

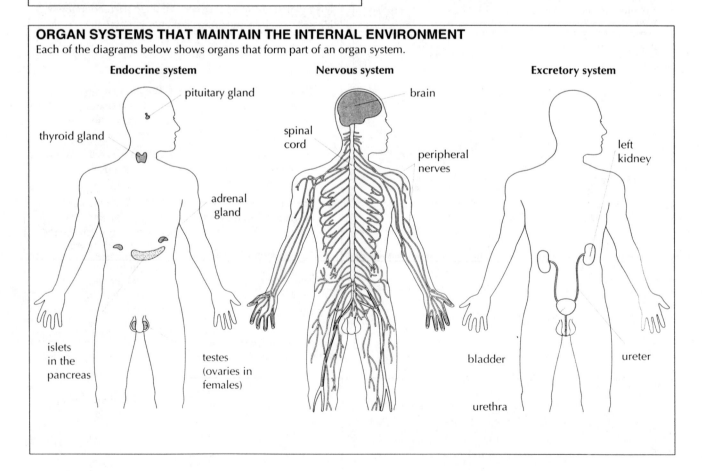

Endocrine system — pituitary gland, thyroid gland, adrenal gland, islets in the pancreas, testes (ovaries in females)

Nervous system — brain, spinal cord, peripheral nerves

Excretory system — left kidney, bladder, ureter, urethra

Negative feedback and homeostasis

CONTROLLING LEVELS BY NEGATIVE FEEDBACK

1. Feedback

In feedback systems, the level of a product feeds back to control the rate of its own production.

Level of product feeds back to affect the rate of production.

Product, e.g. heat or blood glucose.

Processes that cause the production of something.

2. Negative feedback

Negative feedback has a stabilizing effect because a change in levels always causes the opposite change. A rise in levels feeds back to decrease production and reduce the level. A decrease in levels feeds back to increase production and raise the level. These are both negative feedback.

3. Monitoring levels

Levels above set point

Set point

Levels below set point

When the level falls significantly below the set point, it is increased by negative feedback.

TIME

Small fluctuations above and below the set point do not cause a response.

When the level rises significantly above the set point, it is reduced by negative feedback.

CONTROL OF BODY TEMPERATURE

The brain monitors the temperature of the blood and compares it with a set point, usually close to 37 °C. If the blood temperature is lower or higher than the set point, the brain sends messages to parts of the body to make them respond and bring the temperature back to the set point – negative feedback. These messages are carried by neurones. The responses all affect the rate at which heat is produced or the rate at which it is lost from the body.

Responses to chilling

Skin arterioles become narrower and they bring less blood. The blood capillaries in the skin do not move, but less blood flows through them. The temperature of the skin falls, so less heat is lost from it to the environment

Skeletal muscles do many small rapid contractions to generate heat. This is called shivering

Sweat glands do not secrete sweat and the skin remains dry

Responses to overheating

Skin arterioles become wider, so more blood flows through the skin. This blood transfers heat from the core of the body to the skin. The temperature of the skin rises, so more heat is lost from it to the environment

Skeletal muscles remain relaxed and resting, so that they do not generate heat

Sweat glands secrete large amounts of sweat making the surface of the skin damp. Water evaporates from the damp skin and this has a cooling effect

CONTROL OF BLOOD GLUCOSE LEVEL

Cells in the pancreas monitor blood glucose level and compare it with a set point, usually close to 90 mg glucose per 100 ml of blood. If the blood glucose is higher or lower than the set point the pancreas sends messages to target organs to make them respond and bring the blood glucose level back to the set point – negative feedback. These messages are carried by hormones. The responses all affect the rate at which glucose is loaded or unloaded to and from the blood.

Responses to high blood glucose levels

ß cells in the pancreatic islets produce insulin.

Insulin stimulates the liver and muscle cells to absorb glucose from the blood and convert it to glycogen. Granules of glycogen are stored in the cytoplasm of these cells. Other cells are stimulated to absorb glucose and use it in cell respiration instead of fat. These processes lower the blood glucose level

Responses to low blood glucose levels

α cells in the pancreatic islets produce glucagon.

Glucagon stimulates liver cells to break glycogen down into glucose and release the glucose into the blood. This raises the blood glucose level

Reproductive systems

THE FEMALE REPRODUCTIVE SYSTEM

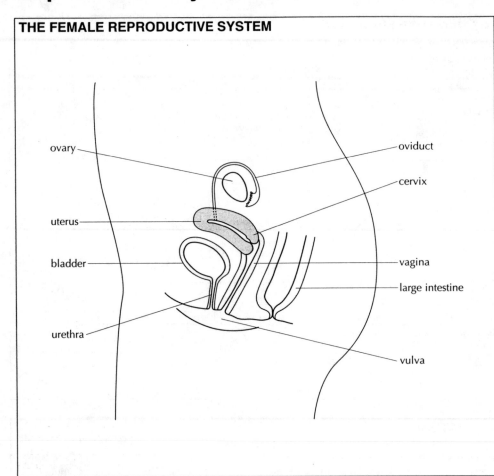

ovary
oviduct
cervix
uterus
bladder
vagina
large intestine
urethra
vulva

PUBERTY IN GIRLS

Over a period of about 4 years in the life of a girl, the amount of estrogen secreted by the ovaries rises. Estrogen causes girls to develop the female secondary sexual characteristics.

- The vagina and uterus grow larger.
- The vagina begins to secrete fluid.
- The breasts grow larger.
- Pubic and armpit hair start to grow.
- The pelvis grows larger.
- Fat is deposited under the skin of the buttocks and thighs.

Girls also start to follow the menstrual cycle and to release eggs.

These changes in the body and the life of girls are called puberty.

PUBERTY IN BOYS

Over a period of about 4 years in the life of a boy, the amount of testosterone secreted by the testes rises. Testosterone causes boys to develop the male secondary sexual characteristics.

- The penis and testes grow larger.
- The prostate gland and the seminal vesicles begin to secrete fluid.
- The larynx grows larger causing the voice to become deeper.
- Pubic and armpit hair start to grow.
- Facial hair starts to grow.
- Skeletal muscles grow larger.

Boys also start to produce sperm and release them during ejaculation.

These changes in the body and the life of boys are called puberty.

THE MALE REPRODUCTIVE SYSTEM

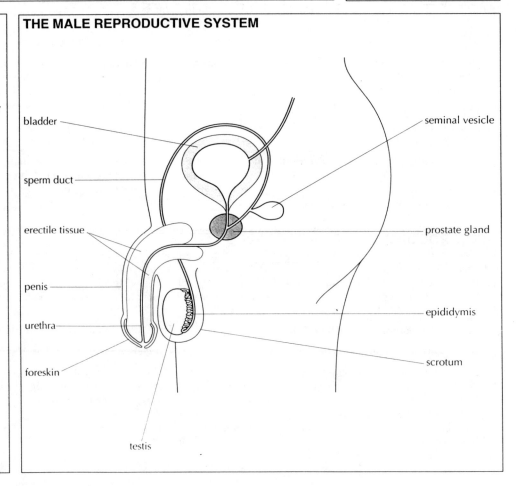

bladder
sperm duct
erectile tissue
penis
urethra
foreskin
testis
seminal vesicle
prostate gland
epididymis
scrotum

Pregnancy and childbirth

FERTILIZATION AND EARLY EMBRYO DEVELOPMENT

If a couple want to have a child, they have sexual intercourse without using any method of contraception. The biological term for sexual intercourse is **copulation**. During copulation, semen is ejaculated into the vagina. Sperm swim through the cervix, up the uterus and into the oviducts. If there is an egg in the oviducts, a sperm can fuse with it to produce a zygote. The fusion of an egg with a sperm is called **fertilization**.

The zygote produced by fertilization in the oviduct is a new human individual. It starts to divide by mitosis to form a 2-cell embryo, then a 4-cell embryo (right) and so on until a hollow ball of cells called a blastocyst is formed. While these early stages in the development of the embryo are happening, the embryo is transported down the oviduct to the uterus. When it is about 7 days old, the embryo implants itself into the wall of the uterus, where it continues to grow and develop.

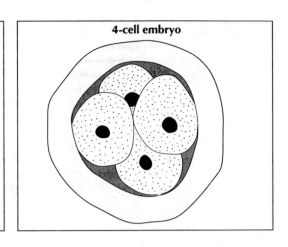

4-cell embryo

Female reproductive system during pregnancy

amniotic sac

developing fetus

uterus wall

vagina

amniotic fluid

placenta

umbilical cord

cervix

DEVELOPMENT OF THE FETUS

By the time that embryo is about 8 weeks old, it starts to develop bone tissue and is known from then onwards as a **fetus**. The fetus develops a placenta and an umbilical cord (left). The placenta is a disc-shaped structure, with many projections called placental villi embedded in the uterus wall. In the placenta the blood of the fetus flows close to the blood of the mother in the uterus wall. Materials are exchanges between maternal and fetal blood. For example, oxygen passes from maternal to fetal blood and carbon dioxide passes from fetal to maternal blood. The fetus also develops around itself an amniotic sac containing amniotic fluid. The fetus floats in this amniotic fluid and is supported by it. The delicate tissues of the fetus are protected from injury by the amniotic fluid, which acts as a shock absorber. This is needed if an everyday event or an accident causes an impact to the mother's abdomen.

A sample of fluid can be taken from the amniotic sac by inserting a hypodermic needle through the abdomen wall. This procedure is known as **amniocentesis**. The fluid contains fetal cells which can be cultured to make them divide. The chromosomes of the dividing cells can be examined to test for chromosomal abnormalities such as Down's syndrome.

CHILDBIRTH

Through the 9 months of pregnancy, the hormone progesterone ensures that the uterus develops and sustains the growing fetus. The level of progesterone in the mother becomes increasingly high. The end of pregnancy is signalled by a fall in progesterone level. This allows the mother's body to secrete another hormone – oxytocin. Oxytocin causes the muscle in the uterus wall to contract. Uterine contractions stimulate the secretion of more oxytocin. The uterine contractions therefore become stronger and stronger. This is an example of **positive feedback**.

While the muscle in the wall of the uterus is contracting, the cervix relaxes and becomes wider. The amniotic sac bursts and the amniotic fluid is released. Finally, often after many hours of contractions, the baby is pushed out through the cervix and the vagina. The umbilical cord is cut and the baby begins its independent life. Contractions continue for a time until the placenta is expelled as the afterbirth.

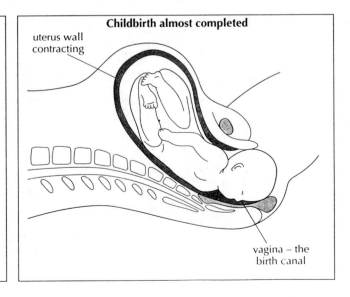

Childbirth almost completed

uterus wall contracting

vagina – the birth canal

The menstrual cycle

Between puberty and the menopause, women who are not pregnant follow a cycle called the menstrual cycle. This cycle is controlled by hormones -FSH and LH produced by the pituitary gland and oestrogen and progesterone produced by the ovary. The figure below shows the levels of these hormones during the menstrual cycle. It also shows the changes in the ovary and in the uterus. Control of the cycle involves negative and positive feedback mechanisms, indicated by arrows.

(- - - - - - - -> = negative feedback   + + + + + + + + +> = positive feedback)

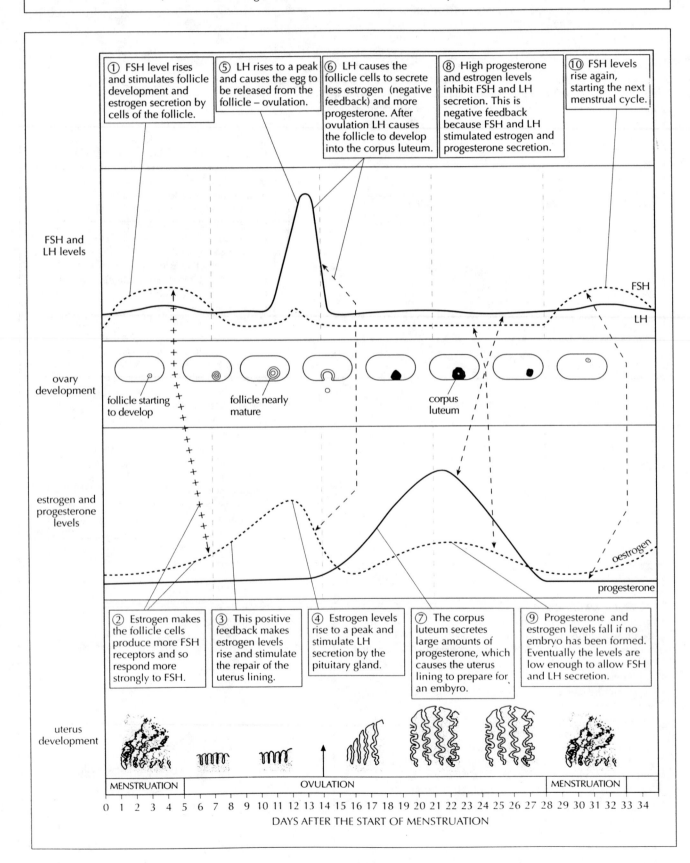

① FSH level rises and stimulates follicle development and estrogen secretion by cells of the follicle.

⑤ LH rises to a peak and causes the egg to be released from the follicle – ovulation.

⑥ LH causes the follicle cells to secrete less estrogen (negative feedback) and more progesterone. After ovulation LH causes the follicle to develop into the corpus luteum.

⑧ High progesterone and estrogen levels inhibit FSH and LH secretion. This is negative feedback because FSH and LH stimulated estrogen and progesterone secretion.

⑩ FSH levels rise again, starting the next menstrual cycle.

FSH and LH levels

FSH

LH

ovary development

follicle starting to develop

follicle nearly mature

corpus luteum

estrogen and progesterone levels

oestrogen

progesterone

② Estrogen makes the follicle cells produce more FSH receptors and so respond more strongly to FSH.

③ This positive feedback makes estrogen levels rise and stimulate the repair of the uterus lining.

④ Estrogen levels rise to a peak and stimulate LH secretion by the pituitary gland.

⑦ The corpus luteum secretes large amounts of progesterone, which causes the uterus lining to prepare for an embryo.

⑨ Progesterone and estrogen levels fall if no embryo has been formed. Eventually the levels are low enough to allow FSH and LH secretion.

uterus development

MENSTRUATION

OVULATION

MENSTRUATION

0 1 2 3 4 5 6 7 8 9 10 11 12 13 14 15 16 17 18 19 20 21 22 23 24 25 26 27 28 29 30 31 32 33 34

DAYS AFTER THE START OF MENSTRUATION

Controlling fertility

Human fertility can now be controlled in many ways. There are also many different views about whether it is right or wrong to control it. It is important to understand the views of different people and also how the methods work.

IN VITRO FERTILIZATION

Some couples do not achieve fertilization and pregnancy as a result of sexual intercourse. One possible cause is blocked oviducts. Many of these couples can be helped to have a child by in vitro fertilization – IVF.

Timetable for IVF

① A drug is injected once a day for three weeks, to stop the women's normal menstrual cycle.

② Large doses of FSH are injected once a day for 10–12 days to stimulate the ovaries to develop many follicles.

③ HCG (another hormone) is injected 36 hours before egg collection, to loosen the egg in the follicles and to make them mature.

④ The man provides semen by ejaculating into a jar. The sperm are processed to concentrate the healthiest ones.

⑤ The eggs are extracted from the follicles using a device inserted through the wall of the vagina.

⑥ Each egg is mixed with sperm in a shallow dish. The dishes are kept overnight in an incubator.

⑦ The dishes are checked to see if fertilization has worked.

⑧ Two or three embryos are selected and placed, via a long plastic tube into the uterus.

⑨ A pregnancy test is done to see if any embryos have implanted.

⑩ A scan is done to see if the pregnancy is continuing normally. The heart should be visible beating.

21.00
9.00
10.00
14.00
8.00
14.00

FAMILY PLANNING AND CONTRACEPTION

Family planning involves couples discussing and deciding how many children they want to have and when, if possible, they want to have them. During one period in the menstrual cycle, sexual intercourse can lead to fertilization and pregnancy. Couples who want a child should have sexual reproduction during this period.

Family planning may also involve contraception. Contraception is used when a couple want to have sexual intercourse but do not want to have a child. Methods of contraception can be divided into two types, mechanical methods such as condoms, diaphragms and IUDs and chemical methods such as the contraceptive pill. The ways in which four methods work are shown below, but not detailed instructions for their use.

The rhythm method – behavioural
The couple try to develop a rhythm of having intercourse only on 'safe' days in the cycle. Temperature changes or tests for the LH surge help to find when ovulation has occurred, which helps to identify safe days.

Condom – a mechanical method
A thin but strong sheath of rubber or plastic is unrolled over the erect penis before intercourse. A teat at the end catches the semen. The condom is removed after withdrawing the penis from the vagina.

IUD – a mechanical method
Intra-uterine devices are plastic or metal objects, which are placed in the uterus and left there for months or years. It causes irritation in the uterus and prevents implantation of embryos into the uterus lining.

Contraceptive pills – a chemical method
Female contraceptive pills contain hormones, that either prevent ovulation or prevent implantation of embryos by altering the uterus lining. One contraceptive pill is taken orally each day.

ETHICAL ARGUMENTS FOR IVF
- Some childless couples are able to have children.
- Suffering due to genetic disease could be reduced if embryos were screened before being transferred to the uterus.

ETHICAL ARGUMENTS AGAINST IVF
- Inherited forms of infertility might be passed on the children, which means that the suffering the parents endured is repeated in their offspring.
- More embryos are often produced than are needed and the spare embryos are sometimes killed although they are new human lives.
- Embryologists select the embryos that are transferred to the uterus, so humans are deciding whether new individuals survive or die.
- Multiple births, which carry the risk of various health problems for the children, are more likely with IVF than with natural conception.

ETHICAL ARGUMENTS FOR FAMILY PLANNING AND CONTRACEPTION
- Prevents the birth of unwanted and therefore probably unloved children
- Couples who are carriers of genetic disease can avoid the birth of children with the genetic disease.
- Women should have the right to choose whether they become pregnant or not.
- Human impacts on the biosphere will be reduced if fewer human babies are born.

ETHICAL ARGUMENTS AGAINST FAMILY PLANNING AND CONTRACEPTION
- Many religions teach that it is wrong to intervene in a natural process.
- Promiscuity might be encouraged.
- Some methods involve the death of zygotes or young embryos, which are new human lives.

1   Respiration in humans and other mammals generates heat which can be used to keep the body temperature above that of the surroundings.

Many mammals found in the southern hemisphere, including marsupials, vary their body temperature according to a daily cycle. The mouse lemur (*Microcebus myoxinus*) is an example of such a mammal. To investigate this daily cycle, *M. myoxinus* was studied in its native habitat in Madagascar. Data-loggers which recorded body temperature (T_b) over 24-hour periods were implanted in the bodies of several of these mammals. Air temperature (T_a) was recorded at the same time. A typical set of results is shown in the graph below.

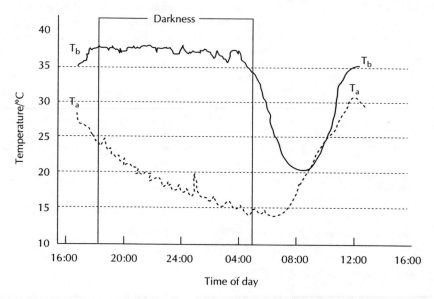

[Source: Cossins and Barnes, Nature (1996), 384, page 582]

a) Using only the data in the graph, state two differences between T_a and T_b during the hours of darkness.   [2]

b) T_b rises from 08:00 to 12:00. Explain briefly how this temperature rise occurs.   [2]

c) Predict, with a reason, whether *M. myoxinus* is active in the hours of daylight or the hours of darkness.   [1]

2   a) (i)   State the function of phagocytic leukocytes.   [1]

    (ii) Outline where in the body phagocytic leukocytes carry out their function.   [2]

    b) Explain briefly the need for small numbers of many types of B-lymphocyte in the body.   [2]

3   The diagram right shows part of the human gas exchange system.

    a) State the name of the parts labelled I and II.   [2]

    b) I and II allow the lungs to be ventilated. Explain briefly the need for ventilation.   [2]

    c) Suggest the name of **one** health problem concerned with gas exchange.   [1]

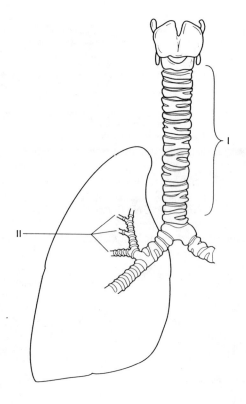

DNA structure and replication

DNA STRUCTURE

At one end of each strand is a phosphate linked to carbon atom 5 of deoxyribose. This is the 5' terminal.

Hydrogen bonds (shown as ■) link the bases. Two bonds form between adenine and thymine and three bonds between guanine and cytosine. Only these pairs can form hydrogen bonds.

Two of the bases in DNA are purines: adenine and guanine. They have two rings in their molecules. Two of the bases in DNA are pyrimidines. Cytosine and thymine are pyrimidines. They have one ring in their molecule. Only a purine plus a pyrimidine will fit in the space between the sugar–phosphate backbones.

Adjacent nucleotides are linked by a bond between the phosphate group of one nucleotide and carbon atom 3 of the other nucleotide.

At one end of each strand is a hydroxyl group attached to carbon atom 3 of deoxyribose. This is therefore the 3' terminal.

adenine   thymine
guanine   cytosine

The two strands have their 3' and 5' terminals at opposite ends – they are anti-parallel. DNA replication can only occur in a 5' ⟶ 3' direction so a different method is needed for the two strands.

THE ROLE OF ENZYMES IN DNA REPLICATION

① The cell produces many free nucleotides for DNA replication. Each has three phosphate groups – they are deoxyribonucleoside triphosphates. Two phosphates are removed during replication to release energy.

③ DNA polymerase III adds nucleotides in a 5' ⟶ 3' direction. On one strand it moves in the same direction as the replication fork, close to helicase. On the other template strand it moves in the opposite direction.

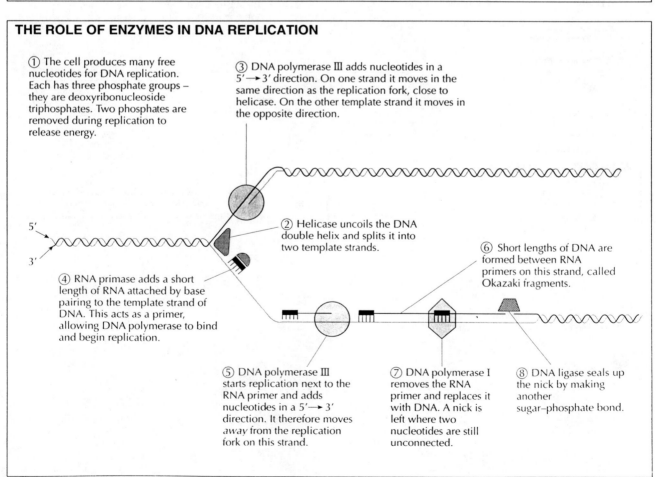

② Helicase uncoils the DNA double helix and splits it into two template strands.

⑥ Short lengths of DNA are formed between RNA primers on this strand, called Okazaki fragments.

④ RNA primase adds a short length of RNA attached by base pairing to the template strand of DNA. This acts as a primer, allowing DNA polymerase to bind and begin replication.

⑤ DNA polymerase III starts replication next to the RNA primer and adds nucleotides in a 5' ⟶ 3' direction. It therefore moves *away* from the replication fork on this strand.

⑦ DNA polymerase I removes the RNA primer and replaces it with DNA. A nick is left where two nucleotides are still unconnected.

⑧ DNA ligase seals up the nick by making another sugar–phosphate bond.

DNA in eukaryotes and prokaryotes

All eukaryotes and prokaryotes use DNA as their genetic material and use the same genetic code, but there are differences in the way that the DNA is used.

- In eukaryotes the DNA is associated with proteins to form **nucleosomes**, whereas in prokaryotes the DNA is naked.
- Replication of DNA begins at special initiation points. Eukaryotes have many of these initiation points along each chromosome. Most prokaryotes have only one point on their DNA molecule where replication is initiated.
- Much of the DNA in eukaryotes consists of repetitive base sequences. These sequences are not genes. They are useful for DNA profiling, but their role in eukaryotes is uncertain. Prokaryotes do not usually have repetitive sequences.
- Many genes in eukaryotes contain **introns**. These are non-coding sequences that are transcribed but not translated. They are found in newly transcribed mRNA but are removed. Mature mRNA does not contain introns. The sequences that are not removed are called exons. Prokaryotes do not usually have introns in their genes.
- Gene expression involves transcription of a gene and translation of the mRNA produced by transcription. In any cell, at a particular time, some genes are being expressed and some genes are not. This is called gene regulation. Each gene in a eukaryote is regulated separately. In prokaryotes some genes are arranged in groups and are regulated together. These groups are called **operons**. The Lac Operon, found in *Escherichia coli* is an example. Figures (below) show how it works.

NUCLEOSOME STRUCTURE

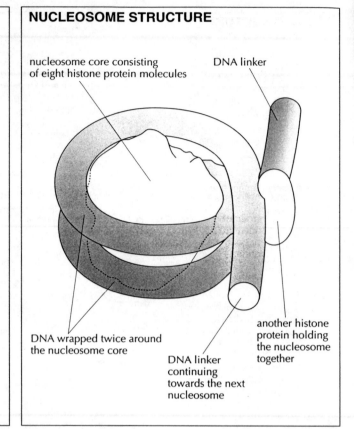

nucleosome core consisting of eight histone protein molecules

DNA linker

DNA wrapped twice around the nucleosome core

DNA linker continuing towards the next nucleosome

another histone protein holding the nucleosome together

THE LAC OPERON

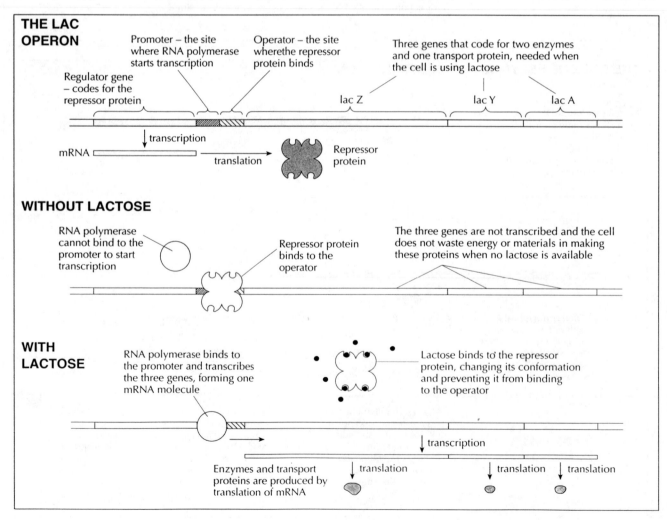

Regulator gene – codes for the repressor protein

Promoter – the site where RNA polymerase starts transcription

Operator – the site wherethe repressor protein binds

Three genes that code for two enzymes and one transport protein, needed when the cell is using lactose

lac Z   lac Y   lac A

transcription

mRNA

translation

Repressor protein

WITHOUT LACTOSE

RNA polymerase cannot bind to the promoter to start transcription

Repressor protein binds to the operator

The three genes are not transcribed and the cell does not waste energy or materials in making these proteins when no lactose is available

WITH LACTOSE

RNA polymerase binds to the promoter and transcribes the three genes, forming one mRNA molecule

Lactose binds to the repressor protein, changing its conformation and preventing it from binding to the operator

transcription

Enzymes and transport proteins are produced by translation of mRNA

translation   translation   translation

Gene expression in eukaryotes

CONTROL OF GENE EXPRESSION IN EUKARYOTES

The process by which a gene has an effect on a cell is called **gene expression**. Although a cell in a multicellular organism contains all of the organism's genes, only some of them will be expressed in that cell. This is the key to controlling the development and differentiation of cells. For example, in humans only ß islet cells in the pancreas express the genes for making insulin.

Gene expression involves several stages:
- Transcription of the gene
- Processing of the mRNA to remove introns (post-transcriptional modification)
- Translation of the mRNA to produce a protein
- Modification of the protein inside the endoplasmic reticulum or inside the Golgi apparatus (post-translational modification)

At any of these stages, the expression of genes can be regulated. Regulation of transcription is the most important stage.

REGULATION OF TRANSCRIPTION

The regulation of transcription in eukaryotic cells is a complex process.
- Genes are only transcribed if RNA polymerase binds to a region of DNA close to the start of the gene called the **promoter**.
- Most genes have several sequences of bases in their promoter that encourage binding of RNA polymerase. There is a wide variety of these sequences.
- Some base sequences always encourage binding, to allow continuous expression of genes.
- Other base sequences are sites where a regulatory protein can bind to the promoter. RNA polymerase only binds if the regulatory protein is present. Many different regulatory proteins are involved in the regulation of transcription.
- Some regulatory proteins only become active if a steroid hormone or other chemical messenger binds to them.

ROLE OF THE PROMOTER AND TERMINATOR IN TRANSCRIPTION

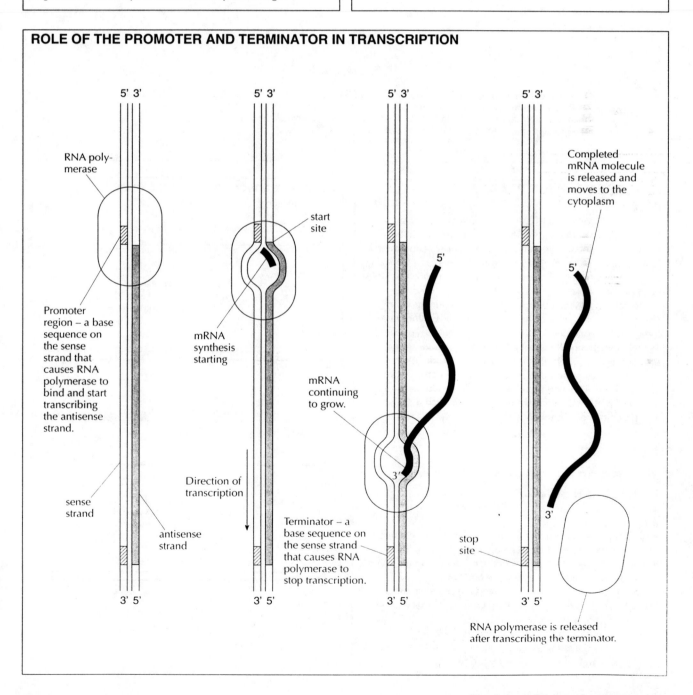

RNA polymerase

Promoter region – a base sequence on the sense strand that causes RNA polymerase to bind and start transcribing the antisense strand.

sense strand

antisense strand

Direction of transcription

start site

mRNA synthesis starting

mRNA continuing to grow.

Terminator – a base sequence on the sense strand that causes RNA polymerase to stop transcription.

stop site

Completed mRNA molecule is released and moves to the cytoplasm

RNA polymerase is released after transcribing the terminator.

Transcription and reverse transcription

RNA POLYMERASE AND TRANSCRIPTION

DNA is split into two strands by RNA polymerase. One of these strands forms the template for transcription. The base sequence of the mRNA is complementary to it. The other strand has the same base sequence as the mRNA (except for T instead of U) and is therefore called the **sense strand**. The strand that forms the template and is transcribed is called the **antisense strand**.

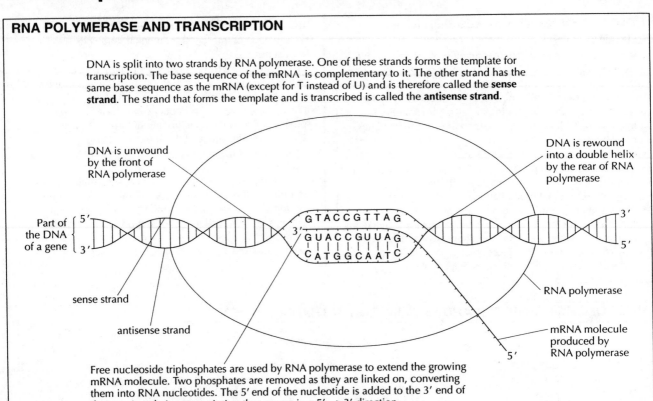

DNA is unwound by the front of RNA polymerase

DNA is rewound into a double helix by the rear of RNA polymerase

Part of the DNA of a gene

```
5'                                    3'
         GTACCGTTAG
       3'
         GUACCGUUAG
         |||||||||
         CATGGCAATC
3'                                    5'
```

sense strand

antisense strand

RNA polymerase

mRNA molecule produced by RNA polymerase

5'

Free nucleoside triphosphates are used by RNA polymerase to extend the growing mRNA molecule. Two phosphates are removed as they are linked on, converting them into RNA nucleotides. The 5' end of the nucleotide is added to the 3' end of the growing chain– transcription thus moves in a 5'⟶3' direction.

REVERSE TRANSCRIPTASE AND ITS ROLE IN HIV

HIV and other retroviruses contain an enzyme that catalyses the production of DNA from RNA. This enzyme is called reverse transcriptase. The genome of retroviruses consists of RNA. When a retrovirus enters a host cell it makes a DNA copy of the genome using reverse transcriptase (see right). The DNA copy becomes inserted into the host cell's chromosomes. In the case of HIV, the viral DNA can remain inactive in chromosomes of lymphocytes for long periods. When the lymphocyte replicates its DNA, the viral DNA is replicated as well. Eventually the viral DNA starts to be transcribed, producing viral mRNA that is translated to produce viral proteins. Transcription also produces copies of the entire viral genome. The viral proteins and RNA assemble to form new HIV particles, which are released from the lymphocyte. The lymphocyte subsequently dies.

THE USE OF REVERSE TRANSCRIPTASE IN MOLECULAR BIOLOGY

Reverse transcriptase is used to obtain copies of genes for use in gene transfer.
• Cells that are transcribing the required gene are obtained and mRNA copies of the gene are extracted from them.
• Single stranded DNA copies of the mRNA are made using reverse transcriptase. This is called cDNA.
• DNA polymerase is used to convert the single-stranded DNA into double stranded DNA – the genes that can be transferred into another organisms.

The genes produced do not have introns, so if they are transferred to bacteria, which do not edit out introns, the correct protein will nonetheless be produced.

Reverse transcription in cells infected by retroviruses

3' ————————————————————— 5'

↓ Reverse transcriptase synthesizes a DNA strand complementary to the RNA

3' ▭▭▭▭▭▭▭▭▭▭▭▭▭▭▭▭▭▭▭ 5'
5' ▭▭▭▭▭▭▭▭▭▭▭▭▭▭▭▭▭▭▭ 3'

↓ Reverse transcriptase breaks down the strand of RNA

5' ————————————————————— 3'

↓ Reverse transcriptase synthesizes a DNA strand complementary to the other DNA strand

3' ▭▭▭▭▭▭▭▭▭▭▭▭▭▭▭▭▭▭▭ 5'
5' ▭▭▭▭▭▭▭▭▭▭▭▭▭▭▭▭▭▭▭ 3'

The RNA of the retrovirus is the sense strand, therefore the first DNA strand is the antisense strand and the second one synthesized is the sense strand

Translating the genetic code

Messenger RNA carries the information needed for making polypeptides out from the nucleus to the cytoplasm of eukaryotic cells. The information is in a coded form, which is decoded during translation. Ribosomes, tRNA molecules and tRNA activating enzymes are needed to carry out this decoding.

THE STRUCTURE OF RIBOSOMES
Ribosomes have a complex structure, with the following principal features.
- Proteins and ribosomal RNA molecules both form part of the structure.
- There are two subunits, one large and one small.
- There are binding sites for tRNA on the surface of the ribosome. Two tRNA molecules can bind at the same time to the ribosome.
- There is a binding site for mRNA on the surface of the ribosome.
The figure (right) shows the shape of a ribosome in outline and two tRNA binding sites.

Ribosome shape

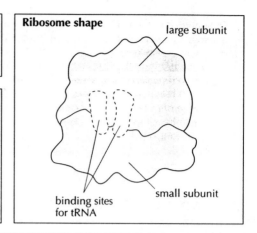

tRNA AND tRNA ACTIVATING ENZYMES
Transfer RNA has a vital role in translating the genetic code. There are many different types of tRNA in a cell.
All tRNA molecules have:
- a triplet of bases called the anticodon, in a loop of seven nucleotides
- two other loops
- the base sequence CCA at the 3′ terminal, which forms a site for attaching an amino acid
- sections that become double stranded by base pairing.
These features allow all tRNA molecules to bind to the binding sites on the ribosome and to mRNA. The base sequence of tRNA molecules varies and this causes some variable features in its structure.
- An extra small loop is sometimes present
- The base paired sections are sometimes helical.
The variable features give each type of tRNA a distinctive three-dimensional shape and distinctive chemical properties. This allows the correct amino acid to be attached to the 3′ terminal by an enzyme called a **tRNA activating enzyme**. There are 20 different tRNA activating enzymes – one for each of the 20 different amino acids. Each of these enzymes attaches one particular amino acid to all of the tRNA molecules that have an anticodon corresponding to that amino acid. The tRNA activating enzymes recognize these tRNA molecules by their shape and chemical properties. The figure (right) is a two-dimensional view of the structure of tRNA molecules and the figure (lower right) shows an example of the three-dimensional structure formed when a tRNA molecule folds up.
Energy from ATP is needed for the attachment of amino acids. A high-energy bond is created between the amino acid and the tRNA. Energy from this bond is later used to link the amino acid to the growing polypeptide chain during translation.

tRNA structure

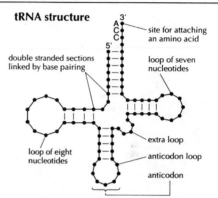

Three-dimensional view of tRNA

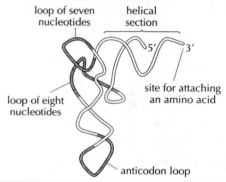

THE GENETIC CODE
Although the genetic code appears completely random at first, there are some rules, which are always or nearly always followed (right).

First base of codon (5′ end)	Second base of codon on messenger RNA				Third base of codon (3′ end)
	U	C	A	G	
U	Phenylalanine	Serine	Tyrosine	Cysteine	U
	Phenylalanine	Serine	Tyrosine	Cysteine	C
	Leucine	Serine	STOP	STOP	A
	Leucine	Serine	STOP	Tryptophan	G
C	Leucine	Proline	Histidine	Arginine	U
	Leucine	Proline	Histidine ·	Arginine	C
	Leucine	Proline	Glutamine	Arginine	A
	Leucine	Proline	Glutamine	Arginine	G
A	Isoleucine	Threonine	Asparagine	Serine	U
	Isoleucine	Threonine	Asparagine	Serine	C
	Isoleucine	Threonine	Lysine	Arginine	A
	Methionine / START	Threonine	Lysine	Arginine	G
G	Valine	Alanine	Aspartic acid	Glycine	U
	Valine	Alanine	Aspartic acid	Glycine	C
	Valine	Alanine	Glutamic acid	Glycine	A
	Valine	Alanine	Glutamic acid	Glycine	G

Polysomes and polypeptide elongation

The figure (right) is an electron micrograph showing groups of ribosomes called **polysomes** (or polyribosomes). A polysome is a group of ribosomes moving along the same mRNA, as they simultaneously translate it. Each ribosome follows a series of steps that is repeated many times to translate the mRNA. One amino acid is added to the elongating polypeptide each time the cycle of steps is repeated (see below). As ribosomes move along the mRNA towards the 3' end, the polypeptide is gradually elongated.

(× 180 000)

POLYPEPTIDE ELONGATION

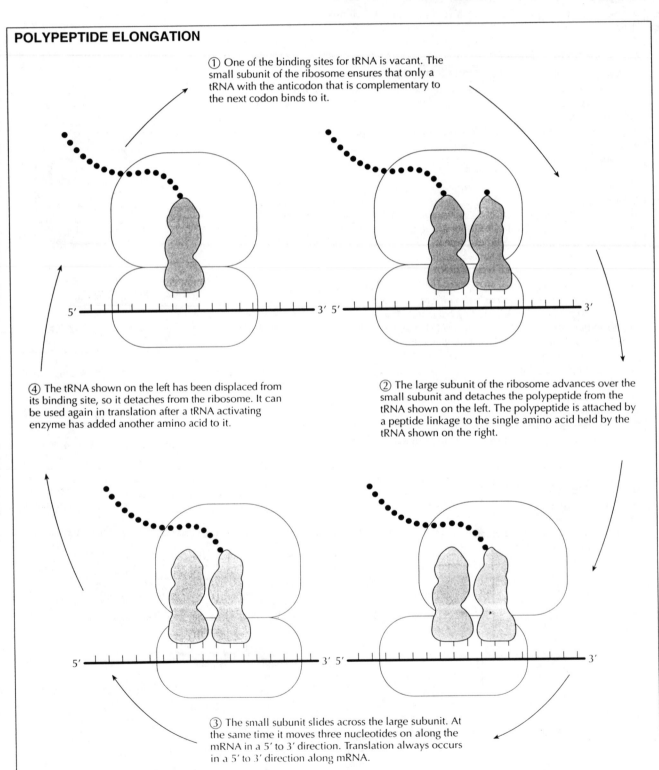

① One of the binding sites for tRNA is vacant. The small subunit of the ribosome ensures that only a tRNA with the anticodon that is complementary to the next codon binds to it.

② The large subunit of the ribosome advances over the small subunit and detaches the polypeptide from the tRNA shown on the left. The polypeptide is attached by a peptide linkage to the single amino acid held by the tRNA shown on the right.

③ The small subunit slides across the large subunit. At the same time it moves three nucleotides on along the mRNA in a 5' to 3' direction. Translation always occurs in a 5' to 3' direction along mRNA.

④ The tRNA shown on the left has been displaced from its binding site, so it detaches from the ribosome. It can be used again in translation after a tRNA activating enzyme has added another amino acid to it.

Starting and stopping translation

Special steps are needed to start the process of translation and to stop it. These steps are called **initiation** and **termination**. The three stages of translation are thus initiation, elongation and termination.

INITIATION OF TRANSLATION

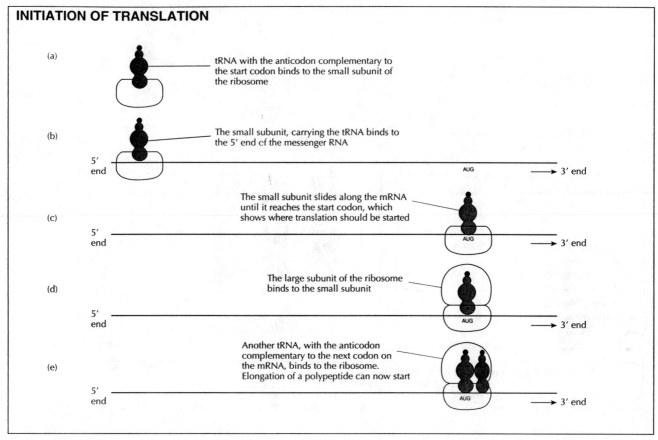

(a) tRNA with the anticodon complementary to the start codon binds to the small subunit of the ribosome

(b) The small subunit, carrying the tRNA binds to the 5′ end of the messenger RNA

5′ end — AUG — 3′ end

(c) The small subunit slides along the mRNA until it reaches the start codon, which shows where translation should be started

5′ end — AUG — 3′ end

(d) The large subunit of the ribosome binds to the small subunit

5′ end — AUG — 3′ end

(e) Another tRNA, with the anticodon complementary to the next codon on the mRNA, binds to the ribosome. Elongation of a polypeptide can now start

5′ end — AUG — 3′ end

TERMINATION OF TRANSLATION

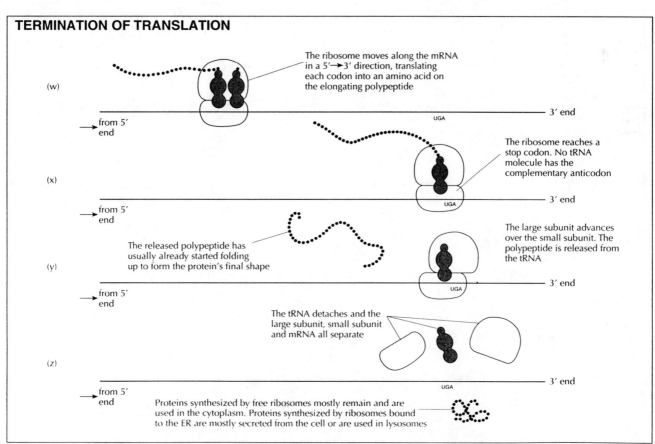

(w) The ribosome moves along the mRNA in a 5′→3′ direction, translating each codon into an amino acid on the elongating polypeptide

from 5′ end — UGA — 3′ end

(x) The ribosome reaches a stop codon. No tRNA molecule has the complementary anticodon

from 5′ end — UGA — 3′ end

(y) The released polypeptide has usually already started folding up to form the protein's final shape

The large subunit advances over the small subunit. The polypeptide is released from the tRNA

from 5′ end — UGA — 3′ end

(z) The tRNA detaches and the large subunit, small subunit and mRNA all separate

from 5′ end — UGA — 3′ end

Proteins synthesized by free ribosomes mostly remain and are used in the cytoplasm. Proteins synthesized by ribosomes bound to the ER are mostly secreted from the cell or are used in lysosomes

Nucleic acids and proteins 65

Intramolecular bonding in proteins

Polypeptides have a main chain consisting of a repeating sequence of covalently bonded carbon and nitrogen atoms: N – C – C – N – C – C, and so on. Each nitrogen atom has a hydrogen atom bonded to it (N – H). Every second carbon atom has an oxygen atom bonded to it (C = O).

$$
\begin{array}{ccccccc}
 & & H & & O \\
 & & | & & \| \\
N & - C & - C & - N & - C & - C \\
 & | & & | \\
 & H & & O
\end{array}
$$

Hydrogen bonds can form between N – H and C = O groups, if they are brought close together. For example, if sections of polypeptide run parallel, hydrogen bonds can form between them. The structure that develops is called a ß-pleated sheet. If the polypeptide is wound into a helix, hydrogen bonds can form between adjacent turns of the helix. The structure that develops is called an α-helix. Because the groups forming hydrogen bonds are regularly spaced, secondary structures always have the same dimensions.

In addition to the hydrogen bonding in ß-pleated sheets and α-helices, there are many other types of bonding. Most of these involve the R groups of the amino acids. The figure (below) shows some of these bonds.

β-PLEATED SHEET

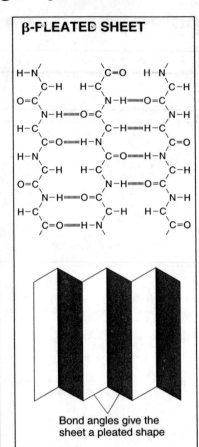

Bond angles give the sheet a pleated shape

α -HELIX

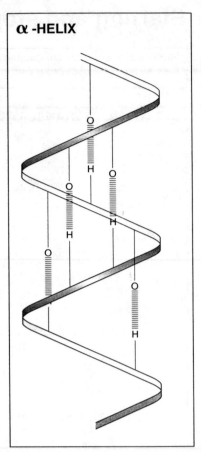

Types of intramolecular bond in proteins

Ionic bonds can form between positively and negatively charged R groups

Acidic amino acids have R groups that can lose an H⁺ ion and so become negatively charged

Basic amino acids have R groups that can accept an H⁺ ion and so become positively charged

Disulfide bridges, which are strong covalent bonds can form between pairs of cysteines

Hydrophobic interactions, which are weak bonds, can form between R groups that are non-polar including all those projecting inwards here

Hydrogen bonds can form between some R groups

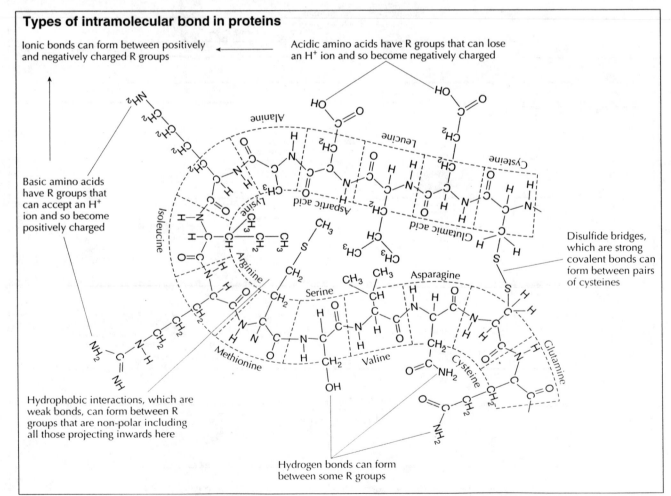

Protein structure

Proteins have a complex structure, which can be explained by defining four levels of structure, primary, secondary, tertiary and quaternary structure.

PRIMARY STRUCTURE

Primary structure is the number and sequence of amino acids in a polypeptide. Most polypeptides consist of between 50 and 1000 amino acids. The primary structure is determined by the base sequence of the gene that codes for the polypeptide. The figure (below) shows the primary structure of ß-endorphin, a protein consisting of a single polypeptide of 31 amino acids that acts as a neurotransmitter in the brain.

Primary structure of ß-endorphin

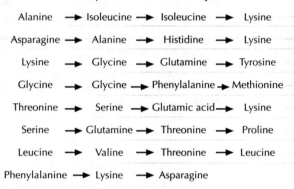

Alanine → Isoleucine → Isoleucine → Lysine

Asparagine → Alanine → Histidine → Lysine

Lysine → Glycine → Glutamine → Tyrosine

Glycine → Glycine → Phenylalanine → Methionine

Threonine → Serine → Glutamic acid → Lysine

Serine → Glutamine → Threonine → Proline

Leucine → Valine → Threonine → Leucine

Phenylalanine → Lysine → Asparagine

TERTIARY STRUCTURE

Tertiary structure is the three-dimensional conformation of a polypeptide. It is formed when the polypeptide folds up after being produced by translation. The conformation is stabilized by intramolecular bonds that form between amino acids in the polypeptide, especially between their R groups. These include ionic bonds, hydrogen bonds, hydrophobic interactions and disulfide bridges. The intramolecular bonds are often formed between amino acids that are widely separated in the primary structure of the polypeptide, but which are brought together during the folding process. The figure below shows the tertiary structure of lysozyme using the sausage model.

Sausage model of lysozyme

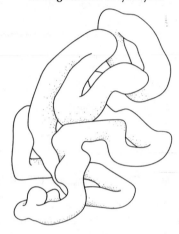

SECONDARY STRUCTURE

Secondary structures are regular repeating structures, including ß-pleated sheets and α-helices stabilized by hydrogen bonds between groups in the main chain of the polypeptide. In many proteins, parts of the polypeptide form secondary structures and other parts do not. In some proteins secondary structures do not form at all. In a few proteins almost all of the polypeptide forms secondary structures. For example almost all of myosin molecules is α-helix and almost all of fibroin (silk protein) is β-pleated sheet.
The figure (below) shows the position of secondary structures in lysozyme, using the ribbon model. Sections of α-helix are represented by helical ribbons and sections of β-pleated sheet are represented by arrows.

Ribbon model of lysozyme

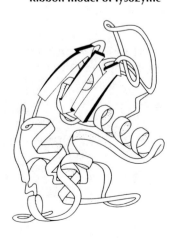

QUATERNARY STRUCTURE

Quaternary structure is the linking together of two or more polypeptides to form a single protein. For example, insulin consists of two polypeptides linked together, collagen consists of three polypeptides and hemoglobin consists of four. In some cases proteins also contain a non-polypeptide structure called a **prosthetic group**. Each of the four polypeptides in hemoglobin is linked to a heme group, which is not made of amino acids. Proteins with a prosthetic group are called **conjugated proteins**.
The figure (below) shows the quaternary structure of hemoglobin.

Sausage model of hemoglobin

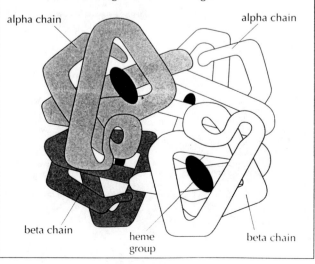

alpha chain

alpha chain

beta chain

heme group

beta chain

Protein functions

Proteins have a huge range of functions in living organisms. Some proteins are located in membranes. The functions of membrane proteins are listed on page 7. Six of the functions of non-membrane proteins are listed (below). Proteins can also be used as food stores, for example, casein in milk, as pigments, for example, opsin in the retina and as toxins as in some snake venom.

Function	Example	Details	Shape
Enzymes	Catalase	The function of catalase is to catalyse the conversion of hydrogen peroxide, a toxic waste product of metabolism, into water and oxygen	Globular
Structural	Collagen	The function of collagen is to strengthen bone, tendon and skin. These tissues all produce tough collagen fibres in the spaces between their cells	Fibrous
Transport	Hemoglobin	The function of hemoglobin is to bind oxygen in the lungs and to transport it to respiring tissues	Globular
Movement	Myosin	The function of myosin (with another protein called actin) is to cause contraction in muscle fibres and as a result cause movement in animals	Fibrous
Hormones	Insulin	The function of insulin is to bind to receptors in the plasma membranes of target cells and stimulate them to remove glucose from the blood	Globular
Defence	Immunoglobulin	The function of immunoglobulin is to act as antibodies. Part of the immunoglobulin molecule can be varied, so that an almost endless variety of different antibodies can be produced	Globular

FIBROUS AND GLOBULAR PROTEINS

The table (above) indicated the shape of each of the named proteins. Proteins can be divided into two types according to their shape, fibrous or globular. Fibrous proteins have a long and narrow shape. They are mostly insoluble in water. Globular proteins have a rounded shape. They are mostly soluble in water.

POLAR AND NON-POLAR AMINO ACIDS IN PROTEINS

Amino acids can be divided into two types according to the chemical characteristics of their R group. Polar amino acids have hydrophilic R groups and non-polar amino acids have hydrophobic R groups. The distribution of polar and non-polar amino acids in a protein molecule influence where the protein is located in a cell and what function it can carry out. The figures (below) show examples of this.

Superoxide dismutase – an enzyme found in all aerobic organisms

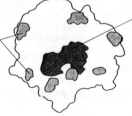

A ring of amino acids with negatively charged R-groups repel the negatively charged superoxide ions and help to direct them to the active site.

The active site is a cleft containing amino acids with positively charged R-groups which attract the negatively charged superoxide ions that are the substrate of the enzyme.

Positions of proteins in and out of membranes

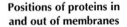
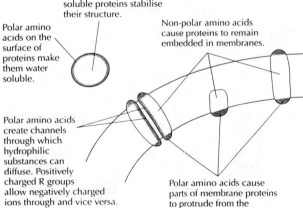

Non-polar amino acids in the centre of water-soluble proteins stabilise their structure.

Polar amino acids on the surface of proteins make them water soluble.

Non-polar amino acids cause proteins to remain embedded in membranes.

Polar amino acids create channels through which hydrophilic substances can diffuse. Positively charged R groups allow negatively charged ions through and vice versa.

Polar amino acids cause parts of membrane proteins to protrude from the membrane. Transmembrane proteins have two such regions.

Lipase – an enzyme that works in the small intestine

polar region

The active site is a cleft containing amino acids with non-polar R-groups which bind non-polar triglycerides.

Part of the enzyme molecule acts as a hinged lid which can cover the active site when not in use, hiding the non-polar R-groups.

non-polar region

A protein cofactor binds to the enzyme, and helps lipase to bind to the surface of lipid droplets because it has non-polar R-groups on its surface.

Enzymes and activation energy

ENERGY CHANGES DURING CHEMICAL REACTIONS

During chemical reactions, reactants are converted into products. Before a molecule of the reactant can take part in the reaction it has to gain some energy. This is called the **activation energy** of the reaction. The energy is needed to break bonds within the reactant. Later, during the progress of the reaction, energy is given out as new bonds are made. In **exergonic** reactions this amount of energy is greater than the activation energy. In **endergonic** reactions it is less. Enzymes reduce the activation energy of the reactions that they catalyse and therefore make it easier for reactions to occur.

The chemical environment provided by the active site for the substrate causes changes within the substrate molecule, which weakens its bonds. The substrate is changed into a transition state, which is different from the transition state during the reaction when an enzyme is not involved. The transition state achieved during binding to the active site has less energy and this is how enzymes are able to reduce the activation energy of reactions.

THE INDUCED FIT MODEL OF ENZYME ACTIVITY

Biochemists have investigated many enzymes and found that the lock and key model does not fully explain the binding of the substrate to the active site. Until the substrate binds, the active site does not fit the substrate precisely. As the substrate approaches the active site and binds to it, the shape of the active site changes and only then does it fit the substrate. The substrate induces the active site to change, weakening bonds in the substrate during the process and thus reducing the activation energy. The figure (below) shows the induced fit model of enzyme activity.

Some enzymes can have quite broad specificity, for example some proteases. The induced fit model explains this better than the lock and key model – if the shape of an active site alters when substrates bind, several different but similar substrates could easily bind successfully to it.

Energy changes during exergonic reactions

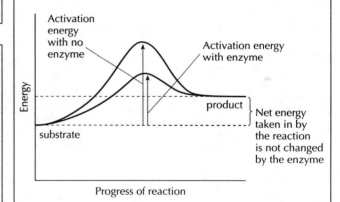

Energy changes during endergonic reactions

In living organisms, endergonic reactions are coupled with exergonic reactions, for example hydrolysis of ATP. The endergonic reaction can then occur more easily.

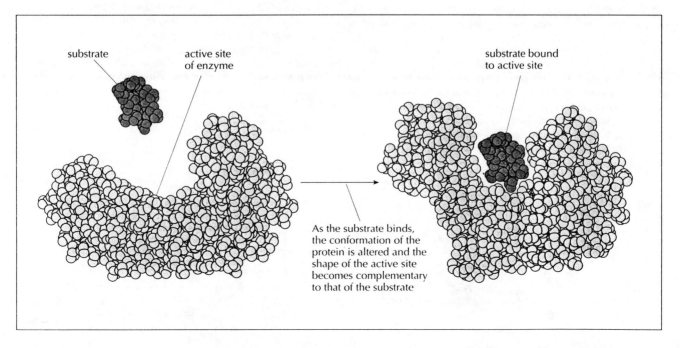

Enzyme inhibition

Some chemical substances reduce the activity of enzymes or even prevent it completely. These substances are called enzyme inhibitors. Some enzyme inhibitors are **competitive** and some are **non-competitive**. Figures below are a comparison of these types of inhibitor, with an example of each.

Competitive inhibition

The substrate and inhibitor are chemically very similar

The inhibitor binds to the active site of the enzyme

While the inhibitor occupies the active site, it prevents the substrate from binding and so the activity of the enzyme is prevented until the inhibitor dissociates

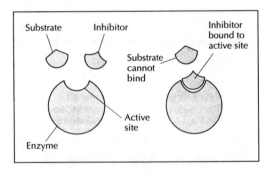

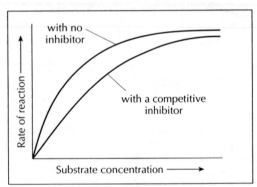

With a fixed low concentration of inhibitor, increases in the substrate concentration gradually reduce the effect of the inhibitor.
The inhibitor and substrate compete for the active site. When the substrate binds to the active site, the inhibitor cannot bind, so the proportion of enzyme molecules that are inhibited becomes less and less. When there are many more substrate molecules than inhibitor molecules, the substrate always wins the competition and binds to the active site. The same maximum enzyme activity rate is then reached as when there is no inhibitor.

EXAMPLE

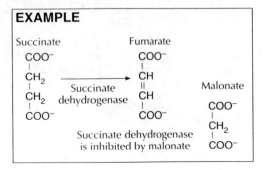

Non-competitive inhibition

The substrate and active site are not similar

The inhibitor binds to the enzyme at a different site from the active site

The inhibitor changes the conformation of the enzyme. The substrate may still be able to bind, but the active site does not catalyse the reaction, or catalyses it at a slower rate

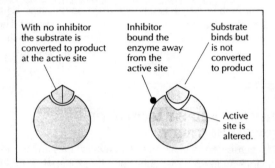

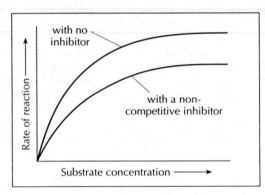

With a fixed low concentration of inhibitor, increases in substrate concentration increase enzyme activity. However, the substrate and inhibitor are not competing for the same site. The substrate cannot prevent the binding of the inhibitor, even at very high substrate concentrations. Some of the enzyme molecules therefore remain inhibited and the maximum enzyme activity rate reached is lower than when there is no inhibitor

EXAMPLE

Metal ions including copper (Cu^{2+}), mercury (Hg^{2+}) and silver (Ag^+) act as non-competitive inhibitors of many enzymes by binding reversibly to the –SH groups of cysteine – the amino acid that forms disulfide bridges. This disrupts the structure of the enzyme:

$$-SH + Ag^+ \longrightarrow -S - Ag + H^+$$

Controlling metabolic pathways

METABOLIC PATHWAYS

Metabolic pathways have these features:
- They consist of many chemical reactions that are carried out in a particular sequence.
- An enzyme catalyses each reaction.
- All the reactions occur inside cells.
- Some pathways build up organic compounds (anabolic pathways) and some break them down (catabolic pathways).
- Some metabolic pathways consist of chains of reactions. Glycolysis is an example of a chain of reactions–a chain of ten enzyme-controlled reactions converts glucose into pyruvate.
- Some metabolic pathways consist of cycles of reactions, where a substrate of the cycle is continually regenerated by the cycle. The Krebs cycle is an example.

The figure (opposite) shows the general pattern of reactions in a chain and a cycle.

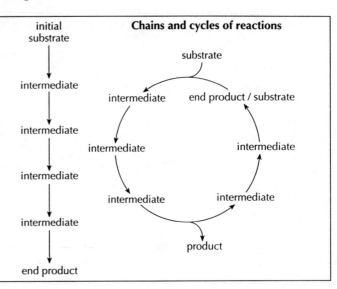

Chains and cycles of reactions

ALLOSTERY AND THE CONTROL OF METABOLIC PATHWAYS

In many metabolic pathways, the product of the last reaction in the pathway inhibits the enzyme that catalyses the first reaction. This is called **end-product inhibition**. The enzyme that is inhibited by the end products is an example of an **allosteric** enzyme. Allosteric enzymes have two non-overlapping binding sites. One of these is the active site. The other is the allosteric site.

In this case the allosteric site is a binding site for the end product. When it binds, the structure of the enzyme is altered so that the substrate is less likely to bind to the active site. This is how the end-product acts as an inhibitor. Binding of the inhibitor is reversible and if it detaches, the enzyme returns to its original conformation, so the active site can bind the substrate easily again (right).

The advantage of this method of controlling metabolic pathways is that if there is an excess of the end-product the whole pathway is switched off and intermediates do not build up. Conversely, as the level of the end-product falls, more and more of the enzymes that catalyse the first reaction will start to work and the whole pathway will become activated. End product inhibition is an example of negative feedback (see example below).

End-product inhibition

Substrate binds to the active site and is converted to the product.

Substrate could bind to the active site as the allosteric site is empty.

Substrate is not likely to bind to the active site as the inhibitor has bound to the allosteric site.

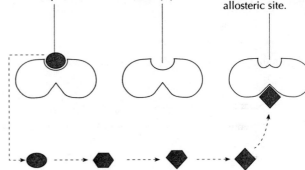

The substrate of the first enzyme in the metabolic pathway is converted by the pathway into an inhibitor of the enzyme.

An example of end product inhibition

isoleucine is the end product of the pathway and inhibits threonine dehydratase which catalyses the first step

1 An enzyme experiment was conducted at three different temperatures. The graph shows the amount of substrate remaining each minute after the enzyme was added to the substrate. W shows the results obtained at a temperature of 40 °C.

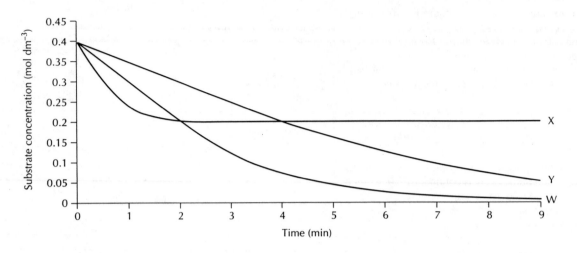

a) (i) Explain whether the temperature used for X was higher or lower than 40 °C. [3]

(ii) Estimate the temperature that was used for Y. [2]

b) Draw a curve on the graph to show the expected results of repeating the experiment at 40 °C with

(i) a fixed low concentration of non-competitive inhibitor. [1]

(ii) a fixed low concentration of competitive inhibitor. [2]

2 Reverse transcriptase is an enzyme found only in cells infected by certain viruses.

a) Outline the process catalysed by this enzyme. [2]

b) (i) State the name of the group of viruses that contain the gene for this enzyme. [1]

(ii) State one example of a virus from this group. [1]

c) Explain briefly why the enzyme is a useful tool for molecular biologists. [3]

3 The diagram below represents the structure of lysozyme, a protein consisting of a single polypeptide, found in egg white.

a) State the name given to the shape of this type of protein. [1]

b) State what is meant by the primary structure of a protein. [1]

c) In the regions labelled X and Y two different types of secondary structure are found.

(i) Identify each type of secondary structure: [2]

(ii) State the type of bonding that is used to stabilize these structures. [1]

d) Explain the importance of the tertiary structure of this protein to its function. [2]

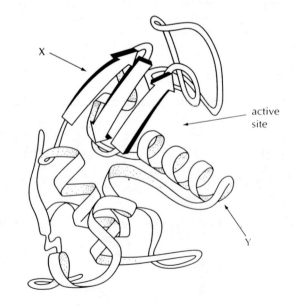

Glycolysis

INTRODUCING GLYCOLYSIS

Cell respiration involves the production of ATP using energy released by the oxidation of glucose, fat or other substrates. If glucose is the substrate, the first stage of cell respiration is a metabolic pathway called **glycolysis**. The pathway is catalysed by enzymes in the cytoplasm. Glucose is partially oxidized in the pathway and a small amount of ATP is produced. This partial oxidation is achieved without the use of oxygen, so glycolysis can form part of both aerobic and anaerobic respiration.

OXIDATION AND REDUCTION IN CELL RESPIRATION

Cell respiration involves many oxidation and reduction reactions. The figure (top right) compares the ways in which chemical substances can be oxidized and reduced. Hydrogen carriers accept hydrogen atoms removed from substrates in cell respiration. The most commonly used hydrogen carrier is NAD^+ (nicotinamide adenine dinucleotide). Hydrogen atoms consist of one proton and one electron. When two hydrogen atoms are removed from a respiratory substrate, NAD^+ accepts the electrons from both atoms and the proton from one of them.

$$NAD^+ + 2H \longrightarrow NADH + H^+$$

The figure (right) shows equations for some of the chemical changes that are part of cell respiration. It is possible to use the information in the figure (top right) to deduce whether each of them is an oxidation, a reduction or both.

CONVERTING GLUCOSE TO PYRUVATE IN GLYCOLYSIS

There are four main stages in glycolysis.
1. Two phosphate groups are added to a molecule of glucose to form hexose biphosphate. Adding a phosphate group is called **phosphorylation**. Two molecules of ATP provide the phosphate groups. The energy level of the hexose is raised by phosphorylation and this makes the subsequent reactions possible.

2. The hexose biphosphate is split to form two molecules of triose phosphate. Splitting molecules is called **lysis**.

3. Two atoms of hydrogen are removed from each triose phosphate molecule. This is an **oxidation**. The energy released by this oxidation is used to link on another phosphate group, producing a 3-carbon compound carrying two phosphate groups. NAD^+ is the hydrogen carrier that accepts the hydrogen atoms.

4. Pyruvate is formed by removing the two phosphate groups and by passing them to ADP. This results in **ATP formation**.

The figure (right) shows the main stages of glycolysis

SUMMARY OF GLYCOLYSIS
• One glucose is converted into two pyruvates.
• Two ATP molecules are used per glucose but four are produced so there is a net yield of two ATP molecules.
• Two NAD^+ are converted into two $NADH + H^+$

Comparison of oxidation and reduction

Oxidation reactions	Reduction reactions
Addition of oxygen atoms to a substance.	Removal of oxygen atoms from a substance.
Removal of hydrogen atoms from a substance.	Addition of hydrogen atoms to a substance.
Loss of electrons from a substance.	Addition of electrons to a substance.

Examples of oxidations and reductions in cell respiration

$$Fe^{3+} + electron \longrightarrow Fe^{2+}$$

$$Fe^{2+} \longrightarrow Fe^{3+} + electron$$

$$succinate + FAD \longrightarrow fumarate + FADH_2$$

$$malate + NAD^+ \longrightarrow oxaloacetate + NADH + H^+$$

$$pyruvate + NADH + H^+ \longrightarrow lactate + NAD^+$$

Stages of glycolysis

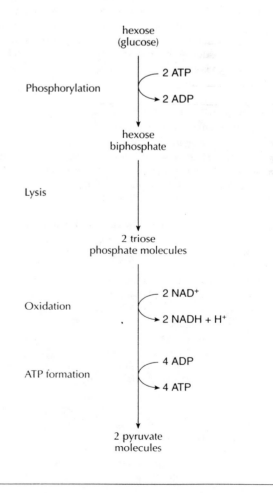

hexose (glucose)

Phosphorylation — 2 ATP → 2 ADP

hexose biphosphate

Lysis

2 triose phosphate molecules

Oxidation — 2 NAD^+ → 2 $NADH + H^+$

ATP formation — 4 ADP → 4 ATP

2 pyruvate molecules

Krebs cycle

Enzymes in the matrix of the mitochondrion catalyse a cycle
of reactions called the **Krebs cycle**. These reactions can only
occur if oxygen is available and so are part of aerobic, but not
anaerobic cell respiration.

THE CENTRAL ROLE OF ACETYL COA IN METABOLISM

Acetyl groups (CH_3CO) are the substrate used in the Krebs
cycle. A carrier called CoA (Coenzyme A) accepts acetyl
groups produced in metabolism and brings them for use in the
cycle.

$$acetyl\ group\ +\ CoA \longrightarrow acetyl\ CoA$$

Acetyl CoA is formed in both carbohydrate and fat metabolism.
- Carbohydrates are converted into pyruvate and the pyruvate
  is converted to acetyl CoA by a reaction that is often called
  the link reaction, as it links glycolysis and the Krebs cycle.
- Fats are broken down into fatty acids and glycerol. The
  hydrocarbon tails of the fatty acids are then broken down
  into two-carbon fragments and oxidized to form acetyl CoA.

Acetyl CoA is therefore the connection between the
metabolism of carbohydrates and fats and the Krebs cycle is
used whether glucose or fats are the substrate for respiration.

Summary of metabolic pathways involving acetyl CoA

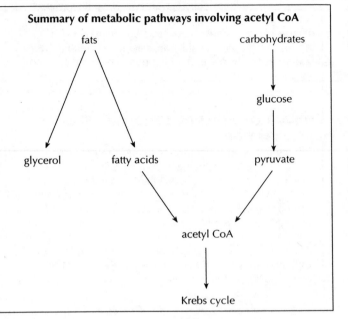

THE LINK REACTION

Pyruvate from glycolysis is absorbed by the mitochondrion.
Enzymes in the matrix of the mitochondrion remove hydrogen
and carbon dioxide from the pyruvate. The hydrogen is
accepted by NAD^+. Removal of hydrogen is oxidation.
Removal of carbon dioxide is decarboxylation. The whole
conversion is therefore **oxidative decarboxylation**. The
product of oxidative decarboxylation of pyruvate is an acetyl
group, which is accepted by CoA (right).

Summary of the link reaction

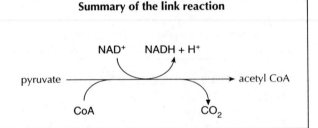

THE KREBS CYCLE

In the first reaction of the cycle an acetyl group is
transferred from acetyl CoA to a four-carbon
compound (oxaloacetate) to form a six-carbon
compound (citrate). Citrate is converted back into
oxaloacetate in the other reactions of the cycle.
Three types of reaction are involved.
- Carbon dioxide is removed in two of the
  reactions. These reactions are **decarboxylations**.
  The carbon dioxide is a waste product and is
  excreted together with the carbon dioxide from
  the link reaction.
- Hydrogen is removed in four of the reactions.
  These reactions are **oxidations**. In three of the
  oxidations the hydrogen is accepted by NAD^+.
  In the other oxidation FAD accepts it. These
  oxidation reactions release energy, much of
  which is stored by the carriers when they
  accept hydrogen. This energy is later released
  by the electron transport chain and used to
  make ATP.
- ATP is produced directly in one of the
  reactions.
  This reaction is **substrate-level
  phosphorylation**.
The figure (right) is a summary of the Krebs cycle.

Summary of the Krebs cycle

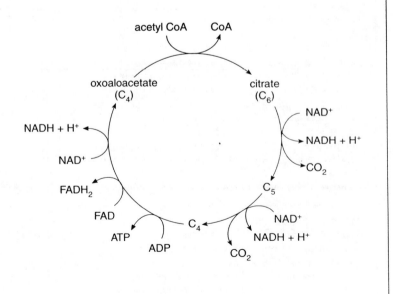

Oxidative phosphorylation

THE ELECTRON TRANSPORT CHAIN

The electron transport chain is a series of electron carriers, located in the inner membrane of the mitochondrion. NADH supplies two electrons to the first carrier in the chain. The electrons come from oxidation reactions in earlier stages of cell respiration. The two electrons pass along the chain of carriers because they give up energy each time they pass from one carrier to the next. At three points along the chain enough energy is given up for ATP to be made by ATP synthase. As this ATP production relies on energy released by oxidation it is called **oxidative phosphorylation**. ATP synthase is also located in the inner mitochondrial membrane. $FADH_2$ also feeds electrons into the electron transport chain, but at a slightly later stage than NADH and at only two stages is sufficient energy released for ATP production by electrons from $FADH_2$.

THE ROLE OF OXYGEN

At the end of the electron transport chain the electrons are given to oxygen. At the same time oxygen accepts hydrogen ions, to form water. This happens in the matrix, on the surface of the inner membrane. This is the only stage at which oxygen is used in cell respiration. If oxygen is not available, electron flow along the electron transport chain stops and $NADH + H^+$ cannot be reconverted to NAD^+. Supplies of NAD^+ in the mitochondrion run out and the link reaction and Krebs cycle cannot continue. Glycolysis can continue because conversion of pyruvate into lactate or ethanol and carbon dioxide produces as much NAD^+ as is used in glycolysis. However, whereas aerobic cell respiration gives a yield of about 30 ATP molecules per glucose, glycolysis produces only two. Oxygen thus greatly increases the ATP yield.

The figure (below) shows the electron transport chain and the role of oxygen as the terminal electron acceptor.

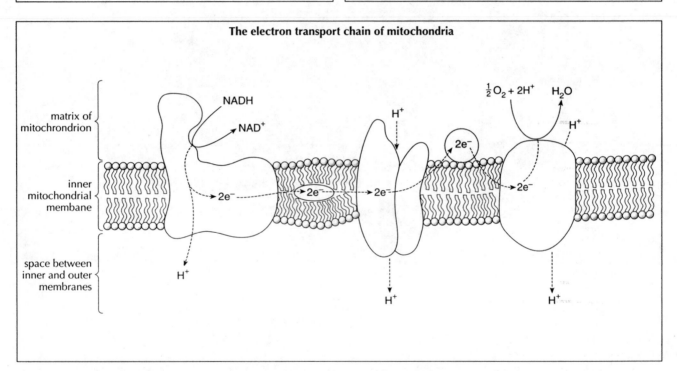

The electron transport chain of mitochondria

THE COUPLING OF ELECTRON TRANSPORT TO ATP SYNTHESIS

Energy released as electrons pass along the electron transport chain is used to pump protons (H^+) across the inner mitochondrial membrane into the space between the inner and outer membranes. A concentration gradient is formed, which is a store of potential energy. ATP synthase, located in the inner mitochondrial membrane, transports the protons back across the membrane down the concentration gradient. As the protons pass across the membrane they release energy and this is used by ATP synthase to produce ATP. The coupling of ATP synthesis to electron transport via a concentration gradient of protons is called **chemiosmosis**.

The figure (right) shows some features of ATP synthase.

Structure of ATP synthase

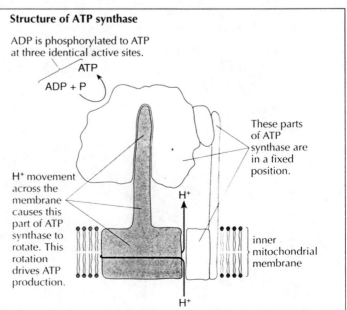

ADP is phosphorylated to ATP at three identical active sites.

ATP

ADP + P

These parts of ATP synthase are in a fixed position.

H^+ movement across the membrane causes this part of ATP synthase to rotate. This rotation drives ATP production.

inner mitochondrial membrane

Mitochondria

The mitochondrion is an excellent example of the relationship between structure and function.

The figure (below) is an electron micrograph of a whole mitochondrion.

The figure (bottom) is a drawing of the same mitochondrion, labelled to show how it is adapted to carry out its function.

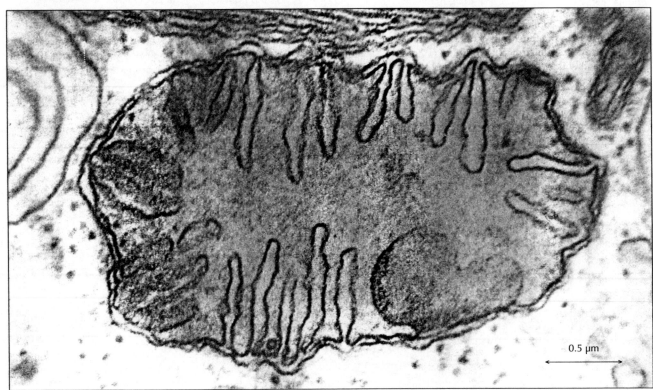

0.5 µm

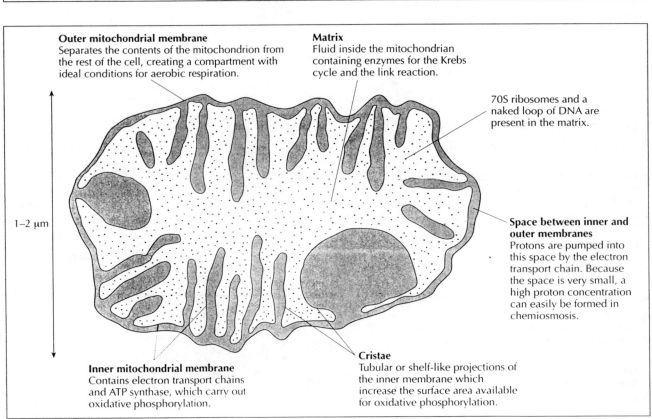

Outer mitochondrial membrane
Separates the contents of the mitochondrion from the rest of the cell, creating a compartment with ideal conditions for aerobic respiration.

Matrix
Fluid inside the mitochondrian containing enzymes for the Krebs cycle and the link reaction.

70S ribosomes and a naked loop of DNA are present in the matrix.

1–2 µm

Space between inner and outer membranes
Protons are pumped into this space by the electron transport chain. Because the space is very small, a high proton concentration can easily be formed in chemiosmosis.

Inner mitochondrial membrane
Contains electron transport chains and ATP synthase, which carry out oxidative phosphorylation.

Cristae
Tubular or shelf-like projections of the inner membrane which increase the surface area available for oxidative phosphorylation.

Light and photosynthesis

Photosynthesis is the process that plants, algae and some bacteria use to produce all of the organic compounds that they need. Photosynthesis involves many chemical reactions. Some of them need a continual supply of light and so are called **light-dependent reactions**. Other reactions need light indirectly, but can carry on for some time in darkness. These are called **light-independent reactions**.

Glucose, amino acids and other organic compounds are produced in the light independent reactions. The light-dependent reactions produce intermediate compounds that are used in the light-independent reactions. In darkness these intermediate compounds are gradually used up.

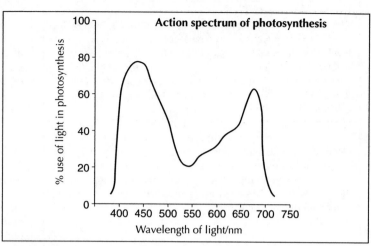

THE ACTION SPECTRUM OF PHOTOSYNTHESIS

A spectrum is a range of wavelengths of electromagnetic radiation. The spectrum of light is the range of wavelengths from 400 nm to 700 nm. Each wavelength is a pure colour of light:

    400–525 violet–blue
    525–625 green–yellow
    625–700 orange–red

The efficiency of photosynthesis is not the same in all wavelengths of light. The efficiency is the percentage of light of a wavelength that is used in photosynthesis. The figure (top right) is a graph showing the percentage use of the wavelengths of light in photosynthesis. This graph is called the **action spectrum** of photosynthesis. The graph shows that violet and blue light are used most efficiently and red light is also used efficiently. Green light is used much less efficiently.

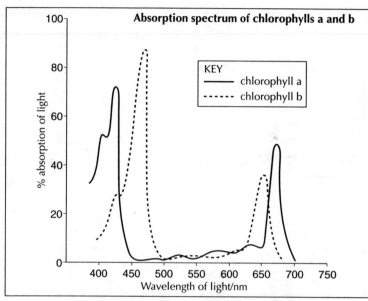

THE ABSORPTION SPECTRA OF PHOTOSYNTHETIC PIGMENTS

The action spectrum of photosynthesis is explained by considering the light-absorbing properties of the photosynthetic pigments. Most pigments absorb some wavelengths better than others. The figure (centre right) shows the percentage of the wavelengths of visible light that are absorbed by two common forms of chlorophyll. This graph is called the **absorption spectrum** of these pigments. The graph shows strong similarities with the action spectrum for photosynthesis.

- The greatest absorption is in the violet–blue range.
- There is a also a high level of absorption in the red range of the spectrum
- There is least absorption in the yellow–green range of the spectrum. Most of this light is reflected.

There are some differences between the action spectrum and the absorption spectra. Whereas little light is absorbed by chlorophylls in the green to yellow range there is some photosynthesis. This is due to accessory pigments, including xanthophylls and carotene, which absorb wavelengths that chlorophyll cannot.

Action and absorption spectra of an alga

Some algae contain large amounts of accessory pigments. For example, kelp (*Laminaria saccharina*) contains carotene and fucoxanthin in addition to chlorophylls and so can absorb and use all wavelengths of light with about the same efficiency in photosynthesis. The graph below shows the action and absorption spectra for kelp. The colour of kelp can be deduced from the data.

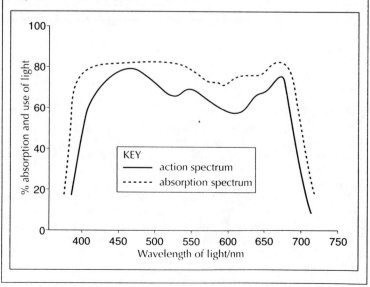

Light-dependent reactions

LIGHT ABSORPTION

Chlorophyll absorbs light and the energy from the light raises an electron in the chlorophyll molecule to a higher energy level. The electron at a higher energy level is an **excited electron** and the chlorophyll is **photoactivated**. In single chlorophyll molecules the excited electron soon drops back down to its original level, re-emitting the energy. Chlorophyll is located in thylakoid membranes and is arranged in groups of hundreds of molecules, called **photosystems**. There are two types of photosystem – photosystems I and II. Excited electrons from absorption of photons of light anywhere in the photosystem are passed from molecule to molecule until they reach a special chlorophyll molecule at the reaction centre of the photosystem. This chlorophyll passes the excited electron to a chain of electron carriers.

PRODUCTION OF ATP

An excited electron from the reaction centre of photosystem II is passed along a chain of carriers in the thylakoid membrane (below). It gives up some of its energy each time that it passes from one carrier to the next. At one stage, enough energy is released to make a molecule of ATP. The coupling of electron transport to ATP synthesis is by chemiosmosis, as in the mitochondrion. Electron flow causes a proton to be pumped across the thylakoid membrane into the fluid space inside the thylakoid. A proton gradient is created. ATP synthase, located in the thylakoid membranes, lets the protons across the membrane down the concentration gradient and uses the energy released to synthesize ATP.

The production of ATP using the energy from an excited electron from Photosystem II is called **non-cyclic photophosphorylation**. An alternative method of photophosphorylation is shown on page 81.

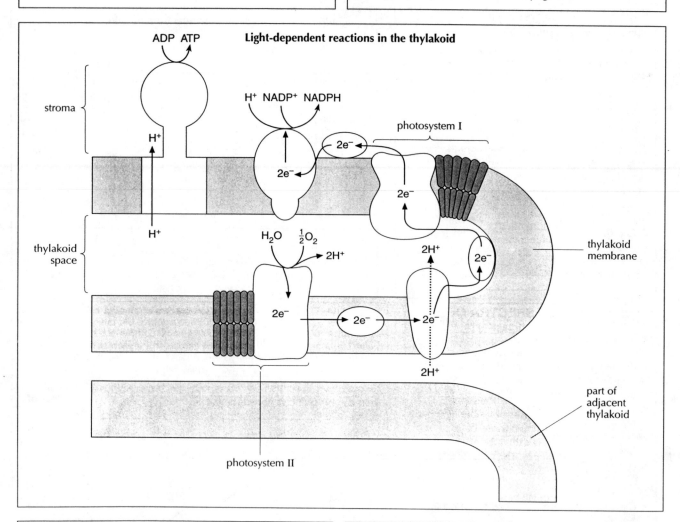

Light-dependent reactions in the thylakoid

PRODUCTION OF NADP

After releasing the energy needed to make ATP, the electron that was given away by photosystem II is accepted by photosystem I. The electron replaces one previously given away by photosystem I. With its electron replaced, photosystem I can be photoactivated by absorbing light and then give away another excited electron. This high-energy electron passes along a short chain of carriers to NADP+ in the stroma. NADP+ accepts two high-energy electrons from the electron transport chain and one H+ ion from the stroma, to form NADPH.

PRODUCTION OF OXY.GEN

Photosystem II needs to replace the excited electrons that it gives away. The special chlorophyll molecule at the reaction centre is positively charged after giving away an electron. With the help of an enzyme at the reaction centre, water molecules in the thylakoid space are split and electrons from them are given to chlorophyll. Oxygen and H+ ions are formed as by-products. The splitting of water molecules only happens in the light, so is called **photolysis**. The oxygen produced in photosynthesis is all the result of photolysis of water. Oxygen is a waste product and is excreted.

Light-independent reactions

THE CALVIN CYCLE

The light-independent reactions take place within the stroma of the chloroplast. The first reaction involves a five-carbon sugar, ribulose bisphosphate (RuBP). RuBP is also a product of the light independent reactions, which therefore form a cycle, called the **Calvin cycle**. There are many alternative names for the intermediate compounds in the Calvin cycle. Glycerate 3-phosphate is sometimes also called 3-phosphoglycerate. Glycerate 3-phosphate is sometimes abbreviated as GP, which could be confused with glyceraldehyde 3-phosphate, which is a form of triose phosphate or with glucose phosphate. The abbreviation GP should therefore be avoided!

CARBON FIXATION

Carbon dioxide is an essential substrate in the light-independent reactions. It enters the chloroplast by diffusion. In the stroma of the chloroplast carbon dioxide combines with ribulose bisphosphate (RuBP), a five-carbon sugar, in a carboxylation reaction. The reaction is catalysed by the enzyme ribulose bisphosphate carboxylase, usually called **rubisco**. Large amounts of rubisco are present in the stroma, because it works rather slowly and the reaction that it catalyses is a very important one. The product of the reaction is a six-carbon compound, which immediately splits to form two molecules of glycerate 3-phosphate. This is therefore the first product of carbon fixation.

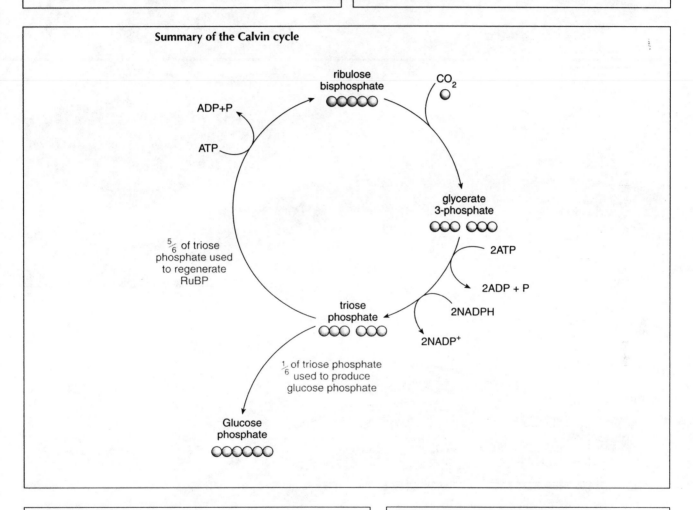

Summary of the Calvin cycle

REGENERATION OF RUBP

For carbon fixation to continue, one RuBP molecule must be produced to replace each one that is used. Triose phosphate is used to regenerate RuBP. Five molecules of triose phosphate are converted by a series of reactions into three molecules of RuBP. This process requires the use of energy in the form of ATP. The reactions can be summarized using equations where only the number of carbon atoms in each sugar molecule is shown.

$$C_3 + C_3 \rightarrow C_6$$
$$C_6 + C_3 \rightarrow C_4 + C_5$$
$$C_4 + C_3 \rightarrow C_7$$
$$C_7 + C_3 \rightarrow C_5 + C_5$$

For every six molecules of triose phosphate formed in the light-independent reactions, five must be converted to RuBP.

SYNTHESIS OF CARBOHYDRATE

Glycerate 3-phosphate, formed in the carbon fixation reaction, is an organic acid. It is converted into a carbohydrate by a reduction reaction. Hydrogen is needed to carry out this reaction and is supplied by NADPH. Energy is also needed and is supplied by ATP. NADPH and ATP are produced in the light-dependent reactions of photosynthesis. Glycerate 3-phosphate is reduced to a three-carbon sugar, triose phosphate (TP). Linking together two triose phosphate molecules together produces glucose phosphate. Starch, the storage form of carbohydrate in plants, is formed in the stroma by condensation of many molecules of glucose phosphate.

Chloroplasts

The chloroplast is another example of close relationship between structure and function. The figure (below) is an electron micrograph of a chloroplast. The figure (bottom) is a drawing of the same chloroplast, labelled to show how it is adapted to carry out its function.

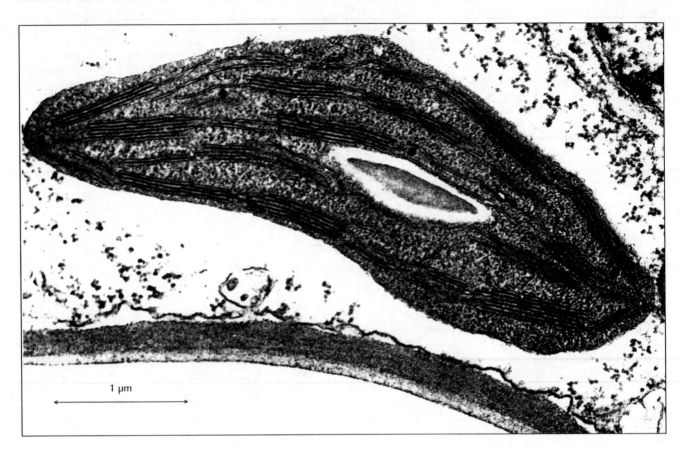

1 µm

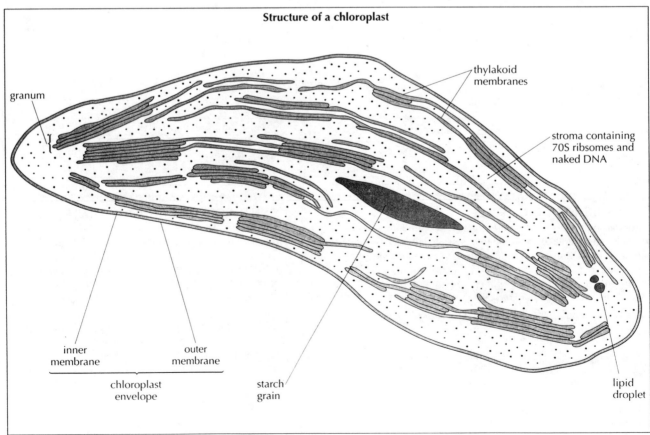

Structure of a chloroplast

granum

thylakoid membranes

stroma containing 70S ribsomes and naked DNA

inner membrane

outer membrane

chloroplast envelope

starch grain

lipid droplet

Limiting factors in photosynthesis

THE CONCEPT OF LIMITING FACTORS

Light intensity, carbon dioxide concentration and temperature are three factors that can determine the rate of photosynthesis. If the level of one of these factors is changed, the rate of photosynthesis changes. Usually, only changes to one of the factors will affect the rate of photosynthesis in a plant at a particular time. This is the factor that is nearest to its minimum and is called the **limiting factor**. Changing the limiting factor increases or decreases the rate, but changes to the other factors have no effect. This is because photosynthesis is a complex process involving many steps. The overall rate of photosynthesis in a plant is determined by the rate of whichever step is proceeding most slowly at a particular time. This is called the **rate-limiting step**. The three limiting factors affect different rate-limiting steps.

The figures on page 17 show the relationship between each of the limiting factors and the rate of photosynthesis.

THE EFFECT OF LIGHT INTENSITY

At low light intensities, there is a shortage of the products of the light-dependent reactions – NADPH and ATP. The rate-limiting step in the Calvin cycle is the point where glycerate 3-phosphate is reduced.
At high light intensities some other factor is limiting.
Unless a plant is heavily shaded, or the sun is rising or setting, light intensity is not usually the limiting factor.

THE EFFECT OF CO_2 CONCENTRATION

At low and medium CO_2 concentrations, the rate-limiting step in the Calvin cycle is the point where CO_2 is fixed to produce glycerate 3-phosphate. RuBP and NADPH accumulate.
At high CO_2 concentrations some other factor is limiting.
Because the level of carbon dioxide in the atmosphere is never very high, carbon dioxide concentration is often the limiting factor.

THE EFFECT OF TEMPERATURE

At low temperatures, all of the enzymes that catalyse the reactions of the Calvin cycle work slowly. NADPH accumulates.
At intermediate temperatures, some other factor is limiting.
At high temperatures, RuBP carboxylase does not work effectively, so the rate-limiting step in the Calvin cycle is the point where CO_2 is fixed. NADPH accumulates.

Results of an investigation into limiting factors

The figure (right) shows the effects of light intensity on the rate of photo-synthesis at two different temperatures and two carbon dioxide concentrations. It is possible to deduce which is the limiting factor at the point marked with an arrow ($①$ – $④$) on each curve.

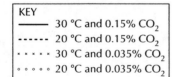

KEY
——— 30 °C and 0.15% CO_2
----- 20 °C and 0.15% CO_2
× × × × 30 °C and 0.035% CO_2
∘ ∘ ∘ ∘ 20 °C and 0.035% CO_2

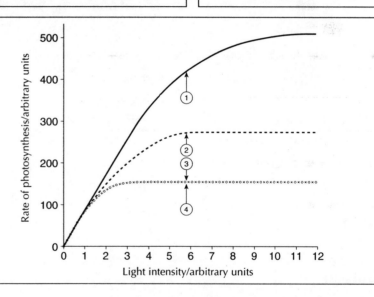

Summary of cyclic photophosphorylation

CYCLIC PHOTOPHOSPHORYLATION

When light is not the limiting factor, NADPH tends to accumulate in the stroma and there is a shortage of NADP⁺. The normal flow of electrons in the thylakoid membranes is inhibited because NADP⁺ is needed as a final acceptor of electrons. An alternative route can be used that allows ATP production when NADP⁺ is not available.
This pathway is called **cyclic photophosphorylation**.
• Photosystem I absorbs light and is photoactivated.
• Excited electrons are passed from photosystem I to a carrier in the chain between photosystem II and photosystem I.
• The electrons pass along the chain of carriers back to photosystem I.
• As the electrons flow along the chain of carriers they cause pumping of protons across the thylakoid membrane.
• A proton gradient is formed and this allows production of ATP by ATP synthase.
The figure (left) shows the pathway used in cyclic photophosphorylation.

1   The electron micrograph below shows part of a plant root cell, including mitochondria.

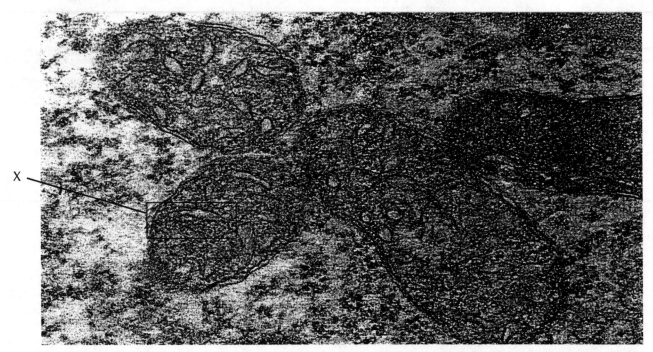

X

[Source: Dr B. E. Juniper, Dept. of Plant Sciences, University of Oxford]

   a) Explain briefly two features that allow the mitochondria in the micrograph to be identified.   [2]

   b) Redraw the structure of the mitochondrion marked X.   [2]

   c) Annotate the micrograph (not your drawing) to show one example of

      (i) a region where the Krebs cycle takes place

      (ii) a location of ATP synthetase

      (iii) a region where glycolysis takes place.

      [3]

2   a) Draw a curve of the action spectrum for photosynthesis on the axis below.   [2]

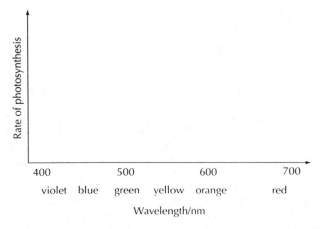

   b) Explain the relationship between the action spectrum and the absorption spectra of photosynthetic pigments.   [3]

3   a) Outline the role of a proton gradient in the process of respiration.   [2]

   b) Explain how the structure of the mitochondrion allows a large proton gradient to be formed.   [2]

Mendel's Law of Independent Assortment

Gregor Mendel discovered the Law of Segregation by doing monohybrid crosses with pea plants. He discovered another law of inheritance by doing crosses in which the parents differed in two characteristics, that are controlled by two different genes. These are called **dihybrid crosses**. Mendel did his dihybrid crosses with pea plants. An example of one of his crosses is shown below. The parents in this cross differ in seed shape, controlled by one gene and in seed colour, controlled by a different gene.

KEY TO SYMBOLS

S = allele for smooth seed.
s = allele for wrinkled seed.
Y = allele for yellow seed.
y = allele for green seed.

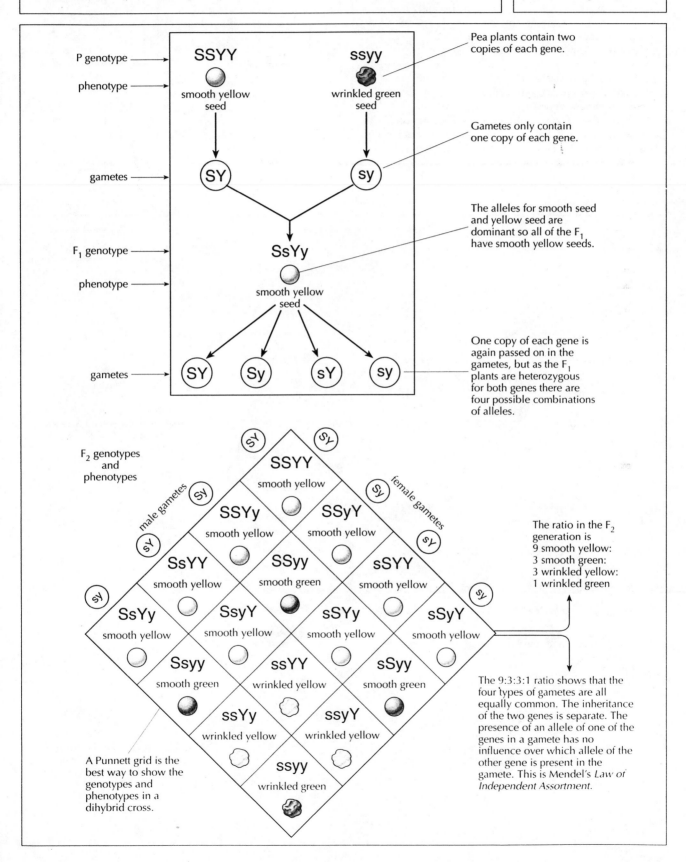

P genotype → SSYY / ssyy

Pea plants contain two copies of each gene.

phenotype → smooth yellow seed / wrinkled green seed

gametes → SY / sy

Gametes only contain one copy of each gene.

The alleles for smooth seed and yellow seed are dominant so all of the F_1 have smooth yellow seeds.

F_1 genotype → SsYy

phenotype → smooth yellow seed

gametes → SY Sy sY sy

One copy of each gene is again passed on in the gametes, but as the F_1 plants are heterozygous for both genes there are four possible combinations of alleles.

F_2 genotypes and phenotypes

male gametes: SY Sy sY sy
female gametes: SY Sy sY sy

SSYY smooth yellow
SSYy smooth yellow
SSyY smooth yellow
SsYY smooth yellow
SSyy smooth green
sSYY smooth yellow
SsYy smooth yellow
SsyY smooth yellow
sSYy smooth yellow
sSyY smooth yellow
Ssyy smooth green
ssYY wrinkled yellow
sSyy smooth green
ssYy wrinkled yellow
ssyY wrinkled yellow
ssyy wrinkled green

The ratio in the F_2 generation is
9 smooth yellow:
3 smooth green:
3 wrinkled yellow:
1 wrinkled green

The 9:3:3:1 ratio shows that the four types of gametes are all equally common. The inheritance of the two genes is separate. The presence of an allele of one of the genes in a gamete has no influence over which allele of the other gene is present in the gamete. This is Mendel's *Law of Independent Assortment*.

A Punnett grid is the best way to show the genotypes and phenotypes in a dihybrid cross.

Dihybrid crosses

PREDICTING RATIOS IN DIHYBRID RATIOS

The 9:3:3:1 ratio is often found when parents that are heterozygous for two genes are crossed together. The ratio is the product of two 3:1 ratios – each of the two genes would give a 3:1 ratio in a monohybrid cross between two heterozygous parents. In a dihybrid cross they follow Mendel's Law of Independent Assortment because they are unlinked.

Dihybrid crosses can give other ratios if:

• either of the genes has codominant alleles,
• either of the parents is homozygous for one/both of the genes,
• either of the genes is sex linked. Sex-linked genes are located on sex chromosomes instead of on autosomes (non-sex chromosomes).

The figure (right) shows ratios that these types of genes could give. Another cause of unusual ratios is interaction between genes. The figure (below) shows an example of a dihybrid cross where there is interaction between genes.

Possible ratios in dihybrid crosses

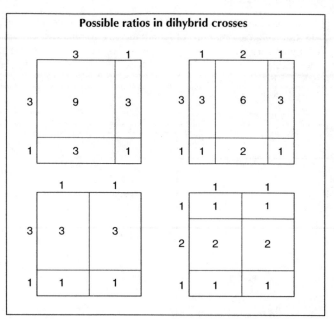

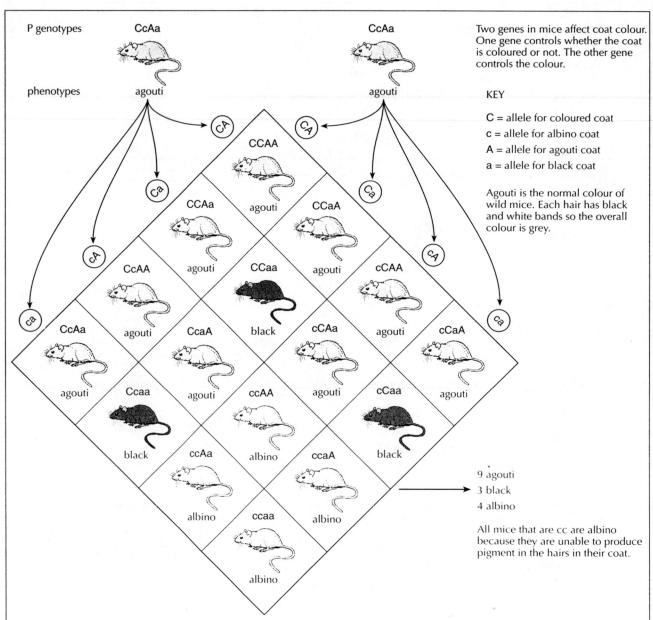

Two genes in mice affect coat colour. One gene controls whether the coat is coloured or not. The other gene controls the colour.

KEY

C = allele for coloured coat
c = allele for albino coat
A = allele for agouti coat
a = allele for black coat

Agouti is the normal colour of wild mice. Each hair has black and white bands so the overall colour is grey.

9 agouti
3 black
4 albino

All mice that are cc are albino because they are unable to produce pigment in the hairs in their coat.

Polygenic inheritance

THE DISCOVERY OF POLYGENIC INHERITANCE

Some characteristics are controlled by more than one gene. This is called polygenic inheritance. Gregor Mendel discovered an example of **polygenic inheritance**, when he crossed a purple-flowered species of bean with a white-flowered species. The F_1 offspring were all purple, so he expected a 3:1 ratio of purple to white flowers in the F_2 offspring. Instead, he found a much smaller proportion of white flowers and a wide variety of shades of purple flower. Mendel suggested that two or three genes might be involved. If these were codominant genes, each with two alleles, one for purple flowers (A^P and B^P) and one for white (A^W and B^W), there could be five shades of flower colour (right).

POLYGENIC INHERITANCE AND CONTINUOUS VARIATION

Most examples of polygenic inheritance involve more than two codominant genes. As the number of genes involved increases, the number of possible phenotypes increases. Eventually, it becomes impossible to divide individuals into discrete groups – the variation is continuous.

Results of a cross between red and white flowered beans

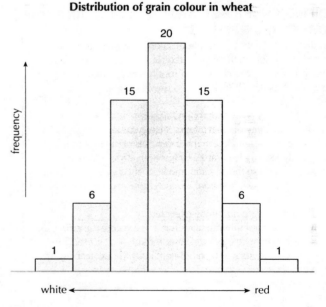

EXAMPLES OF POLYGENIC INHERITANCE

Grain colour in wheat

Wheat grains vary in colour from white to dark red, depending on the amount of a red pigment they contain. Three genes control the colour. Each gene has two alleles, one that causes pigment production and one that does not. Wheat grains can therefore have between 0 and 6 alleles for pigment production. The figure (right) shows the expected distribution of grain colour from a cross between two plants that are heterozygous for each of the three genes.

Skin colour in humans

The colour of human skin depends on the amount of the black pigment melanin in it. There is a continuous distribution of skin colour from very pale (little melanin) to black (much melanin). At least four and possibly more genes are involved, each with alleles that promote melanin production and alleles that do not. There is therefore a wide range of possible genotypes with anything from no alleles promoting melanin production to many.
The figure (below) shows humans with a range of skin colour.

Distribution of grain colour in wheat

white ← → red

Skin colour variation in humans

Statistical testing

USING STATISTICAL TESTS

Biologists use statistical tests to find out whether the differences between sets of results are significant and whether observed results differ significantly from theoretical results. In the chi-squared test and other commonly used tests, the significance level is shown as a percentage. A 5% significance level means that there is a 5% probability (1 in 20) that the differences are due to the samples being, by chance, unrepresentative. Thus there is a 95% probability that the differences are due to a real difference between the populations or treatments being investigated. With less than 95% confidence, there is too great a probability that the differences are due to chance for the conclusion to be trusted. The 5% level of significance is therefore usually the minimum that is accepted in biological investigations.

EXPECTED AND OBSERVED RESULTS IN GENETIC CROSSES

In monohybrid and dihybrid crosses, the **expected results** are calculated using the appropriate genetic ratio. For example, Mendel expected a ratio of 3 round seeds to one wrinkled seed in one of his crosses. 7324 seeds were produced in total in the cross, so the expected results were 5493 round seeds and 1831 wrinkled seeds. The actual numbers of each type of offspring in a cross are the **observed results**. The observed results of Mendel's cross were 5474 round seeds and 1850 wrinkled seeds.

DIFFERENCES BETWEEN EXPECTED AND OBSERVED RESULTS

In the example above, Mendel's observed results are close but not identical to the expected results. In monohybrid and dihybrid crosses the observed results are rarely identical to the expected results. There are two possible reasons for this.

1. *Chance effects of random fertilization*
   The probability of each type of male gamete successfully fertilizing the female gamete depends on the proportions of each type of male gamete. However, each fertilization is random and so the actual numbers of successful male gametes of each type will not usually match these proportions exactly.

2. *The results do not fit the expected ratio*
   If the pattern of inheritance is not as expected, the ratio used to calculate the expected results will not be correct. The observed results will be different from the expected results by a greater amount than differences due to chance.

The chi-squared test can be used to distinguish between these two reasons.

USING THE CHI-SQUARED TEST

An example (above right) shows how the statistic chi-squared is calculated. It is not necessary to memorize the equation for chi-squared, but it is important to understand how the calculated statistic is used. The larger the calculated value of chi-squared, the greater is the difference between the observed and expected results. The calculated value is compared with a table of critical values of chi-squared (right). The correct line of the table must be used, corresponding to the number of degrees of freedom. In the example (above right) there is one degree of freedom and chi-squared is 0.262. This is much lower than the critical value for a significance level of 5%. This shows that there is no significant difference between observed results and the results expected with a ratio of 3:1.

EXAMPLE OF CHI-SQUARED CALCULATION

Total number of offspring = 7324
Expected F_2 ratio = 3 round : 1 wrinkled.

	Round	Wrinkled
Expected results	5493	1831
Observed results	5474	1850

$$\text{Chi-squared} = \sum \frac{(O - E)^2}{E}$$

Where   O = observed result
        E = expected result

$$\text{Chi squared} = \frac{(5474 - 5493)^2}{5493} + \frac{(1850 - 1831)^2}{1831}$$

$$= \frac{361}{5493} + \frac{361}{1831}$$

$$= 0.065 + 0.197$$

$$= 0.262$$

Degrees of freedom = number of classes − 1
= 2 − 1
= 1

CRITICAL VALUES FOR THE CHI-SQUARED TEST

df	Level of significance (P)					df	Level of significance (P)				
	0.05	0.025	0.01	0.005	0.001		0.05	0.025	0.01	0.005	0.001
1	3.84	5.02	6.63	7.88	10.83	20	31.41	34.17	37.57	40.00	45.32
2	5.99	7.38	9.21	10.60	13.81	21	32.67	35.48	38.93	41.40	46.80
3	7.31	9.35	11.34	12.84	16.27	22	33.92	36.78	40.29	42.80	48.27
4	9.49	11.14	13.26	14.86	18.47	23	35.17	38.08	41.64	44.18	49.73
5	11.07	12.83	15.09	16.75	20.52	24	36.42	39.36	42.98	45.56	51.13
6	12.59	14.45	16.81	18.55	22.46	25	37.65	40.65	44.31	46.93	52.62
7	14.07	16.01	18.48	20.28	24.32	26	38.89	41.92	45.64	48.29	54.05
8	15.51	17.53	20.09	21.98	26.13	27	40.11	43.19	46.96	49.64	55.48
9	16.92	19.02	21.67	23.59	27.88	28	41.34	44.46	48.28	50.99	56.89
10	18.31	20.48	23.21	25.19	29.59	29	42.56	45.72	49.59	52.34	58.30
11	19.68	21.92	24.73	26.76	31.26	30	43.77	46.98	50.89	53.67	59.70
12	21.03	23.34	26.22	28.30	32.91	40	43.77	46.98	50.89	53.57	59.70
13	22.36	24.74	27.69	29.82	34.53	50	67.50	71.42	76.16	79.49	86.66
14	23.68	26.12	29.14	31.32	36.12	60	79.08	83.30	88.38	91.95	99.61
15	25.00	27.49	30.58	32.80	37.70	70	90.53	95.02	100.43	104.22	112.32
16	26.30	28.85	32.00	34.27	39.25	80	101.88	106.63	100.43	104.22	112.32
17	27.59	30.19	33.41	35.72	40.79	90	113.15	118.14	124.12	128.30	137.21
18	28.87	31.53	34.81	37.16	42.31	100	124.34	129.56	135.81	140.17	149.44
19	30.14	32.85	36.19	28.58	43.82						

Gene linkage and recombination

In dihybrid crosses some offspring inherit a combination of characteristics that one of the parents had and other offspring inherit new combinations of characteristics. Offspring with a new combination of characteristics are called **recombinants**, because the new combinations are formed by **recombination**. *Recombination is the reassortment of genes or characteristics into different combinations from those of the parents.*

RECOMBINATION OF UNLINKED GENES

Unlinked genes are located on different types of chromosome, so when homologous chromosomes pair up in meiosis they are on different pairs. The pairs of homologous chromosomes are called **bivalents**. The bivalents are orientated randomly on the equator and so the pole to which the each allele of a gene moves is not affected by the pole to which alleles of other unlinked genes are moving. This is how independent assortment of unlinked genes occurs (below). It also allows the recombination of unlinked genes – combinations of alleles inherited from a parent are broken up and new combinations can then be formed by random fertilization.

GENE LINKAGE

Some pairs of genes do not follow the Law of Independent Assortment. The expected 9:3:3:1 ratio is not found when parents that are heterozygous for the two genes are crossed. The figure (right) shows the first example to be discovered. The results show that there were more offspring than expected with the parental character combinations – purple long and red round. There were fewer than expected with the new combinations – purple round and red long. Combinations of genes tend to be inherited together. This is called **gene linkage**. Gene linkage is caused by pairs of genes being located on the same type of chromosome.

RECOMBINATION OF LINKED GENES

Pairs of linked genes are located on the same type of chromosome. Recombination of linked genes involves a special process called **crossing-over**, which happens during the early stages of meiosis.

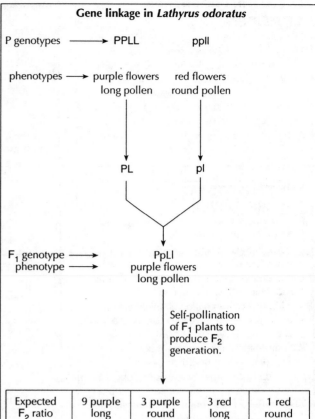

Gene linkage in *Lathyrus odoratus*

P genotypes ⟶ PPLL   ppll

phenotypes ⟶ purple flowers   red flowers
               long pollen   round pollen

PL   pl

F_1 genotype ⟶   PpLl
phenotype ⟶   purple flowers
                     long pollen

Self-pollination of F_1 plants to produce F_2 generation.

Expected F_2 ratio	9 purple long	3 purple round	3 red long	1 red round
Expected results (6952 plants in total)	3910.5	1303.5	1303.5	434.5
Observed results	4831	390	393	1338

Chi-squared = 372 at 3 degrees of freedom
Significance level is less than 0.001
So there is 99.9% confidence of a significant difference between the observed and expected results.

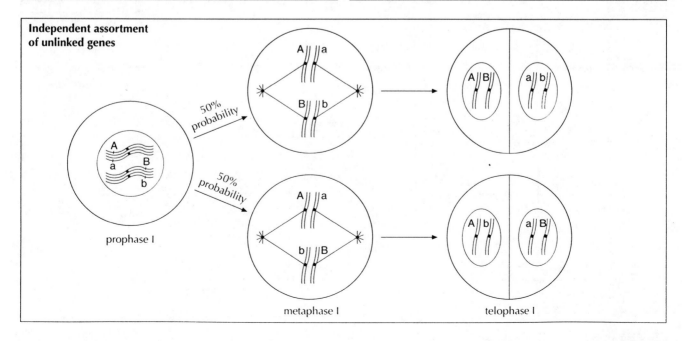

Independent assortment of unlinked genes

prophase I   metaphase I   telophase I

Crossing-over

EVENTS IN PROPHASE I OF MEIOSIS

Homologous chromosomes pair up in prophase I of meiosis. Each homologous chromosome consists of two sister chromatids. Chromatids of different chromosomes are called non-sister chromatids. While the chromosomes are paired, sections of chromatid are exchanged in a process called **crossing-over**.

The figure (right) shows how crossing-over occurs.

THE BENEFITS OF CROSSING-OVER

Crossing-over has two useful consequences.

1. It creates chiasmata which hold homologous chromosomes together in pairs called bivalents, during the later stages of prophase I and metaphase I until microtubules have attached.

2. It allows recombination of linked genes. All of the genes that have their loci on the same chromosome type form a **linkage group**.

Recombination of genes in a linkage group cannot occur without crossing-over. The point where crossing-over occurs along chromosomes is random – it can occur at a vast number of different points. Meiosis can therefore produce an almost infinite amount of genetic variety.

The figure (below) shows how crossing-over can cause recombination of linked genes. The figure (right) shows an example of a cross involving gene linkage, using bars to represent the chromosomes on which the genes are linked. A test cross was done on the F_1 plants.

Recombination of linked genes

Parental gene combinations are AB and ab

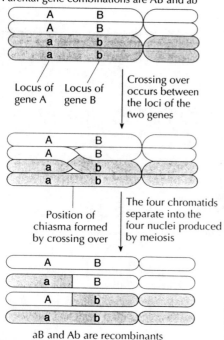

Locus of gene A   Locus of gene B

Crossing over occurs between the loci of the two genes

Position of chiasma formed by crossing over

The four chromatids separate into the four nuclei produced by meiosis

aB and Ab are recombinants

The process of crossing over

At one stage in prophase I all of the chromatids of two homologous chromosomes become tightly paired up together. This is called *synapsis*.

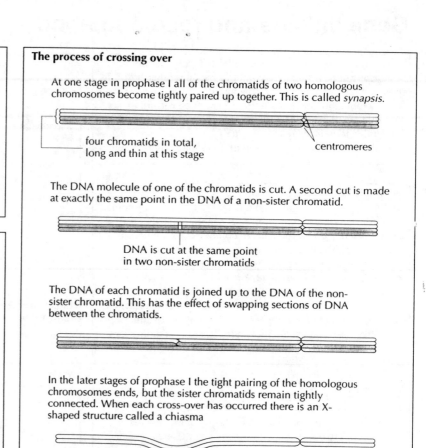

four chromatids in total, long and thin at this stage

centromeres

The DNA molecule of one of the chromatids is cut. A second cut is made at exactly the same point in the DNA of a non-sister chromatid.

DNA is cut at the same point in two non-sister chromatids

The DNA of each chromatid is joined up to the DNA of the non-sister chromatid. This has the effect of swapping sections of DNA between the chromatids.

In the later stages of prophase I the tight pairing of the homologous chromosomes ends, but the sister chromatids remain tightly connected. When each cross-over has occurred there is an X-shaped structure called a chiasma

chiasma

AN EXAMPLE OF GENE LINKAGE AND TEST-CROSSING IN *ZEA MAYS*

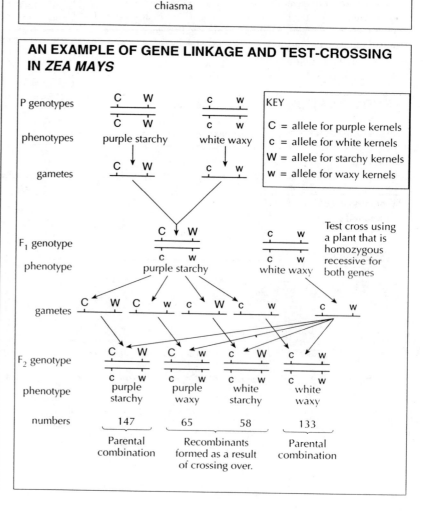

KEY
C = allele for purple kernels
c = allele for white kernels
W = allele for starchy kernels
w = allele for waxy kernels

P genotypes — phenotypes: purple starchy / white waxy

gametes

Test cross using a plant that is homozygous recessive for both genes

F_1 genotype — phenotype: purple starchy / white waxy

gametes

F_2 genotype — phenotype

	purple starchy	purple waxy	white starchy	white waxy
numbers	147	65	58	133
	Parental combination	Recombinants formed as a result of crossing over.		Parental combination

Phases of meiosis

Meiosis involves two divisions. Each division is divided into four phases. The main events of each phase are listed below.

PROPHASE I
- Chromosomes start to coil up and so become shorter and thicker.
- Homologous chromosomes pair up.
- Crossing over occurs.
- Centrioles move to the poles in animal cells.
- Nucleoli break down.
- At the end of prophase I the nuclear membrane breaks down.

METAPHASE I
- Chromosomes continue to shorten and thicken.
- Spindle microtubules attach to the centromeres.
- Bivalents line up on the equator.
- Chiasmata slide towards the ends of the chromosomes, causing the shapes of the bivalents to change.
- At the end of metaphase I the chromosomes start to move.

ANAPHASE I
- The two chromosomes of each bivalent move to opposite poles. This halves the chromosome number. Each chromosome consists of two chromatids. Because of crossing over the two chromatids are not identical.
- At the end of anaphase I the chromosomes reach the poles.

TELOPHASE I
- Nuclear membranes form around the groups of chromosomes at each pole.
- The cell divides to form two haploid cells.
- The chromosomes uncoil partially.
- At the end of telophase I the two cells either enter a brief period of interphase or immediately proceed to the second division of meiosis. The DNA is not replicated.

PROPHASE II
- Chromosomes become shorter and thicken again by coiling.
- Centrioles move to the poles in animal cells.
- At the end of prophase II the nuclear membranes break down.

METAPHASE II
- Spindle microtubules attach to the centromeres.
- Chromosomes line up on the equator
- At the end of metaphase II the centromeres divide.

ANAPHASE II
- The two chromatids of each chromosome move to opposite poles.
- At the end of anaphase II the chromatids reach the poles.

TELOPHASE II
- Nuclear membranes form around the groups of chromatids at each pole. Each chromatid is now considered to be a chromosome.
- The two cells each divide to form to four cells in total.
- The chromosomes uncoil.
- Nucleoli appear.
- In most organisms the cells formed at the end of telophase II develop into gametes.

SUMMARY OF MEIOSIS
1. Meiosis involves two divisions. One cell or nucleus divides to form four cells or nuclei.
2. The chromosome number is halved, from diploid to haploid.
3. An almost infinite amount of genetic variety is produced, as a result of crossing-over in prophase I and the random orientation of bivalents in metaphase I.

The figure (below) shows micrographs of four stages in meiosis in cells from the testis of a locust.

Early prophase I

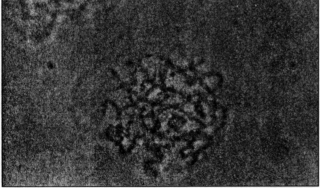

Late prophase I

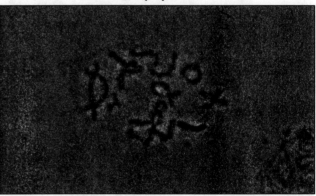

Metaphase I

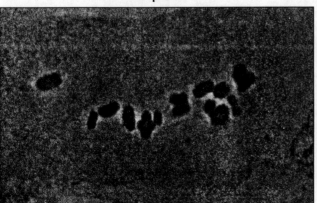

Telophase I

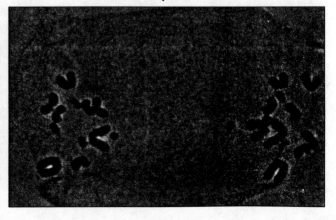

Spermatogenesis

Electron micrograph of adenovirus

Spermatogenesis is the production of spermatozoa.
Spermatozoa are usually simply called sperm.
Spermatogenesis occurs in the testes, in narrow tubes called
seminiferous tubules.
The figures (below and right) show the structure of testis
tissue, including the seminiferous tubules. The figure (bottom)
shows the processes involved in spermatogenesis

Structure of testis tissue

wall of seminiferous tubule   fluid inside seminiferous tubule   basement membrane around the tubule   blood vessel   interstitial (Leydig cells)

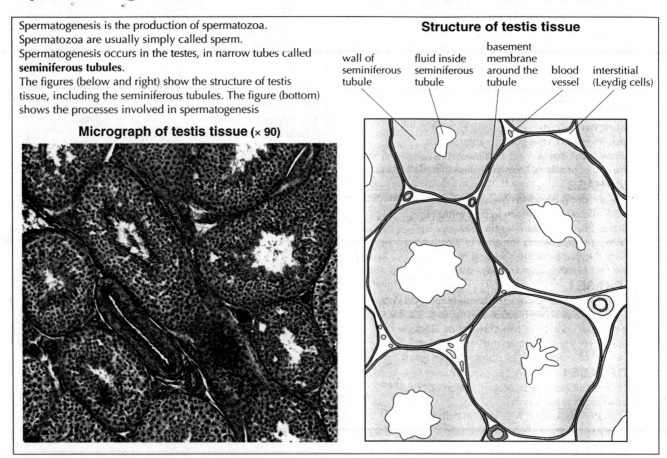

Micrograph of testis tissue (× 90)

STAGES OF SPERMATOGENESIS

basement membrane   spermatogonium

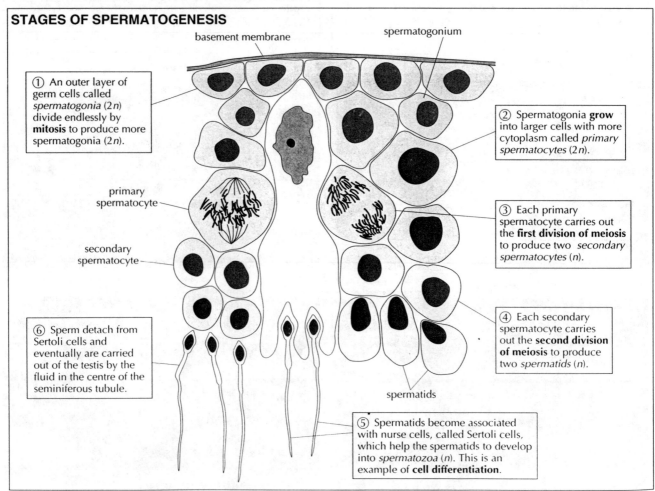

① An outer layer of germ cells called *spermatogonia* (2*n*) divide endlessly by **mitosis** to produce more spermatogonia (2*n*).

② Spermatogonia **grow** into larger cells with more cytoplasm called *primary spermatocytes* (2*n*).

primary spermatocyte

③ Each primary spermatocyte carries out the **first division of meiosis** to produce two *secondary spermatocytes* (*n*).

secondary spermatocyte

④ Each secondary spermatocyte carries out the **second division of meiosis** to produce two *spermatids* (*n*).

⑥ Sperm detach from Sertoli cells and eventually are carried out of the testis by the fluid in the centre of the seminiferous tubule.

spermatids

⑤ Spermatids become associated with nurse cells, called Sertoli cells, which help the spermatids to develop into *spermatozoa* (*n*). This is an example of **cell differentiation**.

Oogenesis

Oogenesis is the production of an ovum. Ova are often simply called eggs. Oogenesis occurs in the ovaries. The figures below show the structure of ovary tissue. The figure (bottom) shows the processes involved in oogenesis.

Micrograph of the ovary of a rabbit

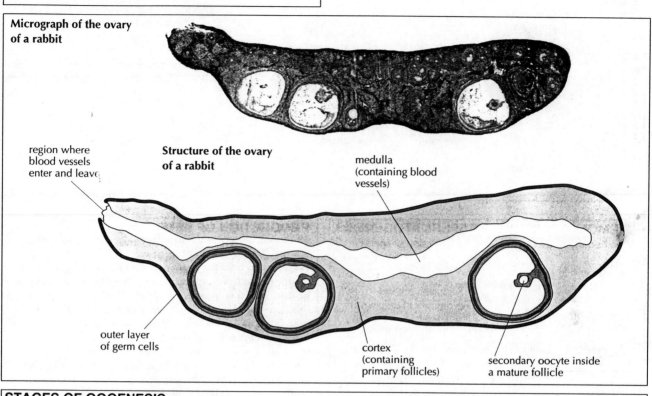

Structure of the ovary of a rabbit

region where blood vessels enter and leave

medulla (containing blood vessels)

outer layer of germ cells

cortex (containing primary follicles)

secondary oocyte inside a mature follicle

STAGES OF OOGENESIS

③ Primary oocytes start the **first division of meiosis** but stop during prophase I. The primary oocyte and a single layer of follicle cells around form a primary follicle.

④ When a baby girl is born the ovaries contain about 400 000 primary follicles.

⑤ Every menstrual cycle a few primary follicles start to develop. The primary oocyte completes the first division of meiosis, forming two haploid nuclei. The cytoplasm of the primary oocyte is **divided unequally** forming a large secondary oocyte (n) and a small polar cell (n).

② Oogonia **grow** into larger cells called *primary oocytes* ($2n$).

① In the ovaries of a female fetus, germ cells called *oogonia* ($2n$) divide by **mitosis** to form more oogonia ($2n$).

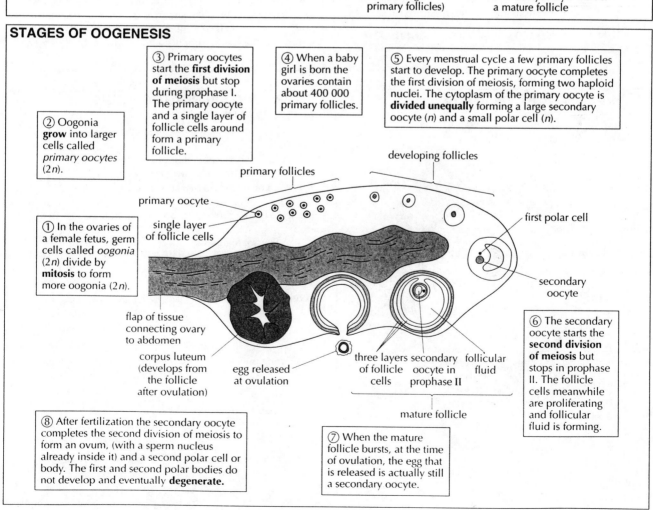

developing follicles

primary follicles

primary oocyte

single layer of follicle cells

first polar cell

secondary oocyte

flap of tissue connecting ovary to abdomen

corpus luteum (develops from the follicle after ovulation)

egg released at ovulation

three layers of follicle cells

secondary oocyte in prophase II

follicular fluid

mature follicle

⑥ The secondary oocyte starts the **second division of meiosis** but stops in prophase II. The follicle cells meanwhile are proliferating and follicular fluid is forming.

⑧ After fertilization the secondary oocyte completes the second division of meiosis to form an ovum, (with a sperm nucleus already inside it) and a second polar cell or body. The first and second polar bodies do not develop and eventually **degenerate.**

⑦ When the mature follicle bursts, at the time of ovulation, the egg that is released is actually still a secondary oocyte.

Gametes

STRUCTURE OF HUMAN SPERM

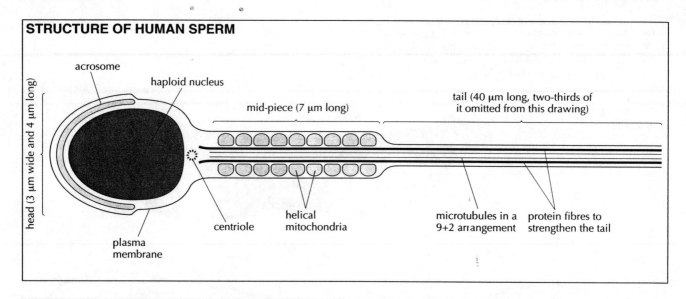

head (3 μm wide and 4 μm long)

acrosome

haploid nucleus

mid-piece (7 μm long)

tail (40 μm long, two-thirds of it omitted from this drawing)

centriole

helical mitochondria

microtubules in a 9+2 arrangement

protein fibres to strengthen the tail

plasma membrane

HORMONAL CONTROL OF SPERMATOGENESIS

Three hormones are involved in the production of sperm.

Hormone	Source	Role
FSH	Pituitary gland	Stimulates primary spermatocytes to undergo the first division of meiosis, to form secondary spermatocytes
Testosterone	Interstitial cells in the testis	Stimulates the development of secondary spermatocytes into mature sperm
LH	Pituitary gland	Stimulates the secretion of testosterone by the testis

PRODUCTION OF SEMEN

Three structures help to produce semen – the epididymis, seminal vesicles and prostate gland

When sperm from the testis arrive in the epididymis, they are unable to swim. The sperm undergo a maturing process while they are stored in the epididymis and become able to swim. The two seminal vesicles and prostate gland produce and store fluids and expel them during ejaculation. The fluid mixes with the sperm and increases the volume of the ejaculate. The fluid from the seminal vesicles contains nutrients for the sperm including fructose. It also contains mucus which protects the sperm in the vagina. The fluid from the prostate gland contains mineral ions and is alkaline so protects the sperm from the acid conditions in the vagina.

STRUCTURE OF A HUMAN EGG

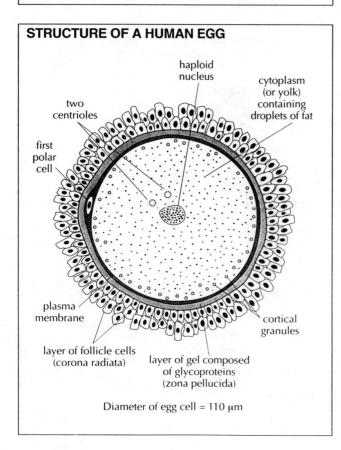

haploid nucleus

cytoplasm (or yolk) containing droplets of fat

two centrioles

first polar cell

plasma membrane

cortical granules

layer of follicle cells (corona radiata)

layer of gel composed of glycoproteins (zona pellucida)

Diameter of egg cell = 110 μm

COMPARING SPERMATOGENESIS WITH OOGENESIS

There are many similarities between the formation of sperm and eggs.
- Both start with proliferation of cells by mitosis.
- Both involve the cell growth before meiosis.
- Both involve the two divisions of meiosis.

The table below shows some of the differences.

Spermatogenesis	Oogenesis
Millions produced daily	One produced every 28 days
Released during ejaculation	Released on about day 14 of menstrual cycle by ovulation
Sperm formation starts during puberty in boys	The early stages of egg production happen during fetal development in females
Sperm production continues throughout the adult life of men	Egg production becomes irregular and then stops at the menopause in women
Four sperm are produced per meiosis	Only one egg is produced per meiosis

Fertilization

Summary of spermatogenesis

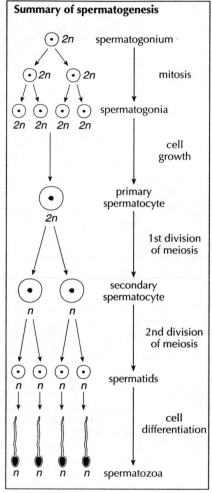

spermatogonium	
	mitosis
spermatogonia	
	cell growth
primary spermatocyte	
	1st division of meiosis
secondary spermatocyte	
	2nd division of meiosis
spermatids	
	cell differentiation
spermatozoa	

Summary of oogenesis

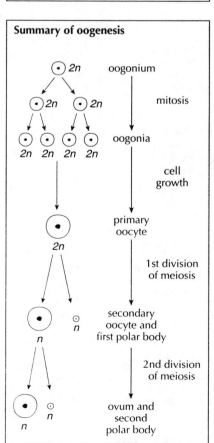

oogonium	
	mitosis
oogonia	
	cell growth
primary oocyte	
	1st division of meiosis
secondary oocyte and first polar body	
	2nd division of meiosis
ovum and second polar body	

Stages in the fertilization of a human egg

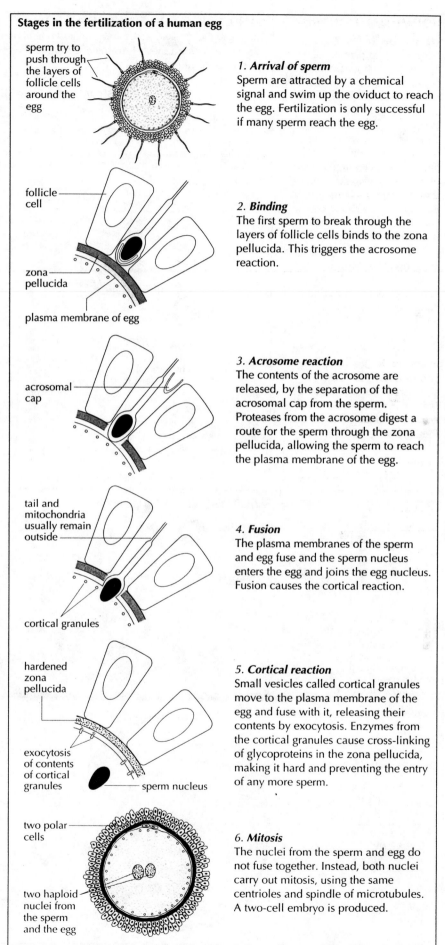

sperm try to push through the layers of follicle cells around the egg

follicle cell

zona pellucida

plasma membrane of egg

acrosomal cap

tail and mitochondria usually remain outside

cortical granules

hardened zona pellucida

exocytosis of contents of cortical granules

sperm nucleus

two polar cells

two haploid nuclei from the sperm and the egg

1. Arrival of sperm
Sperm are attracted by a chemical signal and swim up the oviduct to reach the egg. Fertilization is only successful if many sperm reach the egg.

2. Binding
The first sperm to break through the layers of follicle cells binds to the zona pellucida. This triggers the acrosome reaction.

3. Acrosome reaction
The contents of the acrosome are released, by the separation of the acrosomal cap from the sperm. Proteases from the acrosome digest a route for the sperm through the zona pellucida, allowing the sperm to reach the plasma membrane of the egg.

4. Fusion
The plasma membranes of the sperm and egg fuse and the sperm nucleus enters the egg and joins the egg nucleus. Fusion causes the cortical reaction.

5. Cortical reaction
Small vesicles called cortical granules move to the plasma membrane of the egg and fuse with it, releasing their contents by exocytosis. Enzymes from the cortical granules cause cross-linking of glycoproteins in the zona pellucida, making it hard and preventing the entry of any more sperm.

6. Mitosis
The nuclei from the sperm and egg do not fuse together. Instead, both nuclei carry out mitosis, using the same centrioles and spindle of microtubules. A two-cell embryo is produced.

Pregnancy and the placenta

HORMONAL CONTROL OF PREGNANCY

Estrogen and progesterone are needed throughout pregnancy to stimulate the development of the uterus lining. During the first few days after ovulation the corpus luteum secretes these hormones whether or not there has been fertilization. After implanting in the uterus wall, the embryo starts to secrete a hormone called HCG (human chorionic gonadotrophin). HCG prevents degeneration of the corpus luteum, which would happen at the end of a menstrual cycle. HCG stimulates the corpus luteum to grow and to continue secretion of estrogen and progesterone. This is essential to allow the pregnancy to continue. By the middle of the pregnancy, the corpus luteum starts to degenerate, but by then cells in the placenta are secreting estrogen and progesterone and these cells secrete increasing amounts until the end of the pregnancy.

STRUCTURE AND FUNCTION OF THE PLACENTA

The figure (below) shows the structure and functions of the placenta.
The figure (bottom) shows how materials are exchanged between maternal and fetal blood at the surface of villi in the placenta.

Structure of the placenta

Placenta – a disc-shaped structure, 185 mm in diameter and 20 mm thick when fully grown.

Placental villi – small projections that give a large surface area (14m²) for gas exchange and exchange of other materials. Fetal blood flows through capillaries in the villi.

Inter-villous spaces – maternal blood flows through these spaces, brought by uterine arteries and carried away by uterine veins.

Deoxygenated fetal blood flows from the fetus to the placenta along two umbilical arteries.

Oxygenated fetal blood flows back to the fetus from the placenta along the umbilical vein.

Endometrium – the lining of the uterus, into which the placenta grows.

Myometrium – muscular wall of the uterus, used during childbirth.

EXCHANGE OF MATERIALS ACROSS THE PLACENTA

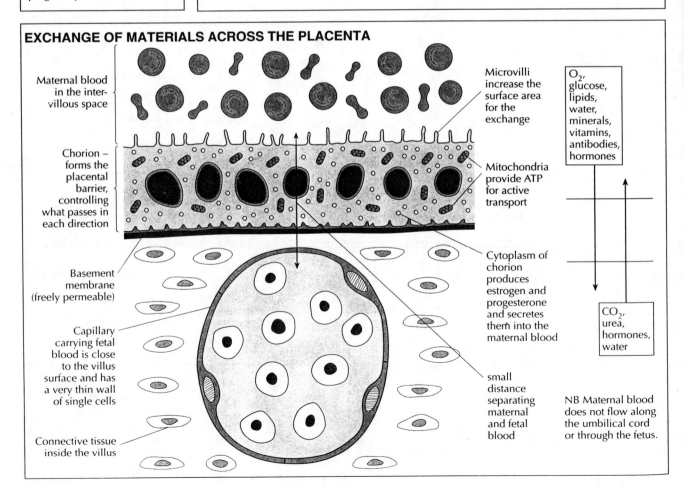

Maternal blood in the inter-villous space

Microvilli increase the surface area for the exchange

Chorion – forms the placental barrier, controlling what passes in each direction

Mitochondria provide ATP for active transport

Basement membrane (freely permeable)

Cytoplasm of chorion produces estrogen and progesterone and secretes them into the maternal blood

Capillary carrying fetal blood is close to the villus surface and has a very thin wall of single cells

Connective tissue inside the villus

small distance separating maternal and fetal blood

O_2, glucose, lipids, water, minerals, vitamins, antibodies, hormones

CO_2, urea, hormones, water

NB Maternal blood does not flow along the umbilical cord or through the fetus.

EXAM QUESTIONS ON TOPICS 8 & 9

1 In some plants two genes control flower colour. [Note: – represents any allele]

Plants with the genotype A_B_ have blue flowers.

Plants with the genotype A_bb have red flowers.

Plants with the genotype aa _ _ have white flowers.

a) State the name given to the type of inheritance where more than one gene controls a single
    phenotypic characteristic.   [1]

A homozygous blue-flowered plant (AABB) is crossed with a homozygous white-flowered plant (aabb).

b) State the genotype and phenotype of the F_1 offspring.   [2]

c) The F_1 plants are allowed to pollinate each other. Deduce, using the Punnett grid below, the genotypes of
    the gametes produced by the F_1 plants and the genotypes and phenotypes of all the possible F_2 offspring.   [5]

gametes ⟶				

d) State the expected ratio of flower colours in the F_2 offspring.   [1]

e) The two genes code for enzymes used to convert a white substance into a red pigment and the red pigment
    into a blue pigment. Deduce the effect of the enzymes produced from gene A and gene B.   [1]

2 a) Define recombination.   [1]

   When grey bodied long winged Drosophila flies were test crossed with black bodied vestigial wing flies the F_1 generation
   was found to contain:

   407 grey bodied long winged flies

   396 black bodied vestigial winged flies

   75 black bodied long winged flies

   69 grey bodied vestigial winged flies

   b) Identify which of the flies were recombinants.   [2]

   c) The F_1 generation does not follow Mendel's Second Law (Law of Independent Assortment).
       Explain how the observed ratio could have arisen.   [3]

   d) Suggest how geneticists could make use of experimental results of the type shown above.   [2]

3 a) Compare the structure of human sperm and eggs.   [4]

   b) Compare the role of FSH in men and women.   [3]

   c) Compare the roles of LH and HCG in women.   [3]

Types of defence

BLOOD CLOTTING

When human tissue is injured and blood escapes from blood vessels, a semi-solid is formed from liquid blood to seal up the wound and prevent entry of pathogens. The semi-solid is called a **blood clot** and the process is called **clotting**.

Platelets have an important role in clotting. Platelets are small cell fragments that circulate with erythrocytes and leukocytes in the blood plasma. The clotting process begins with the release of clotting factors either from damaged tissue cells or from platelets. These clotting factors set off a series of reactions in which the product of each reaction is the catalyst of the next reaction. This system helps to ensure that clotting only happens when it is needed and it also makes it a very rapid process. In the last reaction fibrinogen, a soluble plasma protein is altered by the removal of sections of peptide that have many negative charges. This allows the remaining polypeptide to bind to others, forming long protein fibres called fibrin. Fibrin forms a mesh of fibres across wounds. Blood cells are caught in the mesh and soon form a semi-solid clot. If exposed to air the clot dries to form a protective scab, which remains until the wound has healed.

Final reactions in blood clotting

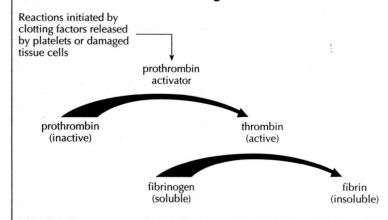

Fibrin and blood cells in a blood clot

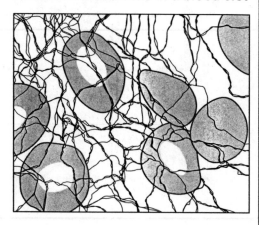

IMMUNITY

The human body is able to resist infection by many types of pathogenic organism. This resistance to infection is called **immunity**. Barriers to infection such as the skin provide immunity to a wide range of pathogens. Phagocytes are able to ingest and kill many organisms and chemicals including acids in the stomach and lysozyme in tears kill others. These defences are not adequate to resist the most potent pathogens. Specific immunity by production of antibodies is needed. Immunity to specific pathogens can be either natural or artificial and either active or passive.

Type of immunity	Definition	Examples
Natural or artificial	*Natural immunity is the result of infection with a pathogen*	The body becomes immune to the measles virus after being infected by it
	Artificial immunity is the result of inoculation with a vaccine	The body becomes immune to the measles virus after being inoculated with measles vaccine
Active or passive	*Active immunity is due to antibodies produced by the body's own immune system, following invasion of the body by pathogens*	Infection with rubella virus causes the development of immunity to rubella and re-infection is very rare
	Passive immunity is due to antibodies received from another organism, which made them as a result of active immunity	During pregnancy, antibodies are passed across the placenta from mother to the fetus and the first milk produced after birth, called colostrum, contains antibodies that can be absorbed into the newborn baby's blood through the stomach wall

THE ROLE OF CYTOTOXIC T-CELLS

Viruses multiply inside body cells, where they are out of reach of antibodies. One class of lymphocyte, called cytotoxic T-cells, can detect viral proteins in the membrane of infected body cells and then destroy these infected cells. Cytotoxic T-cells can also identify and destroy some types of cancer cell in the body.

The figure (right) shows a small cytotoxic T-cell preparing to kill a large cancer cell.

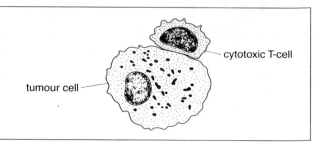

Antibody production

DEVELOPING IMMUNITY

Antibodies are made by lymphocytes called B-cells. The immune system as a whole can make 10^{15} different types of antibody. It would be impossible to make large quantities of all of these antibodies. Instead, a few B-cells that can make each type of antibody are produced and if these cells encounter an antigen to which their antibody binds, they multiply to form a clone of many cells. This is called **clonal selection**. Sometimes, several different types of antibody can bind to the same antigen, so more than one clone of cells is formed. This is called polyclonal selection.

A clone of B-cells can produce large amounts of antibody quickly and so give immunity to the disease with which the antigen is associated. Immunity to a disease is only developed if the immune system is challenged by the disease. This is called the principle of **challenge and response**.

STAGES IN ANTIBODY PRODUCTION

The figure on page 50 summarizes the production of antibodies by B-cells. Antibody production by B-cells usually depends on other types of lymphocyte, including macrophages and helper T-cells.

1. Macrophages take in antigens by endocytosis, process them and then attach them to membrane proteins called MHC proteins. The MHC proteins carrying the antigens are then moved to the plasma membrane by exocytosis and the antigens are displayed on the surface of the macrophage. This is called **antigen presentation**.

2. Helper T-cells have receptors in their plasma membrane that can bind to antigens presented by macrophages. Each helper T-cell has receptors with the same antigen-binding domain as an antibody. These receptors allow a helper T-cell to recognize an antigen presented by a macrophage and bind to the macrophage. The macrophage passes a signal to the helper T-cell changing it from an inactive to an active state. This is called **activation of helper T-cells**.

3. B-cells have antibodies in their plasma membrane. These antibodies recognize an antigen and the antigen binds to the antibody. An activated helper T-cell with receptors for the same antigen binds to the B-cell. The activated helper T-cell sends a signal to the B-cell, causing it to change from an inactive to an active state. This is called **activation of B-cells**. Activated B-cells start to divide by mitosis to form a clone of cells.

4. Plasma cells are active B-cells with a very extensive network of rough endoplasmic reticulum. This is used for synthesis of large amounts of antibody, which is then secreted by exocytosis.

5. Memory cells are B-cells and T-cells that are formed at the same time as activated helper T-cells and B-cells when a disease challenges the immune system. After the activated cells and the antibodies produced to fight the disease have disappeared the memory cells persist and allow a rapid response if the disease is encountered again. Memory cells give long-term immunity to a disease.

The figure (right) shows the events that lead to the production of a clone of plasma cells.

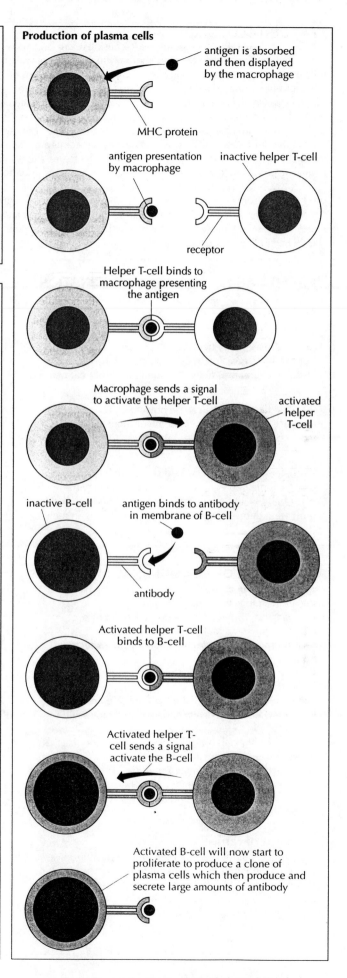

Production of plasma cells

antigen is absorbed and then displayed by the macrophage

MHC protein

antigen presentation by macrophage

inactive helper T-cell

receptor

Helper T-cell binds to macrophage presenting the antigen

Macrophage sends a signal to activate the helper T-cell

activated helper T-cell

inactive B-cell

antigen binds to antibody in membrane of B-cell

antibody

Activated helper T-cell binds to B-cell

Activated helper T-cell sends a signal activate the B-cell

Activated B-cell will now start to proliferate to produce a clone of plasma cells which then produce and secrete large amounts of antibody

Helping to defend the body

VACCINATION

A vaccine is a modified form of a disease-causing micro-organism that stimulates the body to develop immunity to the disease, without fully developing the disease. Vaccines contain weakened forms of the micro-organisms, killed forms or chemicals produced by the micro-organism that act as antigens. The vaccine is either injected into the body or sometimes swallowed. The principle of vaccination is that antigens in the vaccine cause the production of the antibodies needed to control the disease. Sometimes two or more vaccinations are needed to stimulate the production of enough antibodies. The figure (right) shows a typical response to a first and second vaccination against a disease. The first vaccination causes a little antibody production and the production of some memory cells. The second vaccination, sometimes called a booster shot, causes a response from the memory cells and therefore faster and greater production of antibodies. Memory cells should persist to give long-term immunity.

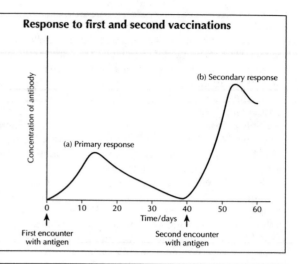

Response to first and second vaccinations

BENEFITS AND DANGERS OF VACCINATION

Vaccination has enormous benefits but also some possible dangers.

Benefits	**Dangers**
1. Some diseases may be completely eradicated. Smallpox, for example, has already been eradicated by vaccination, reducing human suffering and future costs of treatment	1. Excessive amounts of vaccination may reduce the ability of the immune system to respond to new diseases. It has been suggested that having too many vaccinations in a short time harmed soldiers fighting in the Gulf War
2. Deaths due to disease can be prevented. For example measles is a major cause of death of small children in some parts of the world	2. The immunity developed after vaccination may not be as effective as immunity due to actually catching a disease. Vaccination of children might make them vulnerable to more severe infection as adults, for example with measles
3. Long-term disabilities due to disease can be prevented. For example, if pregnant women are infected with rubella their babies can be born with deafness, blindness and heart and brain damage. Mumps can cause infertility in men	3. There is a danger of side effects from some vaccines, which can cause long-term disability. Whooping cough vaccination sometimes causes brain damage and MMR vaccine (combined measles mumps and rubella vaccine) may increase the chance of autism

PRODUCTION OF MONOCLONAL ANTIBODIES

Large quantities of a single type of antibody can be made using an ingenious technique.
- Antigens that correspond to a desired antibody are injected into an animal.
- B-cells producing the desired antibody are extracted from the animal.
- Tumour cells are obtained. These cells grow and divide endlessly.
- The B-cells are fused with the tumour cells, producing hybridoma cells that divide endlessly and produce the desired antibody.
- The hybridoma cells are cultured and the antibodies that they produce are extracted and purified.

The figure (right) shows a factory used for the industrial production of monoclonal antibodies. There are many ways in which monoclonal antibodies can be used. Two examples are described here.

Treatment of rabies

Rabies usually causes death in humans before antibodies produced by the immune system controls it. If a person becomes infected, an effective strategy is to vaccinate against rabies and at the same time inject monoclonal antibodies. These control the rabies virus until antibodies are produced as a result of the vaccination.

Diagnosis of malaria

Tests using monoclonal antibodies have been developed for many diseases, including malaria. Monoclonal antibodies are produced that bind to antigens in malarial parasites. A test plate is coated with the antibodies. A sample is left in the plate long enough for malaria antigens in the sample to bind to the antibodies. The sample is then rinsed off the plate. Any bound antigens are detected using more monoclonal antibodies with enzymes attached that cause a colour change. This is called an ELISA test. It can be used to measure the level of infection and to distinguish between different strains of malaria, either in humans or in mosquitoes.

Nerve impulses

Neurones transmit messages in the form of nerve impulses. These impulses are electrical, but involve the movement of positively charged ions, not electrons. A nerve impulse can travel along a neurone at speeds as high as 100 metres per second.

RESTING POTENTIALS

Neurones pump ions across their plasma membranes by active transport. Sodium is pumped out of the neurone and potassium is pumped in. Concentration gradients of both sodium and potassium are established across the membrane. The inside of the neurone develops a net negative charge, compared with the outside, because of the presence of chloride and other negatively charged ions.
There is therefore an electrical potential or voltage across the membrane. This is called the **resting potential**.
The resting potential is the electrical potential across the plasma membrane of a cell that is not conducting an impulse.

ACTION POTENTIALS

When an impulse passes along the neurone, sodium and potassium ions are allowed to diffuse across the membrane, through voltage-gated ion channels. The electrical potential across the membrane is initially reversed but is then restored. This is called an **action potential**. The figure (right) shows the changes in membrane polarisation that occur during an action potential. The figure (below) shows the net charges inside and outside a neurone and the figure (bottom) shows ion movements.
An action potential is the reversal and restoration of the electrical potential across the plasma membrane of a cell, as an electrical impulse passes along it.

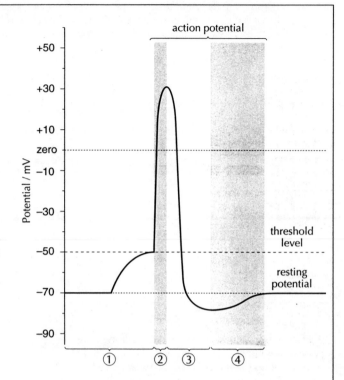

Changes in net charge during an action potential

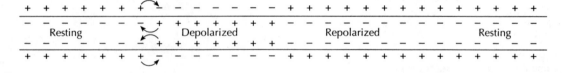

Stages in the passage of a nerve impulse

① An action potential in one part of a neurone causes an action potential to develop in the next section of the neurone. This is due to diffusion of sodium ions between the region with an action potential and the region at the resting potential. These ion movements, local currents, reduce the resting potential. If the potential rises above the threshold level, voltage-gated channels open.

② Sodium channels open very quickly and sodium ions diffuse into the neurone down the concentration gradient. This reduces the membrane potential and causes more sodium channels to open. The entry of positively charged sodium ions causes the inside of the neurone to develop a net positive charge compared to the outside – the potential across the membrane is reversed. This is called **depolarization**.

③ Potassium channels open after a short delay. Potassium ions diffuse out of the neurone down the concentration gradient through the opened channels. The exit of positively charged potassium ions cause the inside of the neurone to develop a net negative charge again compared with the outside – the potential across the membrane is restored. This is called **repolarization**.

④ Concentration gradients of sodium and potassium across the membrane are restored by the active transport of sodium ions out of the neurone and potassium ions into the neurone. This restores the resting potential and the neurone is then ready to conduct another nerve impulse. As before, sodium ions diffuse along inside the neurone from an adjacent region that has already depolarized and initiate depolarization.

Ion movements during an action potential

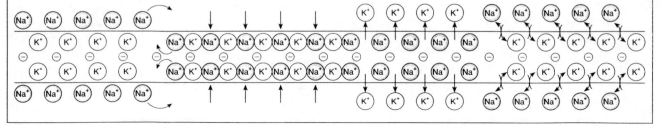

Neurones and synapses

NEURONES

Neurones are nerve cells. The figure below shows the structure of a motor neurone. The function of a motor neurone is to carry impulses from the CNS (brain and spinal cord), to a muscle or to a gland. The cell body is always located in the grey matter of the CNS. The axon is located in one of the nerves of the PNS. The total length of a motor neurone can be more than one metre, most of which is omitted from the figure.

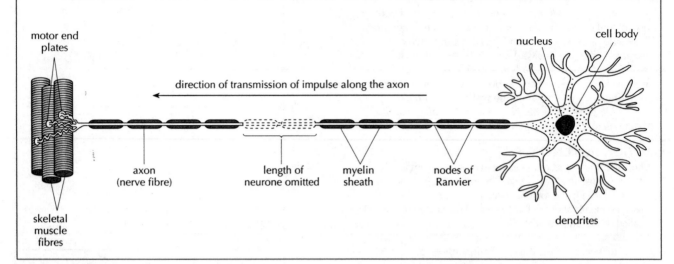

SYNAPSES

A synapse is a junction between two neurones. The plasma membranes of the neurones are separated by a narrow fluid-filled gap called the synaptic cleft. Messages are passed across the synapse in the form of chemicals called **neurotransmitters**. The neurotransmitters always pass in the same direction from the pre-synaptic neurone to the post-synaptic neurone.

Many synapses function in the following way.

(1) A nerve impulse reaches the end of the pre-synaptic neurone.

(2) Depolarization of the pre-synaptic membrane causes voltage-gated calcium channels to open. Calcium ions diffuse into the pre-synaptic neurone

(3) Influx of calcium causes vesicles of neurotransmitter to move to the pre-synaptic membrane and fuse with it, releasing the neurotransmitter into the synaptic cleft by exocytosis.

(4) The neurotransmitter diffuses across the synaptic cleft and binds to receptors in the post-synaptic membrane.

(5) The receptors are transmitter-gated ion channels, which open when neurotransmitter binds. Sodium and other positively charged ions diffuse into the post-synaptic neurone. This causes depolarization of the post-synaptic membrane.

(6) The depolarization passes on down the post-synaptic neurone as an action potential.

(7) Neurotransmitter in the synaptic cleft is rapidly broken down, to prevent continuous synaptic transmission. For example, acetylcholine is broken down by cholinesterase in synapses that use it as a neurotransmitter. Calcium ions are pumped out of the pre-synaptic neurone into the synaptic cleft.

The figure (right) shows the events that occur during synaptic transmission.

Stages in synaptic transmission

(3) Vesicles of neurotransmitter move to the membrane and release their contents

(1) Nerve impulse reaches the end of the pre-synaptic neurone

(7) Calcium is pumped out. Neurotransmitter is broken down in the cleft and reabsorbed into the vesicles

synaptic knob

vesicles of neuro-transmitter

(2) Calcium diffuses in through calcium channels

(4) Neuro-transmitter diffuses across the synaptic cleft and binds to receptors

(5) Sodium ions enter the post-synaptic neurone and cause depolarization

(6) Nerve impulse setting off along the post-synaptic neurone

Muscle contraction

THE STRUCTURE OF SKELETAL MUSCLE

When viewed with a light microscope, skeletal muscle is seen to consist of large multinucleate cells called **muscle fibres**. These have a striated or striped appearance. Electron micrographs show the reason for this. Muscle fibres contain cylindrical structures called **myofibrils**. The myofibrils consist of repeating units called **sarcomeres**, which have light and dark bands. Around each myofibril is a special type of endoplasmic reticulum, called sarcoplasmic reticulum. There are also mitochondria between the myofibrils. The figures are electron micrographs of skeletal muscle and the figure (below right) is a drawing.

Structure of a sarcomere

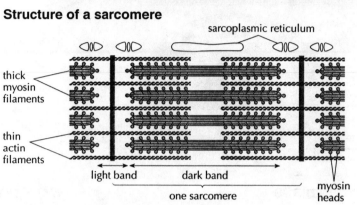

CONTRACTION OF SKELETAL MUSCLE

The contraction of skeletal muscle is due to the sarcomeres in the myofibrils becoming shorter. This is achieved by the sliding of actin and myosin filaments over each other, using ATP to provide the necessary energy. The figure (below) shows how this occurs.

CONTROLLING MUSCLE CONTRACTION

When a skeletal muscle fibre is relaxed, a protein called **tropomyosin** blocks the myosin binding sites on actin. If a motor neurone stimulates the muscle fibre, calcium ions are released from the sarcoplasmic reticulum. These calcium ions bind to another protein called **troponin**. Troponin then causes tropomyosin to move, which exposes the myosin binding sites and allows contraction to begin.

STAGES IN MUSCLE CONTRACTION

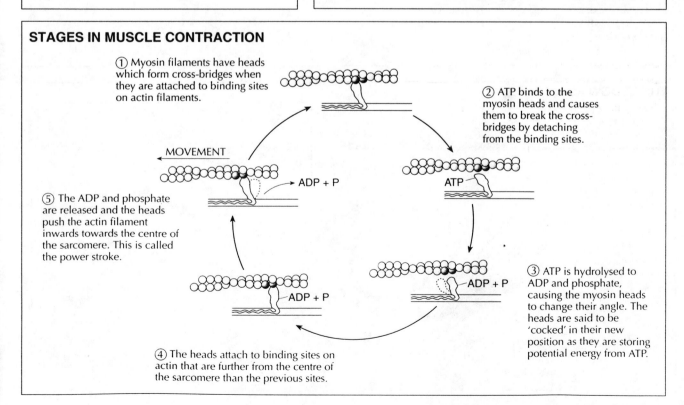

① Myosin filaments have heads which form cross-bridges when they are attached to binding sites on actin filaments.

② ATP binds to the myosin heads and causes them to break the cross-bridges by detaching from the binding sites.

③ ATP is hydrolysed to ADP and phosphate, causing the myosin heads to change their angle. The heads are said to be 'cocked' in their new position as they are storing potential energy from ATP.

④ The heads attach to binding sites on actin that are further from the centre of the sarcomere than the previous sites.

⑤ The ADP and phosphate are released and the heads push the actin filament inwards towards the centre of the sarcomere. This is called the power stroke.

MOVEMENT

Muscles, joints and locomotion

THE ROLES OF MUSCLES IN LOCOMOTION IN ANIMALS

Most animals can move from one place to another. This is called **locomotion**. Animals show a wide diversity of types of locomotion. The figures (below) show four examples of locomotion. When muscles contract they provide the force needed for locomotion. Muscles only do work when they contract so pairs of muscles are needed to carry out opposite movements. These are called **antagonistic pairs**.

THE ROLES OF NERVES AND BONES IN LOCOMOTION

Nerves stimulate muscles to contract. They stimulate each of the different muscles used in locomotion to contract at the correct time, so the movement is coordinated.
Bones provide a firm anchorage for muscles in many animals. They also act as levers, changing the size or direction of forces generated by muscles. Junctions between bones are called **joints**. The figures (bottom) show the structures of the elbow joint and their functions.

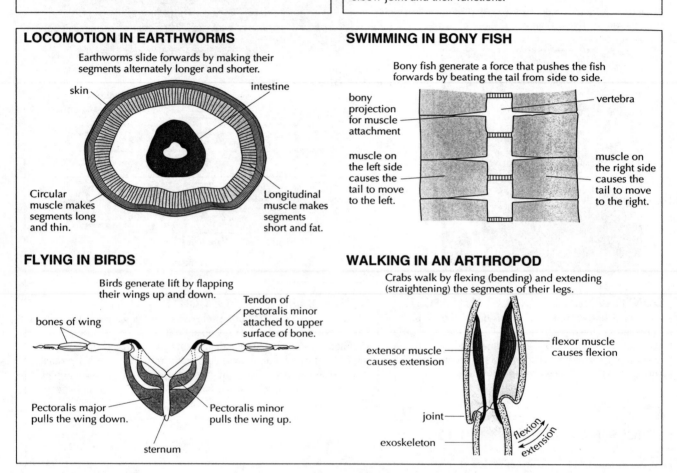

LOCOMOTION IN EARTHWORMS

Earthworms slide forwards by making their segments alternately longer and shorter.

skin

intestine

Circular muscle makes segments long and thin.

Longitudinal muscle makes segments short and fat.

SWIMMING IN BONY FISH

Bony fish generate a force that pushes the fish forwards by beating the tail from side to side.

bony projection for muscle attachment

vertebra

muscle on the left side causes the tail to move to the left.

muscle on the right side causes the tail to move to the right.

FLYING IN BIRDS

Birds generate lift by flapping their wings up and down.

Tendon of pectoralis minor attached to upper surface of bone.

bones of wing

Pectoralis major pulls the wing down.

Pectoralis minor pulls the wing up.

sternum

WALKING IN AN ARTHROPOD

Crabs walk by flexing (bending) and extending (straightening) the segments of their legs.

extensor muscle causes extension

flexor muscle causes flexion

joint

exoskeleton

flexion

extension

THE ELBOW JOINT

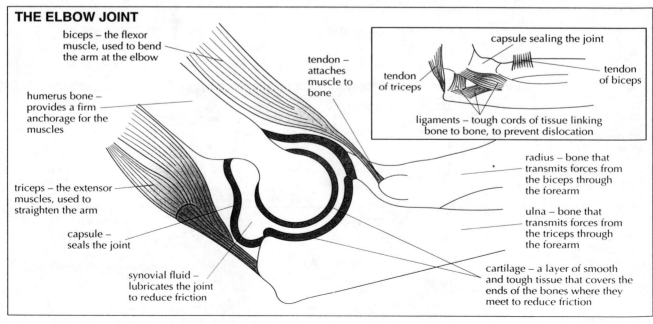

biceps – the flexor muscle, used to bend the arm at the elbow

humerus bone – provides a firm anchorage for the muscles

triceps – the extensor muscles, used to straighten the arm

capsule – seals the joint

synovial fluid – lubricates the joint to reduce friction

tendon – attaches muscle to bone

capsule sealing the joint

tendon of triceps

tendon of biceps

ligaments – tough cords of tissue linking bone to bone, to prevent dislocation

radius – bone that transmits forces from the biceps through the forearm

ulna – bone that transmits forces from the triceps through the forearm

cartilage – a layer of smooth and tough tissue that covers the ends of the bones where they meet to reduce friction

EXAM QUESTIONS ON TOPICS 10 & 11

1 a) State one example of immunity that is

   (I) active and natural [1]

   (ii) passive and natural [1]

   (iii) active and artificial [1]

   b) Outline the process of vaccination against bacterial or viral diseases. [3]

   c) Suggest a reason for rapid production of antibodies after a vaccination is repeated. [1]

2 The electron micrograph below shows part of a myofibril, taken from a skeletal muscle. The parts marked M contain myosin filaments. Three other regions are labelled I, II and III.

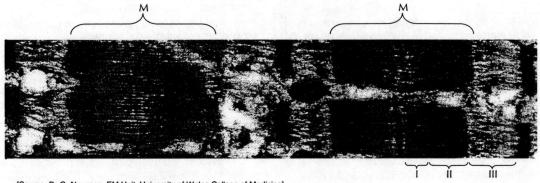

[Source: Dr G. Newman, EM Unit, University of Wales College of Medicine]

   a) (i) State one type of filament, apart from myosin, which is present in myofibrils. [1]

   (ii) Identify in which of the regions labelled I, II and III these other filaments can be found. [1]

   b) The myofibril is partly contracted. Deduce which of the regions would increase in length if

   (i) the myofibril contracted more [1]

   (ii) the myofibril relaxed [1]

3 Animals move using many different methods. The chart below shows how much energy is needed by animals to swim on the surface of the water and underwater. The chart shows the relative amount of energy that is needed to move an equal mass of each animal over an equal distance.

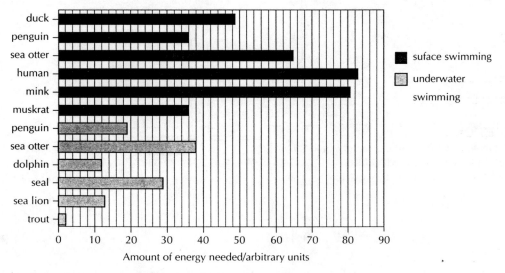

(Source: Alexander, *Nature*, (1999) 397 pages 651–653)

   a) By reference to named animals, compare underwater swimming with surface swimming. [2]

   b) Suggest two reasons for the high energy cost of swimming in humans. [2]

Excretory products and dialysis

EXCRETORY PRODUCTS OF LIVING ORGANISMS

Excretion is one of the basic life processes of all living organisms. The need for excretion is explained on page 48. Living organisms do not all excrete the same substances. Plants excrete oxygen, for example, whereas animals excrete carbon dioxide and nitrogenous waste products. Nitrogenous waste products are formed when surplus amino acids are broken down. The nitrogen compound that is excreted varies between groups of animals and can be related to the habitat and activities of the animal.

- Freshwater fish excrete ammonia, mostly through their gills, but also in urine. Ammonia is very toxic because it is alkaline and so has to be diluted in a large volume of water in urine. Freshwater fish have abundant supplies of water, so water losses can easily be replaced.
- Mammals excrete urea, which is much less toxic and so can be excreted in about a tenth as much water as ammonia. This is an advantage for mammals living in habitats with low water availability. The figure (below) shows a desert rat, with kidneys so efficient at water conservation that it never needs to drink water.
- Birds excrete uric acid, which has low toxicity and is only partially soluble in water. When water is reabsorbed during production of urine in the kidney, uric acid precipitates. The white semi-solid urine of birds contains very little water – only about a fiftieth as much as would be needed to excrete nitrogen as ammonia. Conservation of water and lower body mass are advantages to birds, especially in dry habitats and when flying.

Desert rat – adapted to conserve water

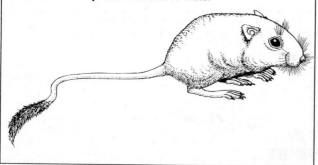

ARTIFICIAL EXCRETION

Humans cannot survive for long without a functioning kidney, unless an artificial method of excretion is used.
The commonest method is dialysis, using a kidney machine.

Patient during a session of dialysis

RENAL DIALYSIS

Dialysis involves the diffusion of solutes from a higher to a lower concentration through a semi-permeable membrane. The dialysis membrane used in kidney machines is usually made of cellulose acetate or nitrate. It has pores that let small solute particles pass through, but not large particles such as plasma proteins or blood cells. Blood flows on one side of the dialysis membrane and dialysis fluid on the other side (above). The formulation of the dialysis fluid ensures that only some substances diffuse into it from the blood.

- It contains no urea or other excretory products, so these waste products diffuse into it rapidly.
- It has the same concentration of glucose, mineral ions and other desirable substances as normal blood plasma, so these substances do not diffuse unless the level in blood plasma is above or below normal.
- It contains dextran, a solute that cannot pass through the dialysis membrane and so causes excess water to move by osmosis from the blood to the dialysis fluid.

During dialysis the patient's blood flows through tubes or between sheets of dialysis membrane. The blood is taken from the patient and returned via needles inserted into a blood vessel in the arm. The dialysis fluid has to be gradually replaced throughout a session to maintain the concentration gradients. A large volume of fluid is used, in contrast to the human kidneys, which can excrete waste products with a very small loss of water.

Structure of a kidney dialysis machine

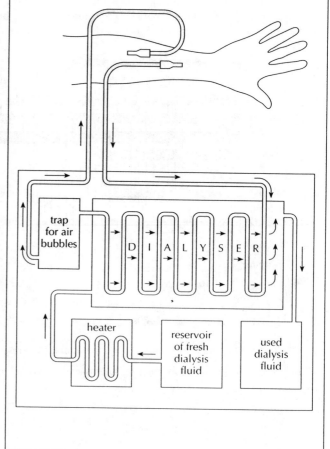

Kidney structure and ultrafiltration

THE STRUCTURE OF THE KIDNEY

The kidneys produce urine. The figure (below) shows the structure of the kidney. The cortex and medulla of the kidney contain many narrow tubes called nephrons. The figure (below) shows the structure of a nephron, together with the associated glomerulus.

Structure of a kidney in vertical section

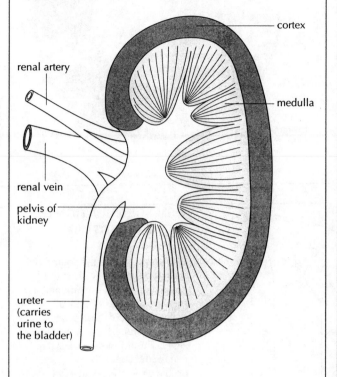

Structure of part of a glomerulus

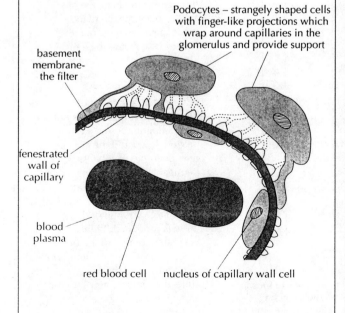

ULTRAFILTRATION IN THE GLOMERULUS

The function of the glomerulus is production of a filtrate from blood by a process called **ultrafiltration**. Part of the blood plasma escapes through the walls of all capillaries, but in the glomerulus 20% escapes, which is much greater than usual. There are two main reasons for this.

- The blood pressure is very high, because the vessel taking blood away from the glomerulus is narrower than the vessel bringing blood.
- The capillaries in the glomerulus are fenestrated – they have many pores through them.

These pores are large enough to allow any molecules through, but on the outside of the capillary wall is a basement membrane, composed of a gel of glycoproteins (below). The basement membrane acts as a filter as it only allows molecules with a molecular mass below 68 000 to pass through. It lets all substances in blood plasma through except plasma proteins. The fluid produced by ultrafiltration is collected by the Bowman's capsule and flows on into the proximal convoluted tubule.

Structure of the nephron

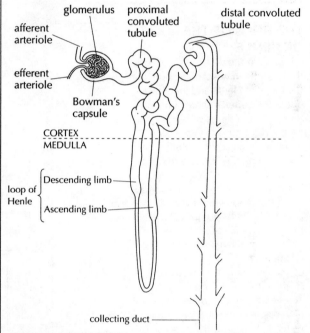

COMPARISON OF THE CONTENTS OF FLUIDS IN THE KIDNEY

The physiology of the kidney can be studied by comparing the content of blood flowing to and from the kidney with the content of glomerular filtrate and urine.

Content (mg per 100 ml of blood)

	Blood plasma in renal artery	Glomerular filtrate	Urine	Blood plasma in renal vein
Glucose	90	90	0	90
Urea	30	30	2000	24
Proteins	740	0	0	740
Sodium ions	900	900	1200	720

Blood in the renal artery has higher oxygen levels and lower CO_2 levels than blood in the renal vein.

Urine production and osmoregulation

SELECTIVE RE-ABSORPTION IN THE PROXIMAL CONVOLUTED TUBULE

Large volumes of glomerular filtrate are produced – about 1 litre every 10 minutes by the two kidneys. As well as waste products, the filtrate contains substances that the body needs, which must be re-absorbed into the blood. Most of this **selective re-absorption** happens in the proximal convoluted tubule. The wall of the nephron consists of a single layer of cells. In the proximal convoluted tubule the cells have microvilli projecting into the lumen (right), giving a large surface area for absorption. Pumps in the membrane re-absorb useful substances by active transport, using ATP produced by mitochondria in the cells. All of the glucose in the filtrate is re-absorbed. About 80% of the mineral ions, including sodium is re-absorbed. Active transport of solutes makes the total solute concentration higher in the cells of the wall than in the filtrate in the tubule. Water therefore moves from the filtrate to the cells and on into the adjacent blood capillary by osmosis. About 80% of the water in the filtrate is re-absorbed, leaving 20% of the original volume to flow on into the loop of Henle.

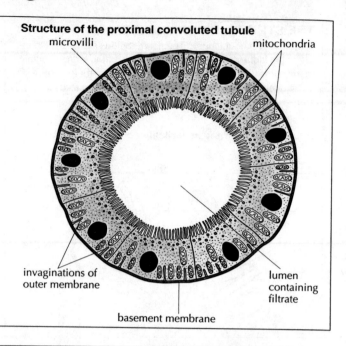

Structure of the proximal convoluted tubule

microvilli — mitochondria — invaginations of outer membrane — lumen containing filtrate — basement membrane

THE ROLE OF THE LOOP OF HENLE

Glomerular filtrate flows deep into the medulla in descending limbs of the loops of Henle and then back out to the cortex in ascending limbs. Descending limbs and ascending limbs are opposite in terms of permeability. Descending limbs are permeable to water but not to sodium ions. Ascending limbs are permeable to sodium ions but not to water (right). Ascending limbs pump sodium ions from the filtrate into the medulla by active transport, creating a high solute concentration in the medulla. As the filtrate flows down the descending limb into this region of high solute concentration, some water is drawn out by osmosis. This dilutes the fluids in the medulla slightly. However the filtrate that leaves the loop of Henle is more dilute than the fluid entering it, showing that the overall effect of the loop of Henle is to increase the solute concentration of the medulla. This is the role of the loop of Henle – to create an area of high solute concentration in the cells and tissue fluid of the medulla.

After the loop of Henle, the filtrate passes through the distal convoluted tubule, where the ions can be exchanged between the filtrate and the blood to adjust blood levels. It then passes into the collecting duct.

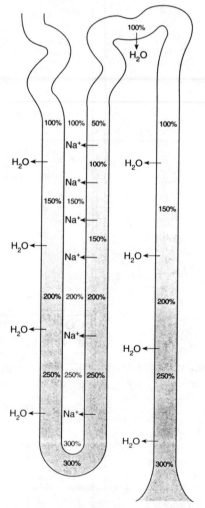

Movements of water and sodium ions in the loop of Henle and the collecting duct. Solute concentrations inside and outside the nephron are shown as a percentage of normal blood solute concentration

OSMOREGULATION IN THE COLLECTING DUCT

Osmoregulation is the control of water and solute levels. The collecting duct has an important role in osmoregulation. If the water content of the blood is too low, the pituitary gland secretes ADH. This hormone makes the cells of the collecting duct produce membrane channels called aquaporins, which makes the collecting duct permeable to water. As the filtrate passes down the collecting duct through the medulla, the high solute concentration of the medulla causes most of the water in the filtrate to be re-absorbed by osmosis. A small volume of concentrated urine is produced.

If the water content of the blood is too high, ADH is not secreted, aquaporins are broken down and the collecting duct becomes much less permeable to water. Little water is reabsorbed as the filtrate passes down the collecting duct and a large volume of dilute urine is produced. In this way the water content of the blood is kept within narrow limits. The urine produced by the collecting ducts drains into the renal pelvis and down the ureter to the bladder.

Plant diversity

Plant classification

There are four main groups of plants, which can be easily distinguished by studying their external structure.

	Roots, leaves and stems	Maximum height	Reproductive structures
Bryophytes – mosses	Bryophytes have no roots, only structures similar to root hairs called **rhizoids** Mosses have simple leaves and stems Liverworts consist of a flattened thallus	**0.5 metres**	Spores are produced in a capsule. The capsule develops at the end of a stalk
Filicinophytes – ferns	Ferns have roots, leaves and short non-woody stems. The leaves are usually curled up in bud and are often **pinnate** – divided into pairs of leaflets	**15 metres**	Spores are produced in sporangia, usually on the underside of the leaves
Coniferophytes – conifers	Conifers are shrubs or trees with roots, leaves and woody stems. The leaves are often narrow with a thick waxy cuticle	**100 metres**	Seeds are produced. The seeds develop from ovules on the surface of the scales of female **cones**. Male cones produce pollen
Angiospermo-phytes – flowering plants	Flowering plants are very variable but usually have roots, leaves and stems. The stems of flowering plants that develop into shrubs and trees are woody	**100 metres**	Seeds are produced. The seeds develop from ovules inside **ovaries**. The ovaries are part of **flowers**. **Fruits** develop from the ovaries, to disperse the seed

ADAPTATIONS OF PLANTS TO THEIR HABITATS

Flowering plants can be found growing in a wide variety of habitats. The structure of each type of plant is closely related to the amount of water available in the habitat. The figure (below left) shows the structure of *Atropa bella-donna*, a dicotyledonous plant that is adapted to growing in a habitat with moderate supplies of water. The figure (below right) shows a plant adapted to grow in deserts and the figure (bottom) shows a plant adapted to grow in water.

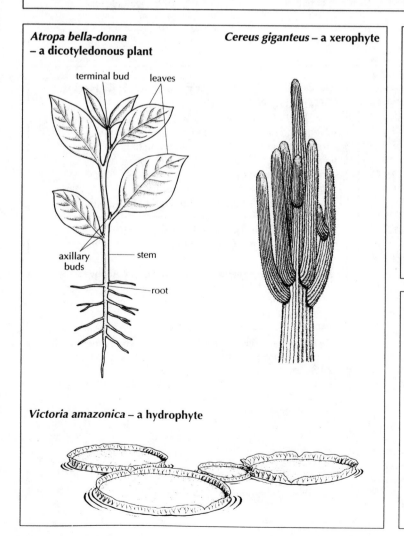

Atropa bella-donna – a dicotyledonous plant

terminal bud
leaves
axillary buds
stem
root

Cereus giganteus – a xerophyte

Victoria amazonica – a hydrophyte

XEROPHYTES

Plants that are adapted to grow in very dry habitats are called **xerophytes**.
Cereus giganteus, the saguaro or giant cactus is a example of a xerophyte. It grows in deserts in Mexico and Arizona and shows many xerophytic adaptations.
- Spines instead of leaves, to reduce transpiration.
- Thick stems containing water storage tissue.
- Very thick waxy cuticle covering the stem.
- Vertical stems to absorb sunlight early and late in the day but not at midday when the light is most intense.
- Very wide-spreading network of shallow roots to absorb water after rains.
- CAM physiology, which involves opening stomata during the cool nights instead of in the intense heat of the day.

HYDROPHYTES

Plants that are adapted to grow either submerged in water or floating on the surface are called **hydrophytes**.
Victoria amazonica, the Amazon water lily is an example of a hydrophyte. It grows on the water surface in shallow pools at the edge of the Amazon River and shows many hydrophyte adaptations.
- Air spaces in the leaf to provide buoyancy.
- Stomata in the upper epidermis of the leaf, which is in contact with the air, but not in the lower epidermis.
- Waxy cuticle on the upper surface of the leaf, but not on the lower surface, which is in contact with water.
- Small amounts of xylem in stems and leaves.

Leaf structure and function

LEAVES AND PHOTOSYNTHESIS

The function of leaves is to produce food for the plant by photosynthesis. The leaf is adapted by its structure to carry out photosynthesis efficiently. On page 3 is a scanning electron micrograph of a leaf. The figure (below) is a plan diagram of tissues in part of a leaf of a dicotyledonous plant to show the adaptations for photosynthesis.

Tissues of the leaf and their functions

LEAVES AND TRANSPIRATION

Photosynthesis depends on gas exchange over a moist surface. Spongy mesophyll cell walls provide this surface. Water often evaporates from the surface and is lost, in a process called transpiration. *Transpiration is the loss of water vapour from the leaves and stems of plants.* The figure (below) shows adaptations to minimize the amount of transpiration.

Palisade mesophyll—consists of densely packed cylindrical cells with many chloroplasts. This is the main photosynthetic tissue and is positioned near the upper surface where the light intensity is highest

Upper epidermis—a continuous layer of cells covered by a thick waxy cuticle. Prevents water loss from the upper surface even when heated by sunlight. Lower epidermis in a cooler position has a thinner waxy cuticle

The main part of the leaf is the leaf blade or lamina. It has a large surface area to absorb sunlight but is very thin—only about 0.3 mm. It is composed of four thin tissue layers with veins at intervals.

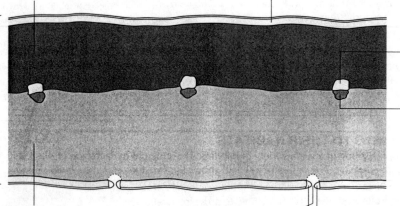

Xylem—brings water to replace losses due to transpiration

Phloem—transports products of photosynthesis out of the leaf.

Vein is centrally positioned to be close to all cells.

Spongy mesophyll—consists of loosely packed rounded cells with few chloroplasts. This tissue provides the main gas exchange surface so must be near the stomata in the lower epidermis.

Stoma—a pore that allows CO_2 for photosynthesis to diffuse in and O_2 to diffuse out.

Guard cells—this pair of cells can open or close the stoma and so control the amount of transpiration.

TRANSPORT IN PHLOEM

Sugars, amino acids and other organic compounds produced in photosynthesis are transported out of the leaf by phloem tissue. Phloem is located in all of the veins of the leaf. The structures within phloem tissue that transport organic compounds are called **sieve tubes**.

Columns of cells develop into sieve tubes by breaking down their nuclei and cytoplasm and making large pores in their end walls to allow a flow of sap. The plasma membranes in sieve tubes remain, and have the important task of pumping organic compounds into the sieve tube by active transport. Transport in phloem is thus an active process, involving the use of ATP. A high solute concentration is created inside the sieve tubes of the leaf, which causes water to enter by osmosis. This creates a high enough pressure to pump the sap inside the sieve tube, containing dissolved organic compounds, to any part of the plant that is using them.

Phloem also transports some spray chemicals if they are absorbed into the leaf after being sprayed onto it.
The transport of any biochemical in phloem whether produced by the plant or not is called **translocation**.

ABIOTIC FACTORS AFFECTING THE RATE OF TRANSPIRATION

Four abiotic factors have an effect on the rate of transpiration.

- Light – guard cells close the stomata in darkness, so transpiration is much greater in the light.
- Temperature – heat is needed for evaporation of water from the surface of spongy mesophyll cells, so as temperature rises the rate of transpiration rises. Higher temperatures also increase the rate of diffusion through the air spaces in the spongy mesophyll, and reduce the relative humidity of the air outside the leaf.
- Humidity – water diffuses out of the leaf when there is a concentration gradient between the air spaces inside the leaf and the air outside. The air spaces are always nearly saturated. The lower the humidity outside the leaf, the steeper the gradient and therefore the faster the rate of transpiration.
- Wind – pockets of air saturated with water vapour tend to form near stomata in still air, which reduce the rate of transpiration. Wind blows the saturated air away and so increases the rate of transpiration.

FOOD STORAGE IN PLANTS

Many perennial plants develop a food storage organ in which food is stored during a dormant season and then used in the next growth season. The food is transported to and from the storage organ in the phloem. Potato tubers are an example of a storage organ. Tubers are swollen underground stems. The figure (right) shows a potato plant with tubers forming.

Food storage in potato tuber

Leaves produce food by photosynthesis.

Phloem in stems transports food to storage organs.

Tuber grows and stores food.

Stems and roots

STRUCTURE AND FUNCTION OF STEMS

Stems connect the leaves, roots and flowers of plants and transport materials between them using xylem and phloem tissue. Stems support the aerial parts of terrestrial plants. Xylem tissue provides support especially in woody stems. Cell turgor also provides support, with both pith and cortex containing many cells that are usually turgid. The figure (below) is a plan diagram to show the position of the tissues in the stem of a young dicotyledonous plant.

Transverse section of a stem

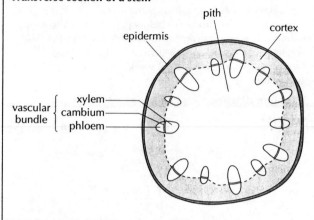

STRUCTURE AND FUNCTION OF ROOTS

Roots absorb mineral ions and water from soil. They anchor the plant in the soil and are sometimes used for food storage. The figure (below) is a plan diagram to show the position of tissues in the root of a young dicotyledonous plant.
The structure of root systems gives them a large surface area for absorption – by branching, by the growth of root hairs and by having a large surface area of cortex cell walls.

Transverse section of a root

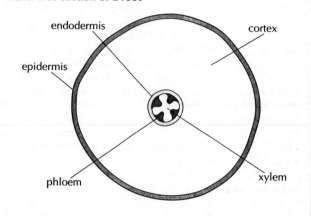

TRANSPORT OF WATER THROUGH THE PLANT

Transpiration causes a flow of water from the roots, through the stems to the leaves of plants. This flow is called the **transpiration stream**.

The process starts with **evaporation** of water from the cell walls of spongy mesophyll cells in the leaf.

The water that evaporates is replaced with water from xylem vessels in the leaf. The water is pulled out of xylem vessels and through pores in spongy mesophyll cell walls by capillary action. Low pressure or suction is created inside xylem vessels when water is pulled out. This is called the **transpiration pull**.

Xylem vessels contain long, unbroken columns of water and the transpiration pull is transmitted down through these columns of water to the roots. The figure (below) shows the structure of a xylem vessel. Mature xylem vessels are dead and the flow of water through them is passive.

The transmission of the transpiration pull through xylem vessels depends on the **cohesion** of water molecules, due to hydrogen bonding.

Structure of xylem vessels

No plasma membranes are present in mature xylem vessels, so water can move in and out freely

Lumen of the xylem vessel is filled with sap, as the cytoplasm and the nuclei of the original cells break down. End walls also break down to form a continuous tube

Helical or ring-shaped thickenings of the cellulose cell wall are impregnated with lignin. This makes them hard, so that they can resist inward pressures

Pores in the outer cellulose cell wall conduct water out of the xylem vessel and into cell walls of adjacent leaf cells

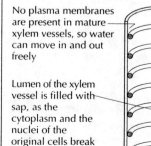

MINERAL ION UPTAKE BY ROOTS

Plants absorb potassium, phosphate, nitrate and other mineral ions from the soil. The concentration of these ions in the soil is usually much lower than inside root cells, so they are absorbed by **active transport**. Root hairs provide a large surface area for mineral ion uptake. The figure (below) shows the structure of root hair cells. Cortex cells can absorb ions that are dissolved in the water that is drawn by capillary action through cortex cell walls.

Vertical section of periphery of root

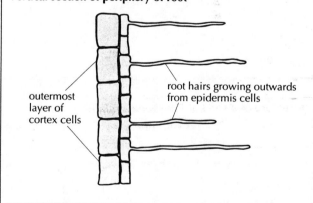

outermost layer of cortex cells

root hairs growing outwards from epidermis cells

WATER UPTAKE BY ROOTS

The cytoplasm of root cells usually has a much higher total solute concentration than water in the surrounding soil, as a result of active transport of mineral ions. Water therefore moves into root cells from the soil by **osmosis**. Most of the water absorbed by roots is eventually drawn by the transpiration pull into xylem vessels in the centre of the root. To reach the xylem, water has to cross the cortex. There are two possible routes. The water could move from cell to cell through the cytoplasm – the **symplastic route**. It could also move by capillary action through cortex cell walls until it reaches the endodermis – the **apoplastic route**.

Reproduction of flowering plants

STRUCTURE AND FUNCTION OF FLOWERS

Flowers are the structures used by flowering plants for sexual reproduction. Female gametes are contained in ovules in the ovaries of the flower. Pollen grains, produced by the anthers, contain the male gametes. A zygote is formed by the fusion of a male gamete with a female gamete inside the ovule. This process is called **fertilization**.

Before fertilization, another process called **pollination** must occur. *Pollination is the transfer of pollen from an anther to a stigma.* Pollen grains containing male gametes cannot move without help from an external agent. Most plants use either wind or an animal for pollination. The structure of a flower is adapted to its method of pollination. The figure (below) shows the structure of a flower of *Lamium album,* which is adapted to bee pollination.

Pollen grains germinate on the stigma of the flower and a pollen tube containing the male gametes grows down the style to the ovary. The pollen tube delivers the male gametes to an ovule, which they fertilize.

Fertilized ovules develop into seeds. The figure (bottom) shows the structure of a seed of *Phaseolus multiflorus.* Ovaries containing fertilized ovules develop into fruits. The function of the fruit is **seed dispersal**.

Structure of *Lamium album* flower

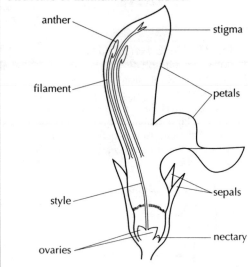

FACTORS NEEDED FOR SEED GERMINATION

Seeds will not germinate unless external conditions are suitable.

- Water must be available to rehydrate the dry tissues of the seed.
- Oxygen must be available for aerobic cell respiration. Some seeds respire anaerobically if oxygen is not available but ethanol produced in anaerobic respiration usually reaches toxic levels.
- Suitable temperatures are needed. Germination involves enzyme activity and at very low and very high temperatures enzyme activity is too slow. Some seeds remain dormant if temperatures are above or below particular levels, so that they only germinate during favourable times of the year.

The figure (below) shows the structure of a seedling of *Phaseolus multiflorus,* about 2 weeks after the start of germination.

Structure of a seedling of *Phaseolus multiflorus*

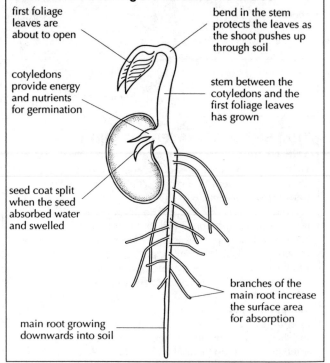

Structure of a seed of *Phaseolus multiflorus*

External structure

Internal structure

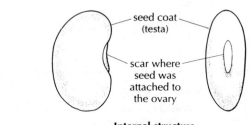
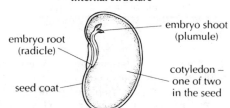

METABOLIC EVENTS DURING GERMINATION

- The first stage in germination is the absorption of water and the rehydration of living cells in the seed. This allows the cells to become metabolically active.
- Soon after absorbing water, a plant growth hormone called **gibberellin** is produced in the cotyledons of the seed.
- Gibberellin stimulates the production of amylase, which catalyses the digestion of starch into maltose in the food stores of the seed.
- Maltose is transported from the food stores to the growth regions of the seedling, including the embryo root and the embryo shoot.
- Maltose is converted into glucose, which is either used in aerobic cell respiration as a source of energy, or is used to synthesize cellulose or other substances needed for growth.

As soon as the leaves of the seedling have reached light and have opened, photosynthesis can supply the seedling with foods and the food stores of the seed are no longer needed.

EXAM QUESTIONS ON TOPICS 12 AND 13

1 a) Outline the need for excretion in living organisms. [2]

b) Compare the composition of the blood in the renal artery and renal vein, by giving two differences in the table below. [2]

Blood in the renal artery	Blood in the renal vein

c) Explain briefly the function of the loop of Henle in the human kidney. [2]

d) Deduce which part of the kidney has been damaged if protein is found in the urine. [1]

2 a) Draw the structure of a nephron. [2]

b) (i) Identify where most active transport occurs in the nephron. [1]

(ii) Identify **one** specific location where active transport occurs in plants. [1]

3 C$_3$ and CAM plants both need CO$_2$ for photosynthesis. They take in CO$_2$ through microscopic pores called stomata. The stomata can very from being fully closed (0% open) to fully open (100% open). The circular graph below shows the width of opening of stomata during a 24 hour period in a C$_3$ plant and a CAM plant.

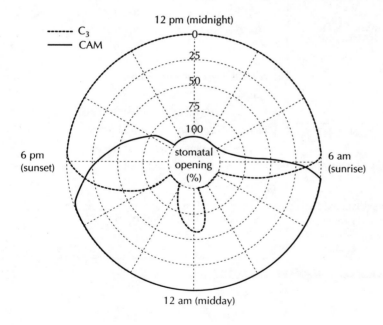

a) Identify the hours during which stomata were fully closed in

(i) the C$_3$ plant [1 mark]

(ii) the CAM plant [1 mark]

b) One of the two plants is a xerophyte.
Use the data in the graph to predict whether the C$_3$ plant or the CAM plant is the xerophyte. [2 marks]

(c) (i) Outline the changes in the stomata of the C$_3$ plant shown in the graph between 11.00 am. and 2.00 pm. [2 marks]

(ii) Suggest a reason for the changes. [1 mark]

Human diets

Humans eat a huge variety of food. Availability, local customs and personal preferences all affect what food each individual includes in his or her diet. *The diet is all the food eaten by an individual during a period of time.*

To remain healthy, humans need a **balanced diet**.
- Balanced diets provide sufficient quantities of all essential nutrients for life processes, including growth and repair.
- Balanced diets also supply enough energy for life processes.

CALCULATING THE NUTRITIONAL CONTENT OF A DIET

The first stage in calculating the nutritional content of a diet is measuring the amount of each food eaten during a period of time. The amount is measured in grams. The nutritional content of each food is obtained, either from labels on packaged foods or from nutritional reference tables. The total amounts of energy and of each nutrient in the diet can then be calculated.

The figure (below) shows the nutritional information given on a package of tofu. The content is given per 100 g of the food. Many packaged foods give nutritional information in a similar way.

NUTRITIONAL INFORMATION

	Per 100g
Energy	258kJ/62Kcal
Protein	6.9g
Carbohydrate	2.4g
of which	
-sugars	1.3g
-polyols	0.0g
-starch	0.0g
Fat	2.7g
of which	
-saturates	0.4g
-mono-unsaturates	0.5g
-polyunsaturates	1.5g
Fibre	100mg
Sodium	36mg
Cholesterol	0.0mg

INGREDIENTS: Water, *Soya Beans, *Isolated Soya Protein Firming Agents (Gulcono-Delta-Lactone, Calcium Chloride) *GM-free Soya beans are used.

Produced by
Morinaga Milk Industry Co., Ltd.
Shiba 5-33-1 Minato-ku.
Tokyo 108-8384 JAPAN

Product of Japan

EVALUATING THE NUTRITIONAL CONTENT OF A DIET

The nutritional content of a diet is evaluated by comparing it with the amounts of each nutrient needed in a balanced diet. It is not always certain how much of a nutrient is needed, but many national governments publish recommendations. The amounts needed vary depending on the age and gender of the person. The figure (below) shows recommended amounts of some nutrients for 15–18-year-old males and females.

Recommended amounts for 15 – 18-year-olds per day

Nutrient	Females	Males
Protein	45.0 g	55.2 g
Iron	14.8 mg	11.3 mg
Calcium	800 mg	1000 mg
Iodine	140 µg	130 µg
Zinc	7.0 mg	9.5 mg
Retinol	600 µg	700 µg
Cyanocobalamin	1.5 µg	1.5 µg
Ascorbic acid	40 mg	40 mg
Fibre	18 g	18 g

EVALUATING THE ENERGY CONTENT OF A DIET

A balanced diet supplies the amount of energy that the body needs but not more. Diets containing excessive amounts of energy cause obesity. Individuals vary considerably in the amount of energy that they need. There are many factors that affect the energy requirement.
- Age – with increasing age up to adulthood more energy is needed, as the body grows larger.
- Gender – females use less energy than males. Two possible reasons are smaller body size and less heat loss because of thicker layers of fat under the skin.
- Activity – physical activity either during work or leisure involves energy expenditure. The more vigorous the activity the more energy is required.
- Condition – extra energy is needed during growth spurts, pregnancy, breast-feeding, illnesses and some other conditions.

The figure (below) shows the energy requirements for some groups.

Mean energy requirement (kJ per day)

Age	Females	Males
0–3 months	2160	2280
1–3 years	4860	5150
4–6 years	6460	7160
7–10 years	7280	8240
11–14 years	7720	9270
15–18 years	8830	11510
19–50 years	8100	10600
75+ years	7610	8770

Activity level and conditions	Difference
Females during pregnancy	+1900
Females – 1st month of breast-feeding	+1900
Females – 2nd month of breast-feeding	+2200
Females – 3rd month of breast-feeding	+2400
Very active 50 kg teenage boys	+1400
Very inactive 50 kg teenage boys	–2500
Very active 50 kg teenage girls	+2500
Very inactive 50 kg teenage girls	–800

Nutritional needs

Food is needed to supply the body with energy and with essential chemical substances, called nutrients. *A nutrient is a substance that must be included in the diet.*
There are five main groups of nutrients – carbohydrate, protein, lipid, minerals and vitamins.
Water and fibre are also needed in the diet.

CARBOHYDRATES

Carbohydrates are divided into three chemical groups.

Chemical group	Examples	Sources
Monosaccharides	Glucose	Honey
	Fructose	Grapes
Disaccharides	Sucrose	Cane sugar
	Lactose	Milk
Polysaccharides	Starch	Bread
	Glycogen	Liver

The main function of carbohydrates is to provide energy that can be released rapidly in cell respiration. If the energy from carbohydrates is not needed immediately, it can be put into stores. Glucose can be converted into glycogen for storage in the liver or into fats for storage in adipose tissue.
Carbohydrates have some other functions.
• They are linked to proteins to form glycoproteins.
• They are subunits of nucleic acids. Ribose, a five-carbon sugar, is needed to make RNA and deoxyribose is needed to make DNA.
• They are needed for the synthesis of some amino acids.

PROTEINS

Many foods contain protein. Foods rich in protein include beef and other meats, salmon and other fish, seeds or nuts such as peanuts and soybean products such as tofu.
Proteins in food are digested into amino acids, which are absorbed and used in **protein synthesis**.
Many processes in the body involve protein synthesis:
• growth
• regeneration of hair, nails, blood and skin
• repair of damaged tissues
If there is an excess of amino acids, some are broken down by removing amino groups from them. This is called **de-amination**. The remainder of the amino acid molecule contains valuable energy, which can be released in cell respiration.
Twenty different amino acids are needed to make human proteins. Some of these can be made converting one type of amino acid into another. Others cannot be synthesized in the human body. These are called **essential amino acids** and must be ingested in food.

WATER

Water is the solvent in which most substances in the body are dissolved and so is a vital component of the body. Losses of water from the body must be replaced by the amount of water in the diet, either in drinks or in food containing water.

LIPIDS

Any food containing oil or fat is a source of lipid. Examples include herrings and other oily fish, walnuts and other nuts and fatty meats such as bacon.
Lipids have important functions in the body.
• Fats and oils provide a source of energy, which is released by cell respiration.
• They contain more energy per gram than carbohydrate and form the body's main energy stores in adipose tissue.
• Fats conduct heat slowly and, because much of the body's adipose tissue is located next to the skin, the fat acts as a thermal insulator.
• Phospholipids are the main constituents of membranes.

MINERALS

Minerals are chemical elements. At least 20 different minerals are essential in the diet. These minerals each have an important function in the body. If they are not ingested in sufficient quantities, a deficiency disease develops. Iodine and zinc are examples of essential minerals.
• Iodine is needed in the thyroid gland for the synthesis of the hormone thyroxin. Each thyroxin molecule contains four iodine atoms. Thyroxin causes the metabolic rate to rise and so has an essential role in homeostasis.
• Zinc is part of some enzymes and also is part of some proteins that control DNA transcription.

VITAMINS

Vitamins are organic compounds that are needed in small quantities, but cannot be synthesized in the body. Each vitamin has a specific function and a deficiency disease develops if insufficient amounts are ingested.
Retinol (vitamin A) is needed to make rhodopsin, the light sensitive pigment in the rod cells of the eye. Rod cells are used for black and white vision in dim light.

Cyanocobalamin (vitamin B_{12}) is needed to make some coenzymes. These coenzymes are part of enzymes that are needed in reactions used to make amino acids and bases for DNA and RNA.

Ascorbic acid (vitamin C) is an anti-oxidant that protects the cytoplasm of cells from attack by free radicals. It is also needed for the synthesis of collagen, a protein that strengthens skin and blood vessel walls.

Calciferol (vitamin D) is needed for calcium absorption in the intestines. It helps to keep calcium levels in the blood within narrow limits and ensures that sufficient calcium is supplied to muscles and bones.

Tocopherol (vitamin E) is an anti-oxidant, especially in membranes, where it helps to prevent damage by hydrogen peroxide and other oxidizing agents.

Malnutrition

In many of the world's countries it is difficult or impossible to eat a balanced diet and the population is malnourished. *Malnutrition is a condition where the diet is not balanced.* Often, the result of malnutrition is a deficiency disease.

IODINE DEFICIENCY DISORDER –AN EXAMPLE OF GLOBAL MALNUTRITION

The main symptom of iodine deficiency disorder (IDD) is swelling of the thyroid gland in the neck. This is called goitre (below).

Appearance of a person with a goitre

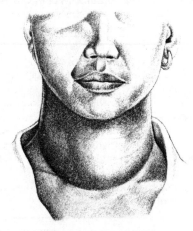

UNICEF has published data on IDD rates in every country, including the percentage of 6 – 11-year-old children with goitre in the ten years from 1985 – 94.
The countries with the highest rates are shown below.

Country	Goitre rate	Country	Goitre rate
Albania	41%	Nepal	44%
Bangladesh	50%	Paraguay	49%
Central African Rep.	63%	Rwanda	41%
Guinea	55%	Syria	73%
Lesotho	43%	Zambia	51%

The data shows that IDD is a very widespread type of malnutrition with high rates in some countries in most parts of the world. Rates are highest in areas where the soil used to grow crops and the drinking water contain little iodine. IDD has very serious consequences. If women are affected during pregnancy, their children are born with permanent brain damage. If children suffer from IDD after birth, their mental development and intelligence are impaired. In 1998, UNICEF estimated that 43 million people worldwide had brain damage due to IDD and 11 million of these had a severe condition called cretinism. 40% of the world's population is estimated to show some mental impairment because of IDD.

At the World Children's Summit in 1990, a campaign was started to eliminate IDD by adding it in small quantities to salt sold for human consumption. This is a highly effective way of preventing IDD at a cost of only about 5 cents per person per year. By the year 2000 iodized salt was reaching more than 3.3 billion people. If the campaign to provide iodized salt throughout the world is successful, no children in the future will suffer from IDD.

CAUSES OF MALNUTRITION

There are many possible causes of malnutrition.
Economic – it may be impossible to buy food because of poverty, or crops grown by farmers might have to be sold instead of eaten. Lack of investment may make farming inefficient.
Environmental – droughts or floods might destroy crops. Mineral deficiencies in soils may cause deficiencies in humans.
Social – population growth may make the food supply insufficient. Disruption to society caused by wars or corrupt government may prevent the production or distribution of food.
Cultural – diet is part of the culture of a population. Cultures that have a diet of mostly maize tend to suffer from pellagra – vitamin B_3 deficiency and cultures that have a diet of mostly polished rice tend to suffer from beri-beri – vitamin B_1 deficiency.

THE IMPORTANCE OF FIBRE IN THE DIET

Although the populations of developed countries do not suffer the severe malnutrition that is common in developing countries, some more minor problems are commonly found. Insufficient fibre in the diet is one example.
Fibre is material that cannot be digested in the small intestine. Cellulose from plant cell walls is the main component of dietary fibre, but there are other components including chitin from fungi and crustaceans.
Many investigations have shown that fibre helps to prevent constipation, by increasing the bulk of material in the large intestine.
There are other possible advantages, but the evidence for these is weaker.
- Fibre may help to prevent obesity by increasing the bulk in the stomach, which reduces the desire to eat more food.
- Fibre may reduce the risk of diseases of the large intestine including appendicitis, cancer and hemorrhoids.
- Fibre may reduce the rate of absorption of sugar and so help in the prevention and treatment of diabetes.

COMPARING THE NUTRITIONAL CONTENT OF FOODS

Some people have special nutritional needs and need to look carefully at food labels. For example doctors sometimes advise a patient to reduce their intake of saturated fats, cholesterol and salt. Protein is needed in the patient's diet and either tofu or beef would supply it. The amounts of protein, saturated fat, cholesterol and sodium in tofu are shown on page 112. Amounts per 100 grams of lean beef are shown below. The amount of saturated fat, cholesterol and sodium in tofu and beef can be compared, either per 100 grams or per gram of protein.

Energy	590 kJ	Fat	4.35 g
Protein	24.35 g	of which	
Carbohydrates	0.0 g	-saturates	1.05 g
		Fibre	trace
		Sodium	220 mg
		Cholesterol	100 mg

Diet and disease

DISEASES ASSOCIATED WITH LIPIDS IN THE DIET

Diets that contain large amounts of lipid tend to cause obesity, because of the high energy content of lipids. Lipids may also increase the risk of coronary heart disease (CHD). Coronary heart disease involves the formation of blood clots, which block the coronary arteries. These arteries supply the wall of the heart with blood. Without a blood supply, the muscle in the wall of the heart cannot contract normally and the result is a heart attack.

Blood clots tend to form in arteries that are affected by atherosclerosis. This is the accumulation of fat and cholesterol in the artery wall. The lumen of the artery becomes narrower, raising blood pressure. The inner surface becomes rough, stimulating clotting.

A high level of lipid in the diet is a possible risk factor for CHD. The figure (below) shows the results of an investigation into the relationship between lipid consumption and rate of death due to CHD.

The data shows a positive correlation, but there are reasons to be cautious about drawing firm conclusions from this data.

- A positive correlation does not show that lipid consumption causes CHD.
- Some countries have a low rate of CHD despite high lipid consumption.
- There are other risk factors for atherosclerosis and CHD, including smoking and high blood pressure.
- There are different types of lipid, including cholesterol and fatty acids, which do not all have the same effects.

Relationship between consumption of fat and CHD

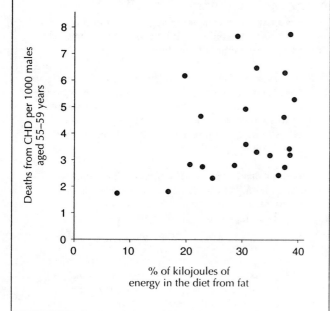

CHOLESTEROL AND HEALTH

Cholesterol is a steroid and is mainly found in animal products. It is an essential component of cell membranes and is used in the synthesis of steroid hormones, including estrogen, progesterone and testosterone.

Many investigations have shown that as the amount of cholesterol in the blood plasma rises, the risk of death from CHD increases. A 10% increase in blood cholesterol level causes a 30% increase in the risk of death from CHD.

There are two sources of cholesterol – it can be synthesized by the liver and it is present in foods containing animal fats. The amount of cholesterol in the diet does not affect the blood cholesterol level much. Other aspects of the diet are more significant. Genetic factors are also very important and in some families high blood cholesterol levels are very common.

EFFECTS OF SATURATED AND UNSATURATED LIPIDS ON HEALTH

- Saturated fatty acids have no double bonds between their carbon atoms and are mainly found in animal products.
- Polyunsaturated fatty acids have two or more double bonds between their carbon atoms and are mainly found in plant products.

There is a positive correlation between the level of saturated fatty acids in the diet and blood cholesterol levels. Increasing the level of some polyunsaturated fatty acids in the diet decreases the blood cholesterol levels. Reducing the amount of saturated fatty acids and increasing the amount of polyunsaturated fatty acids should therefore reduce the risk of CHD. However, because of genetic factors, in some countries the risk of CHD is already very low, so levels of saturated and unsaturated lipids in the diet are not significant.

DISEASES WITH NUTRITIONAL AND NON-NUTRITIONAL CAUSES

Some diseases are difficult to treat because they do not have a single cause.

Rickets is a disease of childhood. The child's legs become bowed, because the bones do not contain enough calcium salts to strengthen them. The cause of rickets is usually calciferol (vitamin D) deficiency, not calcium deficiency. A diet containing insufficient calciferol is an obvious cause, but there is also a non-nutritional cause. Calciferol can be made in the skin if it receives sunlight. This cannot happen if a child stays indoors too much and covers the skin with clothing.

Anemia is a deficiency of red blood cells or hemoglobin. It can be caused by iron deficiency, but there are many other causes. Deficiencies of cyanocobalamin (vitamin B_{12}), folic acid or ascorbic acid are other nutritional causes. There are also non-nutritional causes including genetic diseases, such as sickle cell anemia and infectious diseases such as malaria.

Osteoporosis is progressive loss of mass from bones, making them brittle and easily fractured. It is rare in young people and is commoner in women than in men. Calcium and calciferol deficiencies are not the main cause and non-nutritional factors are more significant. These include the decrease in estrogen levels in women after the menopause, prolonged treatment with some steroid drugs and lack of physical exercise.

Food and health

There are ethical and health issues concerned with food, including the safety of food additives, dangers of food poisoning and nutritional and ethical aspects of vegan and vegetarian diets.

FOOD ADDITIVES

Chemical additives are used in the preparation of many foods. These additives can act as preservatives, antioxidants, colourings, flavourings, stabilizers and acid-regulators. Although food additives are tested to try to ensure that they are safe, some may have harmful effects.

- Nitrite is added to bacon and other cured meat and fish as a preservative. It reacts with amino acids in the gut to produce nitrosamines, which are highly carcinogenic.
- Tartrazine is used in drinks and foods as a yellow colouring. It has a variety of harmful effects including asthma, skin rashes and hyperactivity.
- Monosodium glutamate is added to many foods to act as a flavour enhancer. Some people suffer allergic reactions, including sweating, rapid heartbeat and headache. These may not be due to monosodium glutamate but to impurities added with it .

Most processed foods contain additives, for example savoury snacks (see food label berlow).

CHEESE AND ONION FLAVOUR SNACKS

INGREDIENTS: Potatoes, vegetable oil, cheese and onion flavour [onion powder, lactose, flavour enhancers (monosodium glutamate, sodium guanylate), cheese powder (0.1%) (animal rennet), wheat maltodextrin, flavouring, colour (annatto)], salt.

TYPICAL NUTRITIONAL INFORMATION

	Per 100g	Per 25g Pack
Energy	2210kJ	553kJ
	530kcal	133kcal
Protein	6.5g	1.6g
Carbohydrate	50g	12.5g
of which sugars	2g	0.5g
Fat	33g	8.3g
of which saturates	15g	3.8g
Dietary fibre	4.1g	1g
Sodium	0.7g	0.2g

HYGIENIC METHODS OF FOOD HANDLING AND PREPARATION

If food is not handled and prepared hygienically, bacteria may grow in it, for example, *Salmonella*. When the food is eaten the bacteria grow in the gut and release toxins. This is called food poisoning.

Food poisoning can be fatal in severe cases. It can be prevented if simple rules of hygiene are followed.

- Foods that may contain food poisoning bacteria must be cooked thoroughly to kill the bacteria.
- Frozen food must be fully defrosted before cooking.
- Cooked and uncooked food must be stored separately to avoid bacteria passing from the uncooked to the cooked food.
- Food must not be eaten after its 'use by' date.
- Flies and other animals must not be allowed to contaminate food.

VEGAN AND VEGETARIAN DIETS

Vegans include no animal products in their diet. They do not eat meat, fish, milk, butter, cheese and other dairy products, eggs or honey. Vegans eat products of plants and micro-organisms. Vegetarian diets vary but they usually include eggs, milk, butter, cheese and other dairy products and exclude meat and fish.

There are ethical issues involved in deciding whether to eat animal products.

- Is it right for one animal to take the life of another animal to obtain food?
- Is the pain caused to animals during transport and slaughter justifiable?
- Is the suffering of animals reared for meat or eggs in unnatural and crowded conditions justifiable?
- Is it right to take fish from their habitats and deprive their natural predators of food?
- Is the pollution of natural ecosystems caused by fish farming acceptable?
- Is the suffering caused by keeping egg-laying hens in small cages justifiable?
- Is it right to kill male chicks at 1–3 days old because they do not lay eggs?
- Is the suffering of cows whose calves are taken away from them soon after birth justifiable?
- Is it right to make cows have calves in order to stimulate milk production, knowing that the calves will have to be killed eventually?

POSSIBLE DEFICIENCIES IN VEGAN AND VEGETARIAN DIETS

There is a risk of deficiencies in vegetarian and vegan diets. For vegetarians, the only risk is iron deficiency. For vegans there are four possible deficiencies.

- **Iron** – levels of iron in many plant products are high, but the iron is in a chemical form that is less easy to absorb than iron in animal products. To avoid anemia, vegans need to eat plenty of iron-rich plant products, especially ones containing ascorbic acid, as this vitamin encourages iron absorption.

- **Calcium** – plant products do contain calcium but not as much as animal foods. Vegans usually have a lower calcium intake than meat eaters but calcium deficiency has not been reported.

- **Calciferol (vitamin D)** – plant products do not contain calciferol and, although it can be synthesized in the skin, this only happens with ultra-violet light, which is lacking in northerly latitudes during winter. However, during summer the liver can store enough calciferol to avoid a deficiency in winter. Soya milk, margarine and breakfast cereals usually have calciferol added.

- **Cyanocobalamin (vitamin B_{12})** – plant products do not contain this vitamin, unless they are contaminated with soil bacteria! There have been cases of cyanocobalamin deficiency, especially in babies breast-fed by vegan mothers. Vegans must be careful to ensure that they absorb enough of this vitamin. Yeast extract contains it and soya milk, margarine and breakfast cereals usually have it added. It can also be taken in tablets.

EXAM QUESTIONS ON OPTION A – DIET AND HUMAN NUTRITION

A1 Humans vary considerably in the amount of energy that they need in their diet. Two factors which affect the energy requirement are body mass and physical activity. The amount of physical activity carried out can be measured and expressed as a numerical value, PAL (physical activity level). In teenage boys PAL values range from 1.4 (inactive) to 2.0 (extremely active).

The nomogram below can be used to estimate the energy requirements of boys from 10 to 18 years old if their body mass and PAL are known.

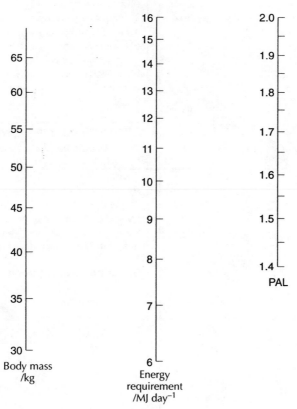

a) Estimate, using the nomogram, the energy requirement for a boy with

   (i) a mass of 60kg and PAL of 1.4   [1]

   (ii) a mass of 40kg and PAL of 1.8   [1]

b) State the relationship between

   (I) physical activity level and energy requirement   [1]

   (ii) body mass and energy requirement   [1]

c) Suggest one reason for the relationship between

   (i) physical activity level and energy requirement   [1]

   (ii) body mass and energy requirement   [1]

d) Predict how the energy requirement of a 10 to 18-year-old boy would differ from that of

   (i) a 10 to 18-year-old girl with the same mass and PAL as the boy   [1]

   (ii) an adult man with the same mass and PAL as the boy   [1]

A2 a) List three sources of protein for humans.   [1]

   b) State two functions of protein in humans.   [2]

   c) Outline how the body responds to an intake of more protein than it needs.   [2]

A3 a) List **two** foods that can be eaten by vegetarians but not by vegans.   [2]

   b) Discuss the possibility of cyanocobalamin deficiency in the diet of vegans.   [3]

Skeleton, joints and muscle

THE AXIAL AND APPENDICULAR SKELETON

The human skeleton consists of bones. It is sometimes divided into two groups of bones.

- The axial skeleton, consisting of the skull and the vertebral column.
- The appendicular skeleton, consisting of the limb bones and the pelvic girdle and shoulder girdle.

The figure (below) shows the axial and appendicular parts of the skeleton.

The human skeleton

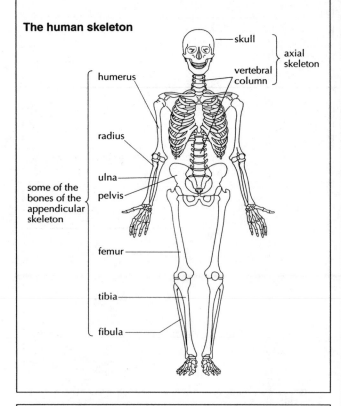

THE STRUCTURE AND FUNCTION OF SKELETAL MUSCLE

Skeletal muscle is composed of muscle fibres. Each muscle fibre is surrounded by a plasma membrane but is larger than a normal cell and contains many nuclei. Muscle fibres also contain cylindrical structures called myofibrils. Myofibrils often extend over the whole length of the muscle fibre. The figure below shows the arrangement of myofibrils within a muscle fibre.

Structure of a muscle

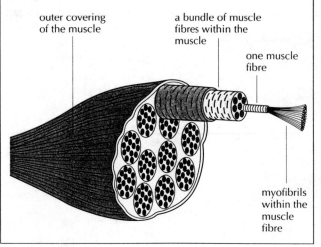

THE STRUCTURE AND FUNCTION OF JOINTS

Joints are junctions between bones. Some joints are fixed, but many allow movement. The hip and the knee are examples of joints that allow movement.

The knee joint allows considerable movement in one plane – flexion and extension. It allows very little movement in the other two planes.

The hip allows movements in all three planes:

- Flexion and extension – forward and backward
- Abduction and adduction – sideways
- Rotation – swivelling.

Figures on page 102 show the structure of the elbow joint and the functions of the different parts of the joint.

STRUCTURE OF BONES

The appendicular skeleton includes some long bones. The figure below shows the structure of a long bone. The shaft of the bone is hollow. This gives much more strength than a solid bone of the same mass, which would be much narrower. The ends of the bone, called heads, form joints with other bones. They are made of spongy bone, which allows them to act as shock absorbers. They can also absorb impacts from different directions without fracturing.

Structure of a long bone

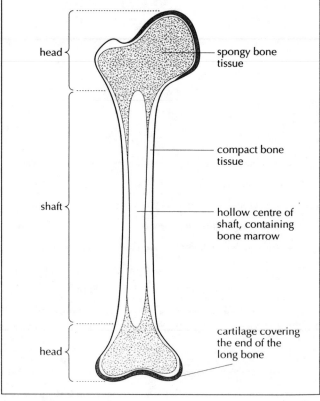

Further information on muscle

The structure of muscle fibres has been investigated using electron microscopes. Figures on page 101 are electron micrographs showing sarcomeres in muscle fibres, with light and dark bands and also actin and myosin filaments. Another figure (also on page 101) shows the arrangement of actin and myosin filaments in a sarcomere.

The contraction of skeletal muscle involves the shortening of sarcomeres, by the sliding of actin and myosin filaments over each other. The figure (bottom) on page 101 shows how this occurs.

Co-ordination of muscle activity

Muscles are controlled by the nervous system. The general organization of the nervous system is described on page 52.

CONTROL OF MUSCLE CONTRACTION

The part of the brain that controls muscle contraction is called the **cerebral cortex**. The regions of the cerebral cortex that are used are called **motor areas**. Each part of the motor areas is used to control one muscle. Instructions are sent to the muscle along motor neurones. The figure (top) on page 100 shows the structure of a motor neurone. The instructions sent along motor neurones are in the form of electrical impulses. The instructions pass across junctions called **synapses**, between motor neurones and muscle fibres. The muscle fibres then contract.

There are cells within muscles which monitor how stretched the muscle is. These cells are called **proprioceptors**. They send messages along sensory neurones to the cerebral cortex. The figure (below) shows the structure of a sensory neurone. The messages indicate whether the muscle is fully contracted, partly contracted or relaxed. The cerebral cortex uses the information to help to decide whether the muscle needs to contract more, less, or not change. The involvement of the proprioceptors is an example of feedback control.

SYNAPTIC TRANSMISSION

Messages are transmitted from one neurone to another at junctions called synapses. Neurones are also connected to receptor cells and muscle fibres by synapses. The transmission of a message across a synapse from one neurone to another involves a series of events.

- A message travels along a neurone, in the form of an electrical impulse.
- The message arrives at the end of the neurone, where it forms a synapse with another neurone.
- A chemical substance, called a neurotransmitter is released from the end of the neurone when the electrical impulse arrives.
- The neurotransmitter substance diffuses across a narrow fluid-filled gap to the neurone beyond the synapse.
- When the neurotransmitter substance reaches the membrane of this neurone it causes an electrical impulse to be propagated.
- The electrical impulse travels on down the neurone beyond the synapse.
- Meanwhile in the synaptic gap an enzyme breaks down the neurotransmitter substance.

The figure (bottom right) shows part of a neurone which has synapses with several other neurones.

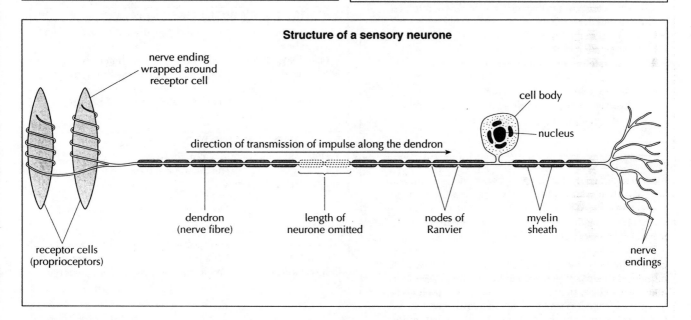

Structure of a sensory neurone

nerve ending wrapped around receptor cell

cell body

nucleus

direction of transmission of impulse along the dendron

dendron (nerve fibre)

length of neurone omitted

nodes of Ranvier

myelin sheath

receptor cells (proprioceptors)

nerve endings

INHIBITORY NEURONES AND ANTAGONISTIC MUSCLE ACTION

Muscles that cause opposite movements at a joint are called **antagonistic muscles**. For example, the biceps and triceps are antagonistic muscles that cause opposite movements of the elbow joint. The biceps causes flexion and the triceps causes extension.

To allow movement at a joint, one muscle of the antagonistic pair must contract and the other must relax. The muscle that relaxes is prevented from contracting by **inhibitory neurones**. Synapses connect these inhibitory neurones with the motor neurones that stimulate the muscle to contract. The inhibitory neurones release a neurotransmitter substance that prevents an electrical impulse being propagated in the motor neurones, even if other neurotransmitter substances are present that normally cause propagation of an impulse.

Synapses between many neurones and one motor neurone

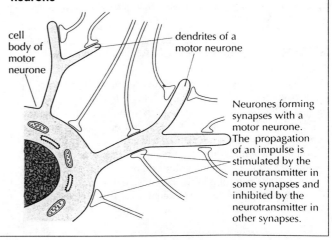

cell body of motor neurone

dendrites of a motor neurone

Neurones forming synapses with a motor neurone. The propagation of an impulse is stimulated by the neurotransmitter in some synapses and inhibited by the neurotransmitter in other synapses.

Muscles and energy

CHANGES IN VENTILATION RATE DURING EXERCISE

During exercise, the rate of aerobic respiration in muscles rises so there is an increase in the CO_2 concentration of the blood. This reduces the pH of the blood and is detected by cells in the walls of arteries, which monitor blood pH. These cells are called **chemosensors**. The chemosensors send nerve impulses to the parts of the brain that control the ventilation rate, called the **breathing centres**. The breathing centres send nerve impulses to the diaphragm and intercostal muscles, causing them to increase the rate at which they contract and relax. This increase in the ventilation rate helps to remove from the body the CO_2 produced in aerobic cell respiration. It also helps to increase the rate of oxygen uptake, which allows aerobic cell respiration to continue in the muscles. After exercise, the level of CO_2 in the blood falls, the pH of the blood rises and the breathing centres cause the ventilation rate to decrease.

THE ROLE OF ADRENALIN

Adrenalin is secreted before or during vigorous exercise. It has many effects on the body, which help to supply muscles with more glucose and oxygen.

- Glycogen stores in the liver are broken down to release glucose into the blood.
- Bronchioles in the lungs widen, to increase ventilation and oxygenation of the blood.
- The heart rate is increased to pump blood faster.
- Blood vessels leading to the skin, gut, kidneys and liver constrict, to reduce flow to these organs.
- Blood vessels leading to muscles dilate, to carry more blood, rich in glucose and oxygen to them.

SOURCES OF ATP IN MUSCLES

Muscle contraction requires a supply of energy in the form of ATP. When muscle fibres start to contract, they usually contain enough ATP for about five seconds of contraction. To go on contracting for longer than this, more ATP must be produced by cell respiration. Aerobic respiration can produce ATP continuously, if oxygen is available. With low intensity exercise, such as walking or jogging, the blood flowing to a muscle maintains oxygen availability. However, with high intensity exercise, such as sprinting, oxygen is used up faster in a muscle than it is supplied by blood. Anaerobic respiration therefore has to be used. Lactate (lactic acid) is produced during anaerobic respiration. Lactate is toxic, so anaerobic respiration can only be used to produce energy for a maximum of 2 minutes during high intensity exercise.

ANAEROBIC RESPIRATION AND THE OXYGEN DEBT

Lactate is carried by blood from muscles to the liver, where it is broken down. Oxygen is needed to do this, so if lactate is present in the body there is an **oxygen debt**. If a large amount of lactate builds up during vigorous exercise, a large amount of oxygen is needed to repay the oxygen debt. This is the reason for the ventilation rate remaining higher than the resting rate for some time after the end of vigorous exercise. The figure (top right) shows the percentage of glycogen broken down and the lactate concentration reached after endurance events of varying duration.

MUSCLE FATIGUE

Some muscle fibres contain a store of carbohydrate, in the form of glycogen. During exercise, the glycogen is gradually converted into glucose for use in cell respiration. When the glycogen in a muscle has been used up, a feeling of tiredness develops called **muscle fatigue**. Accumulation of lactate can also cause muscle fatigue.

Causes of muscle fatigue in races

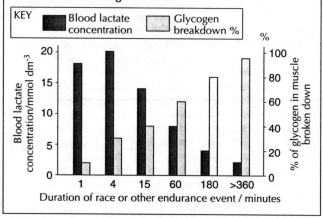

THE ROLE OF MYOGLOBIN IN MUSCLE

Some muscle fibres contain a red pigment called myoglobin. Oxygen binds to myoglobin when the oxygen level in muscle is high. Oxygen is released by myoglobin when the oxygen level in muscle is very low. The role of myoglobin is to act as an oxygen store, allowing muscle fibres to continue aerobic respiration for longer than if myoglobin was not present.

FAST AND SLOW MUSCLE FIBRES

Skeletal muscles contain two main types of muscle fibre, fast fibres and slow fibres. The differences between the two types of fibre are shown below.

	Fast fibres	Slow fibres
Blood supply	Moderate	Very good
Glycogen	Little present	Much present
Myoglobin	Little present	Much present
Cell respiration	Often anaerobic	Always aerobic
Stamina	Low	High
Strength	High	Low

Fast fibres can release large amounts of energy for a short period of time by anaerobic respiration, so are useful in high intensity exercise. Slow fibres release energy more slowly but can continue for longer, so are useful in endurance events. The figure (below) shows the mean percentage of fast and slow fibres in a thigh muscle in five groups of athletes.

Proportions of fast and slow muscle fibres in athletes

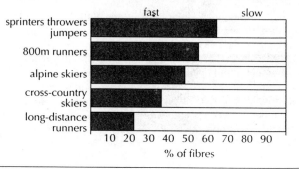

Fitness, training and injuries

PRINCIPLES OF TRAINING

In the world of sport, exercises that are done to change the physical condition of the body are called training. The overall amount of training in a programme depends on three aspects – how often a training session takes place – **frequency**, how long each training session lasts – **duration** and how vigorous the exercise is – **intensity**.

All forms of training have effects on the cardiovascular system and the lungs. Vigorous exercise makes the heart grow larger, so that it has a greater maximum output. The rate at which it beats when the body is at rest becomes slower. Lung capacity can be increased during adolescence by training.

To make any muscle stronger by training, the load on it during training must be greater than during normal daily activities – this is called overload. To go on making a muscle stronger, the overload must periodically be increased. This aspect of training is called **progressive overload**. For example, if training involves lifting weights, the weight lifted must be increased during the training programme.

The types of exercise in a training programme are chosen to develop the body for a particular type of exercise or sport. The muscles that are used in the exercises become stronger, but other muscles are not affected. This aspect of training is called **specificity**.

The result of a successful training programme is a condition called **fitness**. *Fitness is the physical condition of the body that allows it to perform exercise of a particular type.*

MEASURES OF FITNESS

Different measures of fitness are needed for different types of exercise.

- **Flexibility** is the range of movement of the body. It helps prevent injury in any exercise and is very important in gymnastics, swimming or ice skating.
- **Agility** is the ability to perform movements quickly that involve changes of direction. It is particularly important in basketball, tennis, and other racquet games.
- **Speed** is the rate at which a movement is performed. The distance moved and the time taken must be measured. Speed is important in sprinting and football.
- **Stamina** is the ability to continue an exercise for a long time. The maximum duration time is measured. Stamina is important in rowing and in long-distance running.

WARM-UP AND COOL-DOWN ROUTINES

Warm-up routines involve gentle exercises, before a period of vigorous exercise. Many possible advantages have been suggested. Warm muscles may be more supple and less likely to tear. Warm joints may be more mobile and less likely to be strained. Despite this, many animals carry out sudden vigorous exercise without warming up and without injury.

Cool-down (or warm-down) routines involve a gradual decrease in the intensity of exercise at the end of an exercise session. The aim is to disperse lactic acid and allow the cardiovascular system to adjust gradually to the end of vigorous exercise. Athletes have sometimes collapsed after stopping exercise suddenly, because of a reduction in the rate of blood returning to the heart.

INJURIES TO MUSCLES AND JOINTS

Vigorous exercise sometimes causes injuries to joints and muscles.

- **Sprains** – abnormal movement at a joint causes minor tearing of ligaments.
- **Torn ligaments** – large abnormal movements cause ligaments to tear completely.
- **Torn muscles** – excessive stretching of muscles cause them to tear, often at a junction with a tendon.
- **Dislocation** – abnormal movement at a joint causes the bones to move out of alignment.
- **Inter-vertebral disc damage** – abnormal movements or heavy loads cause the soft centre of the disc to bulge out through a tear in the wall of the disc. The figure (below) is a scan showing a damaged disc in the neck.

Scan showing disc bulges in the neck of a patient

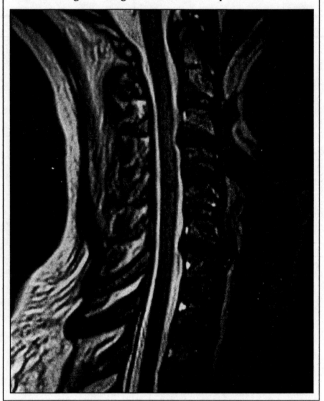

TREATMENT OF SOFT TISSUE INJURIES

If the correct first-aid treatment is given, muscle strains heal more quickly. The aim of the treatment is to reduce the amount of swelling.

Rest – stop the activity that caused the injury and prevent it becoming worse.

Ice – apply an ice pack, to reduce blood flow to the injured area.

Compression – use an elastiç stocking or other tight bandage.

Elevation – lift the injured part to help drain fluid away.

PERFORMANCE-ENHANCING DRUGS

Drugs can be used to enhance performances in sport, but there are strong ethical arguments against their use.

- The long-term health of sportsmen and women who are encouraged to take them may be damaged.
- Drug-users gain an unfair advantage in competitions.
- Criminals profit from the sale of banned drugs.

EXAM QUESTIONS ON OPTION B – PHYSIOLOGY OF EXERCISE

B1 Humans and other mammals can store oxygen in the lungs, in muscles and in the blood. The pie charts below show the volume of oxygen (cm^3) per kilogram of body mass stored in these tissues in humans and in a marine mammal, the Weddell seal.

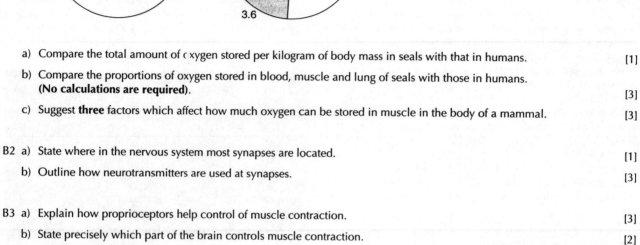

a) Compare the total amount of oxygen stored per kilogram of body mass in seals with that in humans. [1]

b) Compare the proportions of oxygen stored in blood, muscle and lung of seals with those in humans. **(No calculations are required).** [3]

c) Suggest **three** factors which affect how much oxygen can be stored in muscle in the body of a mammal. [3]

B2 a) State where in the nervous system most synapses are located. [1]

b) Outline how neurotransmitters are used at synapses. [3]

B3 a) Explain how proprioceptors help control of muscle contraction. [3]

b) State precisely which part of the brain controls muscle contraction. [2]

EXAM QUESTIONS ON OPTION C – CELLS AND ENERGY

Topics in Option C are covered on pages 66–81.

C1 The rate of photosynthesis in plants can be influenced by many factors. Experiments were carried out to investigate the effect of high and low light intensities on photosynthesis at different temperatures. All other factors were kept constant. A summary of the results is presented in the graph below.

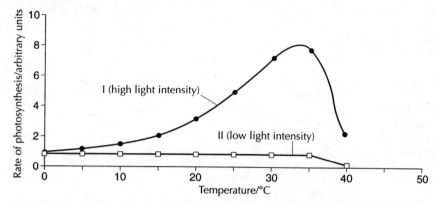

a) State the name of one limiting factor of photosynthesis, apart from temperature and light intensity. [1]

b) (i) Deduce the factor limiting the rate of photosynthesis in experiment I, between 0 and 30°C. Give a reason for your answer. [2]

   (ii) Discuss which factor limits the rate of photosynthesis in experiment I, between 35 and 40°C. [2]

c) Suggest one explanation for the difference between the results of experiments I and II. [2]

C2 Enzymes can be inhibited competitively and non-competitively.

a) State one example of:

   (i) a competitive inhibitor [1]

   (ii) a non-competitive inhibitor [1]

b) Compare competitive and non-competitive inhibition by stating one similarity and one difference in the table below. [2]

	Competitive inhibition	Non-competitive inhibition
Similarity		
Difference		

C3 A simplified version of the Krebs cycle is shown below.

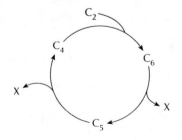

a) Deduce which product of the Krebs cycle is shown as X. [1]

b) Explain which products of the Krebs cycle are useful. [3]

Origin of life on earth

THEORIES FOR THE ORIGIN OF LIFE

Pasteur showed in an experiment in the nineteenth century that spontaneous generation of life from inorganic matter does not now take place – cells can only be formed from other cells. This is not surprising, as even the simplest prokaryotic cells are very complicated. Nevertheless, about 3.5 – 4.0 billion years ago, the first living cells did appear on Earth.

- Followers of many religions believe that God created life. This is called **special creation**.
- Some scientists believe that organic material can travel throughout the universe and so could have arrived on Earth from outer space. This is called **panspermia**.

The scientific method cannot be used to test these theories for the origin of life. Modern experiments can never establish with certainty what did or did not occur billions of years ago. They can only show what is possible, in conditions similar to those when life appeared. Results of experiments simulating conditions in interstellar dust clouds were published in 2001. Simple compounds such as ammonia, carbon dioxide and methanol were mixed with ice crystals in a vacuum at extremely low temperatures and in ultra violet light. Hundreds of complex organic compounds were formed, which self assembled into spherical vesicles. The results greatly strengthen the case for panspermia.

Another theory is that life did evolve here on Earth from inorganic matter, despite the fact that it is not possible today. In 1953, Stanley Miller and Harold Urey investigated this theory by trying to recreate the conditions of pre-biotic Earth. Inside their apparatus (top right) they mixed the gases ammonia, methane and hydrogen to form a reducing atmosphere. Electrical discharges and the boiling and condensing of water simulated lightning and rainfall. After one week, the clear water in the apparatus had turned to a murky brown. Analysis revealed many organic compounds, including 15 amino acids. Miller and Urey concluded that organic compounds could have formed on pre-biotic Earth.

THE ORIGIN OF PROKARYOTIC CELLS

If prokaryotic cells did evolve from inorganic matter billions of years ago, membranes would have been needed. Phospholipids naturally group together to form bilayers. These membranes form into cell-like spheres, enclosing a droplet of fluid. For these small structures to develop into living cells, a working genetic mechanism would have been needed. In modern prokaryotes the parts of the genetic mechanism cannot function without each other. For example, genes cannot be replicated without enzymes and enzymes cannot be made without genes. It seems inconceivable that the whole mechanism could have evolved at once, but gradual evolution would have required simpler intermediate stages. One possibility is the use of RNA instead of both DNA and enzymes. RNA can be replicated and can also act as a catalyst. Some reactions in the ribosome are still catalysed by RNA. Another possible intermediate stage is the use of clay minerals to catalyse reactions. Clay minerals are very variable. Some can divide, grow and catalyse specific reactions, including the formation of polypeptides from amino acids.

Before the origin of life, conditions on Earth were very different from those today. The figure (right) shows conditions on pre-biotic Earth.

Miller and Urey's apparatus

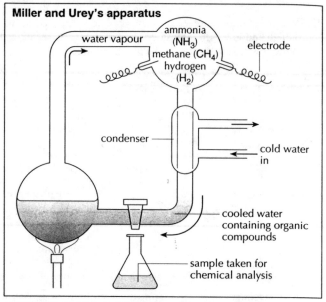

Conditions of pre-biotic Earth

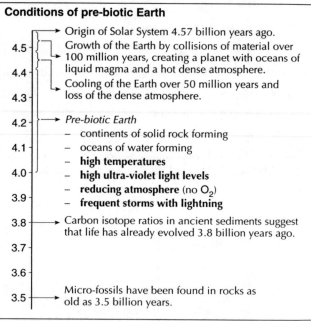

- → Origin of Solar System 4.57 billion years ago.
- Growth of the Earth by collisions of material over 100 million years, creating a planet with oceans of liquid magma and a hot dense atmosphere.
- Cooling of the Earth over 50 million years and loss of the dense atmosphere.
- → *Pre-biotic Earth*
  - – continents of solid rock forming
  - – oceans of water forming
  - – **high temperatures**
  - – **high ultra-violet light levels**
  - – **reducing atmosphere** (no O_2)
  - – **frequent storms with lightning**
- → Carbon isotope ratios in ancient sediments suggest that life has already evolved 3.8 billion years ago.
- → Micro-fossils have been found in rocks as old as 3.5 billion years.

THE ORIGIN OF EUKARYOTIC CELLS

Eukaryotic cells contain mitochondria and chloroplasts, which are not found in prokaryotic cells. If eukaryotic cells evolved from prokaryotic cells, the origin of these organelles must be explained.

According to the **endosymbiotic theory**, both mitochondria and chloroplasts have evolved from independent prokaryotic cells, which were taken into a larger heterotrophic cell by endocytosis. Instead of being digested, the cells were kept alive and continued to carry out aerobic respiration and photosynthesis. The characteristics of mitochondria and chloroplasts support the endosymbiotic theory.

- They grow and divide like cells.
- They have a naked loop of DNA, like prokaryotes.
- They synthesize some of their own proteins using 70S ribosomes, like prokaryotes.
- They have double membranes, as expected when cells are taken into a vesicle by endocytosis.
- Cristae are similar to mesosomes of prokaryotes
- Thylakoids are similar to structures containing chlorophyll in photosynthetic prokaryotes.

Theories of evolution

If species do evolve, a mechanism must exist to cause this evolution. Two theories have been proposed.

LAMARCK'S THEORY OF EVOLUTION

Lamarck, a French naturalist, proposed his theory in 1809. He observed that the characteristics of living organisms can change during their lifetime. For example, if muscles are used, they grow stronger. These are called acquired characteristics. Lamarck proposed that when organisms reproduce, they pass on these acquired characteristics to their offspring – this is called **inheritance of acquired characteristics**.

An example that is often discussed is the neck of the giraffe. The ancestors of the modern giraffe had short necks. According to Lamarck's theory they had to stretch up into trees to reach food, so their necks lengthened slightly. The next generation inherited the lengthened necks and stretched more. Many generations of this process would lead to the long neck of the modern giraffe.

Despite many attempts, no significant cases of inheritance of acquired characteristics have been found. Also, Lamarck's theory does not fit in with our knowledge of inheritance. For the characteristic to be inherited, acquired characteristics would have to cause a mutation in the gene controlling the characteristic. No mechanism for this exists.

DARWIN AND WALLACE'S THEORY OF EVOLUTION

Darwin and Wallace jointly proposed their theory in 1858. Darwin is usually given credit for it as he had been working on it for much longer and he published a detailed account of it in 1859. The Darwin–Wallace theory is evolution by natural selection. An explanation of this theory is given on page 38.

According to this theory, the ancestors of the giraffe varied in neck length. At times of food shortage, when all the lower leaves on trees had been eaten, only the giraffes with longer necks could reach the higher leaves. They survived and passed on their characteristics to their offspring, including longer necks. There is much evidence for evolution by natural selection, including modern examples of observed evolution.

MODERN EXAMPLES OF OBSERVED EVOLUTION

It is not realistic to expect large-scale evolution over a few years, but there are some well-documented examples of small-scale evolution.

1. Evolution of melanism in ladybugs

Adalia bipunctata, the two-spot ladybug (or ladybird) is a small beetle, which usually has red wing cases with two black spots. The red colour warns predators that it tastes unpleasant. Melanic forms also exist, with black wing cases. The melanic form absorbs heat more efficiently than the red form. It therefore has a selective advantage when sunlight levels are low and it is difficult for ladybugs to warm up. The melanic form of *Adalia bipunctata* became common in industrial areas of Britain, but declined again after 1960. The decline correlates with decreases in smoke in the air (below). In air darkened by smoke, the melanic forms will be able to warm up more quickly, but if the smoke is no longer present this advantage is lost and warning colouration is more important.

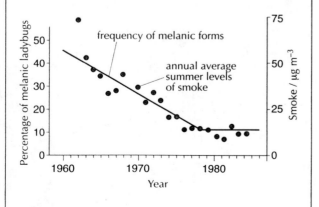

2. Evolution of the beaks of Galapagos Islands finches

Rosemary and Peter Grant have studied the two species of finch that inhabit Daphne Major, in the Galapagos Islands, for many years. One of the finches, *Geospiza fortis* (above), has a short, wide beak and feeds on a variety of seeds, including large hard ones. During 1982 – 3 there was a severe El Niño event, which brought very heavy rain to Daphne Major. With more food available, the population of *G. fortis* rose considerably, reaching a peak in 1983. It dropped back in the drier years following, and in 1987 was only 37% of its peak. The period of heavy rain changed the vegetation and until 1991 there were fewer plants producing large, hard seeds and more producing small soft ones. The diet of *G. fortis* therefore changed. The 37% of the population remaining in 1987 were not a random sample of the population. They had longer, narrower beaks than the average in 1983. There had been a significant change in the beaks of the population. The conclusion that this change was caused by natural selection due to the change in diet is supported by evidence from the other finch on the island, *Geospiza scandens*. Its population rose and fell in the same way as *G. fortis*, but neither its diet nor the size of its beak changed.

Evidence for evolution

Although it is not possible to prove using the scientific method that the organisms on Earth today are the result of evolution, there is much evidence that makes it very likely.

EVIDENCE FROM BIOCHEMISTRY

There are remarkable similarities between living organisms in their biochemistry.
- All use DNA (or RNA) as their genetic material.
- All use the same universal genetic code, with only a few insignificant variations.
- All use the same 20 amino acids in their proteins.
- All use left, and not right-handed amino acids.

These similarities suggest that all organisms have evolved from a common ancestor that had these characteristics.

EVIDENCE FROM HOMOLOGOUS ANATOMICAL STRUCTURES

There are also remarkable similarities between some groups of organisms in their structure.

1. At an early stage, vertebrate embryos are very similar, despite huge differences in the structure of the adults (below).

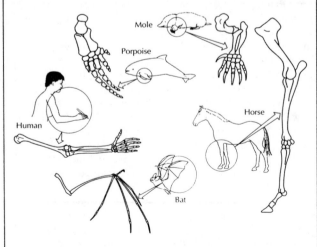

Tortoise

Chick

Rabbit

2. The limbs of vertebrates show striking similarities in their bones, despite being used in many different ways (below). The structure is called the pentadactyl limb.

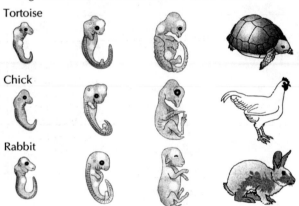

Mole

Porpoise

Human

Horse

Bat

The most likely explanation for these structural similarities is that the organisms have evolved from a common ancestor. Structures that have developed from the same part of a common ancestor are called homologous structures.

EVIDENCE FROM PALAEONTOLOGY

The existence of fossils is very difficult to explain without evolution. An example of this is *Acanthostega*. The figure (below) is a drawing of a 365-million-year-old fossil of *Acanthostega*. It has similarities to other vertebrates, with a backbone and four limbs, but it has eight fingers and seven toes, so is not identical to any existing organism. This suggests that vertebrates and other organisms change over time. *Acanthostega* is an example of a 'missing link'. Although it has four legs, like most amphibians, reptiles and mammals, it also had a fish-like tail and gills and lived in water. This shows that land vertebrates could have evolved from fish via an aquatic animal with legs.

Fossil of *Acanthostega*

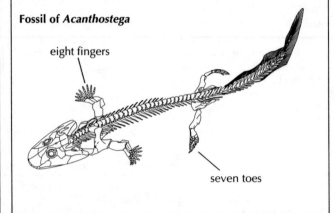

eight fingers

seven toes

EVIDENCE FROM GEOGRAPHICAL DISTRIBUTIONS

Wallace's Line is an example of geographical distribution of organisms that is difficult to explain without evolution. The figure (below) shows its position. There are huge differences in the types of land animal that are found on either side of Wallace's Line. For example, placental mammals are found on the Asian side and mainly marsupial and monotreme mammals are found on the Australasian side. The landmasses on the two sides of the boundary separated about 100 million years ago and came together again by continental drift about 15 million years ago. The mammals on the separated landmasses followed different evolutionary paths, so different types evolved. In similar habitats where natural selection acts in the same way on different organisms, the results are sometimes strikingly similar, for example the marsupial mole of Australia and the golden moles of Africa.

Map showing Wallace's Line

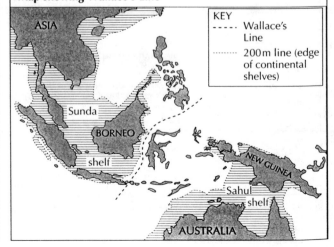

KEY	
- - - - -	Wallace's Line
....	200 m line (edge of continental shelves)

Evolutionary history

The study of the evolutionary history of groups of organisms is called **phylogeny**. Two useful approaches in phylogeny are study of the fossil record and comparison of molecular structure.

DATING FOSSILS

To place fossils into a sequence, it is necessary to know their dates. Fossils, or the rocks containing fossils can be dated using radioisotopes – radioactive isotopes of chemical elements. When an atom of a radioisotope decays, it changes into another isotope and gives off radiation. The rate of decay varies between different radioisotopes and is expressed as the **half-life**. *The half-life is the time taken for the radioactivity to fall to half of its original level.* The figure (right) shows a decay curve for radioisotopes.

The two radioisotopes that are most commonly used are ^{14}C and ^{40}K. In radiocarbon dating the percentage of surviving ^{14}C atoms in the sample is measured. In potassium–argon dating, the proportions of parent ^{40}K atoms and daughter ^{40}Ar atoms are measured. In both methods the age in half-lives can then be deduced from the decay curve. The half-life of ^{14}C is 5730 years, so it is useful for dating samples that are between 1000 and 100 000 years old. The half-life of ^{40}K is 1250 million years so it is useful for dating samples older than 100 000 years.

METHODS OF PRESERVATION

The remains of past living organisms can be trapped and preserved in various ways.
- In resins, which turn to amber.
- Frozen, in ice or snow.
- In acid peat, which prevents decay.
- In sediments that turn to rock.

The last method is the most important. Sediments accumulate in layers in parts of the sea and sometimes on land. The weight of sediments compresses those beneath until they become rock. If hard parts of animals such as shells or bones form part of the sediment, they will be preserved in the rock. Sometimes, the shape of an organism is preserved as a cast. Minerals sometimes seep into the soft parts of an organism as it decays and harden to form a petrified replica of the organism.

Decay curves for radioisotopes

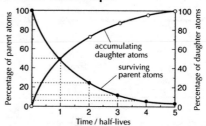

COMPARING MOLECULAR STRUCTURE

The phylogeny of many groups of organisms has been studied by comparing the structure of a protein or other biochemical that they contain. For example, the amino acid sequence of the polypeptide of hemoglobin has been compared in many vertebrates. The figure (below) shows the numbers of differences in the amino acid sequence in ten vertebrates. Differences in amino acid sequence accumulate gradually over long periods of time. There is evidence that differences accumulate at a roughly constant rate. They can therefore be used as an evolutionary clock. The number of differences in amino acid sequence can be used to deduce how long ago species split from a common ancestor. Using this information and the details of what the amino acid differences are, the probable phylogeny of groups of organisms can be deduced. The figure (bottom) shows the probable phylogeny of the ten organisms.

The phylogeny of many groups has been studied in this way. Usually, the results fit in with earlier studies of fossils, or anatomical studies, but sometimes there are surprises. For example, figure (bottom) shows larger than expected differences between the various types of fish.

Numbers of differences in the amino acid sequence of hemoglobin in ten vertebrates

	Elephant	Platypus	Ostrich	Starling	Crocodile	Lungfish	Coelacanth	Goldfish	Shark
Human ⟶	26	40	43	41	47	83	70	68	71
Elephant ⟶		45	45	48	50	84	72	63	74
Platypus ⟶			54	52	51	89	74	70	76
Ostrich ⟶				26	36	91	75	68	73
Starling ⟶					47	91	77	67	70
Crocodile ⟶						85	78	70	77
Lungfish ⟶							90	94	86
Coelacanth ⟶								83	78
Goldfish ⟶									88

Phylogenetic tree diagram for ten vertebrates

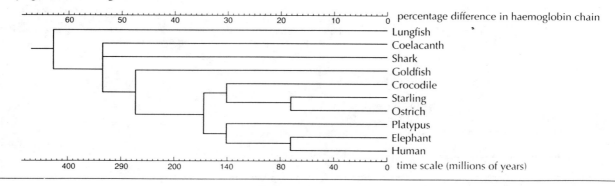

Human origins

HUMAN CLASSIFICATION

Humans are classified as primates (see below), because they show primate characteristics:

- grasping limbs, with long fingers and a separated opposable thumb
- mobile arms, with shoulder joints allowing movement in three planes and the bones of the shoulder girdle allowing weight to be transferred via the arms
- stereoscopic vision, with forward facing eyes on a flattened face, giving overlapping fields of view
- skull modified for upright posture

These characteristics are sometimes described as adaptations for tree life, though many other tree-living mammals do not show them.

Human classification

Kingdom	Animalia
Phylum	Chordata
Class	Mammalia
Order	Primate
Family	Hominidae
Genus	Homo
Species	sapiens
Subspecies	sapiens

SEARCHING FOR THE EARLIEST HUMAN ANCESTORS

The question of where the earliest hominid ancestors lived has still not been answered with certainty. The closest existing relatives of humans are chimpanzees and gorillas from Africa and orang-utans of South East Asia. Research into the differences between these primates in the amino acid sequences of their proteins, including hemoglobin, myoglobin and fibrinogen shows that humans are more closely related to chimpanzees and gorillas than orang-utans. The oldest human ancestors therefore probably lived in Africa.

Mitochondrial DNA from humans and related primates has been sequenced. The differences in base sequence have been used to construct a hypothetical phylogeny (shown below). The data confirms that humans are closer to African apes than Asian ones and therefore supports the theory that human ancestors split from the ancestors of chimps and gorillas in Africa. The data also allows approximate dating of the splits between African and other humans – 140 000 ago and the split between Europeans and Japanese – 70 000 years ago. The conclusion for this and other studies is that the ancestors of modern humans migrated out of Africa less than half a million years ago. Other fossil hominids found out of Africa must have been the result earlier migrations and modern hominids are not descended from these hominids.

Phylogenetic tree for humans and closely related apes

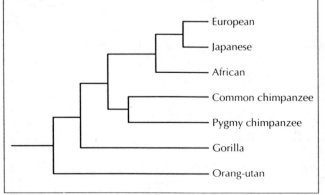

NEOTENY

Neoteny is keeping juvenile characteristics as an adult. Adult humans show similarities in appearance to baby apes (below), with flat faces, large brain to body size ratio, upright heads and little body hair. This suggests that human evolution from an ape ancestor might have involved a slowing down of development, with a long childhood, delayed puberty and retention of juvenile characteristics in adulthood.

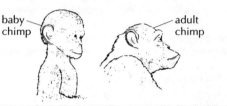

BIPEDALISM

Bipedalism is walking on two legs. Even the oldest fossils, of *Australopithecus afarensis*, show at least partial bipedalism, so it was a very early development in human evolution. The change to bipedalism involved many adaptations.

- The foramen magnum, a hole in the skull through which the spinal cord and brain connect, moved forwards. This allows the head to balance on the backbone.
- The arms became shorter and less powerful.
- The legs became longer and stronger.
- The knee changed to allow the leg to straighten fully.
- The foot became more rigid, with a longer heel, shorter toes and a non-opposable big toe.

There are many consequences of bipedalism. Collecting food from bushes is easier and also walking long distances while carrying food, water, infants, tools or weapons. It makes tree climbing more difficult.

INCREASED BRAIN SIZE

The brains of early hominids (*Australopithecus*) were only slightly larger in relation to body size than the brains of apes. The brains of later hominids (*Homo*) were larger (below). This was due to continued rapid brain growth after birth. In apes and earlier hominids brain growth slows after birth. There are many consequences of increased brain size. Capacities for learning, complex thought and memory are increased. Language and more complex tool manufacture and use are possible. However, the larger brains take longer to develop and more energy to use.

Brain sizes of *Homo* and *Australopithecus*

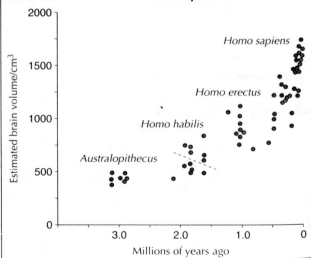

Human evolution

TRENDS IN HOMINID FOSSILS

Many hominid fossils have been found, dated and assigned to a species. These fossils show evolutionary trends, including increasing adaptation to bipedalism and increasing brain size. Other trends and dates of emergence are shown in the figure (right). *Australopithecus and Homo habilis* fossils were all found in Southern or Eastern Africa. *Homo erectus* fossils were found in Eastern Africa, but also in Asia, indicating that there was migration out of Africa. *Homo neanderthalensis* fossils were found in Europe and *Homo sapiens* in many parts of the world indicating further migrations.

EVOLUTIONARY RELATIONSHIPS BETWEEN HOMINIDS

There are many gaps in the hominid fossil record and so it is far from clear how species of hominid evolved. Three hypotheses for the origin of *Homo* are shown below.

Hypotheses for human origins

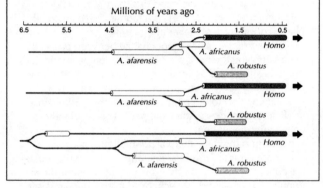

ECOLOGICAL CHANGES AND HOMINID ORIGINS

Five million years ago Africa became drier. Dense forest was replaced by thinner woodland with clearings. This may have prompted the evolution of bipedalism, although the earliest hominids probably still lived partly in trees. The powerful jaws and teeth of *Australopithecus* indicate a mainly vegetarian diet. About 2.5 million years ago Africa became much cooler and drier. Savannah grassland replaced forest. This change of habitat may have prompted the evolution of the first species of *Homo*, with the development of increasingly sophisticated tools and a change to a diet that included meat obtained by hunting and killing large animals. *Homo erectus* and later species developed the use of fire and were able to colonise colder areas and survive during ice ages.

CULTURAL EVOLUTION

The large brains of *Homo sapiens* and other species of *Homo* allow much to be learned, both during the long period of childhood and during adulthood. Language, tool-making skills, hunting techniques, methods of agriculture, religion, art and many other forms of behaviour are passed on from one generation of a tribe or other group to the next by teaching and learning. These things are the culture of the group. New methods, inventions or customs can be incorporated into what is passed on. This is called cultural evolution. Cultural evolution has been very important in the recent evolution of humans and has allowed much more rapid change than genetic evolution could alone.

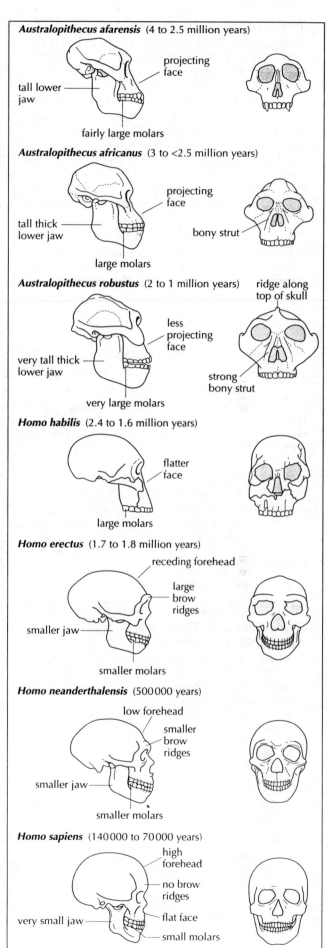

Population genetics

THE HARDY–WEINBERG PRINCIPLE

If there are two alleles of a gene in a population, there are three possible genotypes – homozygous for each of the two alleles and heterozygous. The frequency of the two alleles in the population is usually represented by the letters p and q. The total frequency of the alleles in the population is 1.0, so

$$p + q = 1.$$

If there is random mating in a population, the chance of inheriting two copies of the first of the two alleles is $p \times p$. The chance of inheriting two copies of the second of the two alleles is $q \times q$. The expected frequency of the two homozygous genotypes is therefore p^2 and q^2. The expected frequency of the heterozygous genotype is $2pq$. The sum of all of these frequencies is 1.

$$p^2 + 2pq + q^2 = 1$$

This is called the **Hardy–Weinberg equation**. If the allele frequencies and the genotype frequencies in a population are known, this equation can be tested. The example below shows the results of a survey of the MN blood group gene in Japanese town. The two alleles of this gene are codominant.

Allele frequencies in the parental generation

M allele: $p = 0.525$ N allele: $q = 0.475$

Genotype frequencies in the offspring

	Predicted	Actual
MM	$p^2 = 0.276$	0.274
MN	$2pq = 0.499$	0.502
NN	$q^2 = 0.225$	0.224

The results of the survey show that the actual genotypes fit those predicted by the Hardy–Weinberg equation very closely. They therefore follow the **Hardy–Weinberg Principle**. Various conditions can prevent the Hardy–Weinberg Principle from operating – non-random mating, natural selection, mutation, small population size, immigration or emigration.

CALCULATING ALLELE AND GENOTYPE FREQUENCIES USING THE HARDY–WEINBERG EQUATION

If a population is known to be following the Hardy–Weinberg Principle, the Hardy–Weinberg equation can be used to calculate unknown frequencies. An example of this is a gene with two alleles that controls the ability to taste phenylthiocarbamide (PTC). The ability to taste PTC is due to the dominant allele (T) and non-tasting is due to the recessive allele (t).

1600 people were tested in a survey. 461 were non-tasters – a frequency of 0.288. Their genotype was homozygous recessive ($t\,t$).

If $q =$ frequency of t allele, $q^2 = 0.288$ so $q = 0.537$
If $p =$ frequency of T allele, $p = (1 - q) = 0.463$
The frequency of homozygous dominants (TT) and heterozygotes (Tt) can be calculated.

$\quad p^2 =$ frequency of homozygous dominants
$\quad p^2 = 0.463 \times 0.463 = 0.214$
$\quad 2pq =$ frequency of heterozygotes
$\quad 2pq = 2\,(0.463 \times 0.537) = 0.497$

GENE POOLS

A new individual, produced by sexual reproduction inherits genes from its two parents. If there is random mating, any two individuals in an interbreeding population could be the two parents, so the individual could inherit any of the genes in the interbreeding population. These genes are called the **gene pool.**

A gene pool is all the genes in an interbreeding population.

NATURAL SELECTION AND CHANGES TO THE GENE POOL

If an allele increases the chances of survival and reproduction of individuals that possess it, the frequency of the allele in the gene pool will tend to increase. Conversely, if the allele reduces chances of survival and reproduction, it will decrease in frequency. These changes are due to natural selection. The Hardy–Weinberg Principle can be used to test for natural selection. If allele and genotype frequencies in a population show that the Hardy–Weinberg Principle is being followed for a particular gene, this indicates that there is no natural selection. Members of the population all have an equal chance of survival whatever alleles of the gene they possess. The allele frequencies will not change between one generation and the next. If allele and genotype frequencies do not follow the Hardy–Weinberg Principle, a possible reason is that natural selection favours one allele over another. Adaptations develop in populations as a result of changes in allele frequencies in the gene pool. This is sometimes called **microevolution.**

A population in which there are two alleles of a gene in the gene pool is **polymorphic**. If one allele is gradually replacing the other, the population shows **transient polymorphism**. Populations of ladybug that changed from having red wing cases with black spots to black wing cases are an example of transient polymorphism.

MACROEVOLUTION, GRADUALISM AND PUNCTUATED EQUILIBRIUM

Over long periods of time, many advantageous alleles will appear and spread through a species. These micro-evolutionary steps together constitute **macroevolution**. Eventually the amount of evolution becomes so great that the species is no longer the same – one species has evolved into another.

There has been much discussion among biologists about rates of evolution. One idea, called gradualism, is that evolution proceeds very slowly, but over long periods of time large changes can gradually take place. This does not fit in with the fossil record, which shows periods of stability, with fossils showing little evolution, followed by periods of sudden major change. The periods of stability may be due to equilibrium where living organisms have become well adapted to their environment so natural selection acts to maintain their characteristics. The periods of sudden change that punctuate the equilibrium may correspond with rapid environmental change, caused for example by volcanic eruptions or meteor impacts. New adaptations would be necessary to cope with new environmental conditions, hence strong directional selection and rapid evolution.

Mutations and polymorphisms

CHROMOSOME MUTATIONS

Mutations are changes to genes or chromosomes. Although mutations occur by chance, the rate at which they occur can be predicted. Down's syndrome and Klinefelter's syndrome are examples of conditions caused by chromosome mutations. Down's syndrome is described on page 21. Klinefelter's syndrome is caused by males having one or more extra X chromosomes (XXY). Although recognizably male, those with the syndrome have low testosterone levels and so are infertile and do not fully develop the male secondary sexual characteristics. Chromosome mutations often cause infertility and so the variation that they cause is not inherited. They are therefore not usually significant in evolution.

GENE MUTATIONS

Sickle cell anemia, cystic fibrosis and phenylketonuria (PKU) are diseases caused by gene mutations. PKU is caused by mutations of an autosomal gene that codes for phenylalanine hydroxylase. This enzyme converts the amino acid phenylalanine into tyrosine. Without it, phenylalanine accumulates in the blood to a harmful level that can cause mental retardation and death in young children. Over 30 different alleles cause PKU. Natural selection has kept them at low frequencies in the human population because, until screening and treatment for the disease recently became possible, children homozygous for PKU alleles died at an early age. In a similar way, natural selection will keep alleles that cause other genetic diseases at a low level in the human population.

CYSTIC FIBROSIS

Cystic fibrosis is the commonest genetic disease in Europe. It is caused by mutations of a gene coding for a chloride channel. This protein transports chloride ions across membranes in epithelium cells. Without the chloride channels in the plasma membrane, mucus secreted by epithelia becomes thick and sticky and tends to block airways of the breathing system, causing respiratory infections. Although various mutations of the chloride channel gene can cause cystic fibrosis, 70% of cases are due to one mutant allele, in which three bases coding for phenylalanine have been deleted. The frequency of this allele in Europe can be estimated using the Hardy–Weinberg equation.

Frequency of cystic fibrosis in Europe is one birth in 2500.

70% of these are due to the commonest allele,
    i.e. one birth in 1750 = 0.00057

If the frequency of this allele is q,
    $q^2 = 0.00057$ so $q = 0.023$

All the other alleles of the gene have a combined frequency of $(1 - 0.023) = 0.977$

The estimated frequency of carriers of this allele can be calculated.
    $2pq = 2(0.977 \times 0.023) = 0.045$

So, in Europe about 1 in 20 people are carriers.

BALANCED POLYMORPHISMS

The allele that causes 70% of cystic fibrosis cases probably originated in one person in Europe about 50 000 years ago. Far from being eliminated by natural selection, it increased in frequency, despite causing a severe disease in homozygous individuals. The figure (below) shows the results of an experiment that shows a possible cause. The experiment involved genetically engineered mouse cells that expressed the cystic fibrosis allele. These cells and two groups of control cells were mixed with strains of *Salmonella enterica typhi*, the bacterium that cause typhoid fever.

Effect of human alleles on infection rates in mice

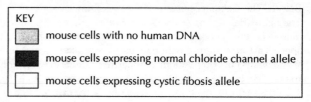

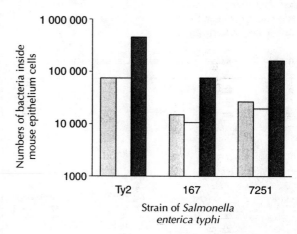

The results show that, in cells where the cystic fibrosis allele is expressed, *S. typhi* infection rates were much lower. Typhoid fever can cause death, so in populations where it is found, the cystic fibrosis allele might increase in frequency. Heterozygotes are the best adapted, because they are resistant to typhoid fever, but are do not develop cystic fibrosis. Natural selection will tend to maintain both the cystic fibrosis allele and the normal allele in the gene pool. It is not therefore a transient polymorphism and instead is called **balanced polymorphism**.

There is another well-known example of balanced polymorphism – sickle cell anemia (see page 28). Heterozygous individuals (Hb^AHb^S) do not develop sickle cell anemia and are resistant to malaria. They are therefore the best adapted in areas where malaria is found. The sickle cell allele has increased in frequency to high levels in some areas. In parts of Africa, as many as 40% of the population are carriers of the sickle cell allele, so show resistance to malaria. The Hardy – Weinberg equation can be used to calculate the frequency of the allele.

The frequency of carriers is 0.4
If the frequency of Hb^A is p
and the frequency of Hb^S is q,
    $2pq = 0.4$.
        So, $p = 0.724$ and $q = 0.276$

SPECIATION

The formation of new species is called speciation. New species are formed when a pre-existing species splits. This usually involves the isolation of a population from the remainder of its species and thus the isolation of its gene pool. The isolated population will gradually diverge from the rest of the species if natural selection acts differently on it. Eventually the isolated population will be unable to interbreed with the rest of the species – it has become a new species.

The most obvious way in which isolation can occur is by migration of members of a species to a new area that is geographically separated from the original territory of the species. This explains why there are often many **endemic species** (species that are found nowhere else) on isolated islands. The Galapagos finches are a well-known example. The figure (below) shows the distribution of another type of animal on the Galapagos archipelago – lava lizards. The seven species are all endemic, although there are related species on the mainland of South America. They are present on twelve of the islands, but on each of these islands only one species is found. Six species are only found on one island. The other species is found on six islands, but there are behavioural differences between these six populations, so they have already begun to diverge.

Distribution of lava lizards on the Galapagos Islands

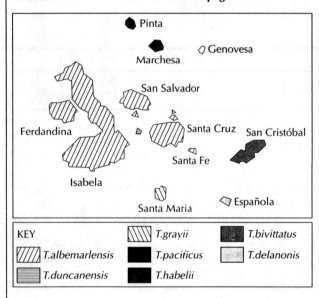

A second possible type of isolation is ecological. If two populations live in the same geographical area, but in different habitats, they may rarely interbreed. The apple maggot fly (*Rhagoletis pomonella*) of North America is an example of this. It originally only laid its eggs on hawthorn fruits, which were the food of its larvae. In the nineteenth century it started to infest non-native apple trees as well. Some strains of this species now prefer to lay their eggs on apple fruit and other strains prefer hawthorn fruit. Because the fruits ripen at different times, adults of the two strains emerge and mate at different times. They also inherit the preference either for apple or hawthorn and tend to remain on their preferred species. In addition to these behavioural differences, clear differences in allele frequencies have been found. If differences continued to build up, the two strains of *Rhagoletis pomonella* would eventually become separate species as a result of ecological isolation.

WHAT IS A SPECIES?

Biologists have been arguing about the exact meaning of the term species for over 200 years. Before the discovery that species can evolve, a species was regarded as a type of living organism with fixed characteristics, which distinguish it from other species. This is known as the morphological definition of a species. It is still a useful idea. Species can usually be distinguished from each other by their characteristics – this is how specimens are identified.

However the morphological definition does not recognize the fact that species evolve. If two populations with similar but not identical characteristics are geographically separated, they may be in the gradual process of splitting from one species into two separate ones. It is not easy for a taxonomist to decide whether to classify them as one or two species and some criterion is needed to decide. The reason for members of a species having common features is that they interbreed with each other. The reason for the characteristics of one species being different from those of another is that the two species do not interbreed and are evolving separately.

Biologists now regard interbreeding as a more important criterion than morphology. The biological definition of a species is *a group of actually or potentially interbreeding populations, with a common gene pool, which are reproductively isolated from other such groups.*

Only if two separated populations can be shown to be capable of interbreeding should they be classified as one species. The biological species definition is widely accepted, but it does cause some problems.

• Many sibling species have been found. These are species that cannot interbreed, but show no significant differences in appearance. Although separate species, they are very difficult for ecologists to identify. For example, the Pipistrelle bat in Britain was recently shown to be two sibling species.

• Some pairs of species that are clearly different in their characteristics will interbreed. Many plant species can hybridize and some animals also can, including ruddy ducks and white-headed ducks (below).

• Some species always reproduce asexually, so the members of a population do not interbreed. The biological species definition is therefore unusable.

• Some species have spread around the Earth to form a series of interlinked but slightly different populations. If the ends of this series overlap, the populations are sometimes so different that they do not interbreed and behave as separate species. An example is herring gulls and lesser black backed gulls in north-west Europe, which are linked by a ring of populations around the northern hemisphere. Examples like this are called ring species.

• Fossils cannot be classified according to the biological species definition, as it is impossible to decide with which organisms they would have been able to interbreed.

Two animal species that can interbreed

White-headed duck   Ruddy duck

EXAM QUESTIONS ON OPTION D – EVOLUTION

D1 The scattergram below shows the relationship between brain size and total body mass in species of mammal. Primate species are shown as solid circles and other species of mammal as open circles.

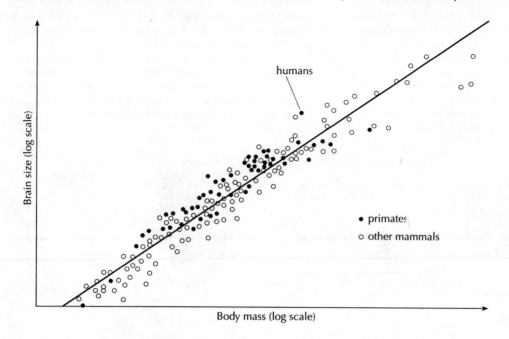

[Source: CUP, Encyclopaedia of Human Evolution,]

a) Using the data in the scattergram,

   (i) state the relationship between body mass and brain size in mammals   [1]

   (ii) compare the brain size in relation to body mass of primates with that of other mammals   [2]

   (iii) explain briefly how the scattergram can be interpreted to show that human brains are larger than those of other primates.   [2]

b) Increases in brain size in relation to body mass could be due either to increases in brain size or decreases in body mass. Suggest one advantage to primates of reduced body mass.

                                                                                                      [1]

D2 a) Outline a theory which explains the origin of eukaryotic cells.   [2]

   b) State two pieces of evidence which support this theory.

                                                                                                      [2]

D3 In Africa, south of the Sahara and north of the Zambezi, the sickle cell allele Hb_s is very common. In some ethnic groups the proportion of newborn babies that are homozygous recessive can be as high as 0.053 (5.3%). These babies suffer from sickle cell anemia.

   a) Calculate the **frequency** of the sickle cell allele in these ethnic groups.   [2]

   b) Calculate the **percentage** of the population that are carriers of the sickle cell allele.   [1]

   c) Outline the reasons for the high frequency of the sickle cell allele in these ethnic groups, despite the serious consequences of sickle cell anemia.   [2]

Examples of behaviour

MATE SELECTION IN *CERVUS ELAPHUS*

Mate selection is choosing partners for reproduction. In the autumn (fall) male red deer (*Cervus elaphus*) try to take possession of as large a group of females as they can and mate with them when they come into estrus. This involves defending the females against rival males by roaring, meeting the challenger at the edge of his territory and threatening him with horns lowered and sometimes by fighting (below left). Fights are usually brief, with the weaker male quickly conceding. If a male is not large or aggressive enough, females tend to move to other males' territories.

GROOMING IN *MACACA MULATTA*

Grooming is using fingers or tongue to clean the coat of a mammal to remove parasites. Rhesus monkeys (*Macaca mulatta*) spend up to 20% of the day grooming. It makes the individual being groomed visibly more relaxed and may be almost addictive because it causes endorphins to be released. Grooming is a submissive gesture, which helps to reinforce rankings within the group, as each monkey grooms higher-ranking individuals (see below). If a group becomes too large for all of its members to groom each other, the group tends to split.

Mate selection in red deer

Grooming in rhesus monkeys

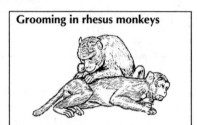

Migration of greater shearwaters

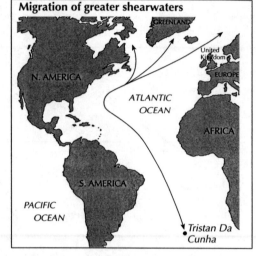

Courtship in green-backed herons

COURTSHIP IN *BUTORIDES VIRESCENS*

Courtship is attracting a mate. Male green backed herons (*Butorides virescens*) establish a territory. If a female enters the territory, a complex courtship begins, involving these stages:
- Flying in circles, with the male trying to attack the female.
- Flying in undulating circles with bent necks.
- The female perches and the male flies at her with slow deep wing flaps and neck plumage displayed.
- The male perches on the nest site and stretches his head and neck upwards, hopping from foot to foot.
- The male stretches his head downwards and snaps his bill (see figure above).

If the male is successful, the female is enticed to the nest and coitus and nest building soon follows.

MIGRATION IN *PUFFINUS GRAVIS*

Migration is movement between different areas according to the seasons of the year. Greater shearwaters (*Puffinus gravis*) use migration to avoid cold winters with low food availability. Between June and August they live and feed over a wide ocean area from Newfoundland to Scandinavia. In the fall (autumn) very large flocks form and fly more than 10 000 kilometres to Tristan da Cunha, a tiny group of islands in the South Atlantic. About four million birds nest there between January and March, feeding in the surrounding ocean. In the fall they migrate north again. The figure (above) shows the migration routes. Shortening day length triggers the migration and navigation involves using the Sun's position and also distinctive smells.

COMMUNICATION IN *FRINGILLA COELEBS*

Communication is the transfer of information from one animal to another. Bird-song is an example of communication. Male chaffinches (*Fringilla coelebs*) use their song to keep other males out of their territory and attract females. The song varies a little between males, allowing identification of individuals. It also has recognizable features to show that it is a chaffinch singing. There are also 13 calls that male and females make in different situations. The figure (below) shows the song and two calls. When a group of chaffinches is trying to drive away a bird of prey sitting on a perch, they make the 'chink' alarm call. The wide frequency range and repetition allows the source of the call to be located, encouraging other chaffinches to join the attack. When an individual chaffinch is hiding from a flying bird of prey it makes the 'seeet' alarm call to warn other chaffinches. The small frequency range, gradual start and finish and lack of repetition make the source of the call hard to locate.

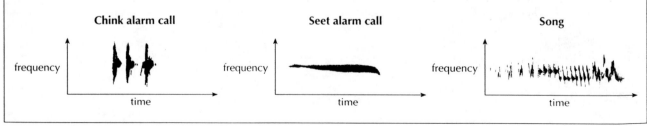

Perception of stimuli

SENSORY RECEPTORS

Animal behaviour is the response that an animal makes to the environment around it. This involves perception of stimuli from the environment, using sensory receptors. There are many different types of sensory receptor. They can be classified into groups. The figure (right) shows four groups of sensory receptor, with an example of each from the human body. All sensory receptors convert energy from the stimulus into the electrical energy of a nerve impulse. They are thus all **energy transducers**.

PHOTORECEPTORS IN THE EYE

The photoreceptors of the eye are contained in the retina. The figure (right) shows the structure of the eye including the position of the retina.

There are two types of photoreceptor cell – rod cells and cone cells. Rod cells are more sensitive to light than cone cells, so they function better in dim light. Rod cells become bleached in bright light, but cone cells function well. Rod cells absorb all wavelengths of visible light, so they give monochrome vision, whereas the three types of cone cell, sensitive to red, green and blue light, give colour vision. Groups of up to two hundred rod cells pass impulses to the same sensory neurone of the optic nerve, whereas cone cells have their own individual neurones through which messages can be sent to the brain. Cone cells therefore give greater visual acuity than rod cells. Rod cells are more widely dispersed through the retina so they give a wider field of vision.

Nerve pathways between retina and visual cortex

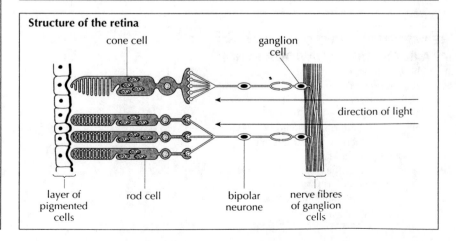

CLASSIFICATION OF SENSORY RECEPTORS

Group	Stimulus perceived	Example
Mechano-receptors	Mechanical energy in the form of movement, sound, pressure or gravity.	Hair cells in the inner ear send nerve impulses to the brain when sounds make them vibrate. Different hair cells respond to different frequencies of sound.
Chemo-receptors	Chemical substances	Nerve cells in the nostrils send impulses to the brain when specific chemicals bind to receptors in their membranes.
Thermo-receptors	Temperature	Warm and cold nerve endings in the skin send messages to the brain or spinal cord at a rate determined by skin temperature.
Photo-receptors	Electromagnetic radiation, usually in the form of light.	Rods and cones in the eye send messages to the brain, when they absorb light.

Structure of the eye (in horizontal section)

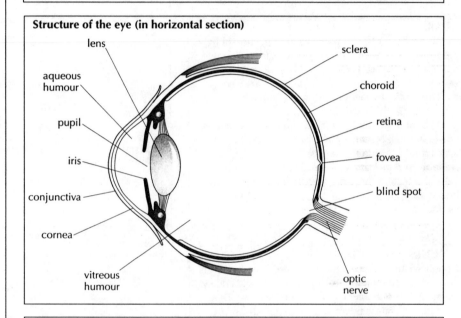

PROCESSING OF VISUAL STIMULI

Bipolar cells in the retina combine the impulses from rod or cone cells and pass them on to sensory neurones of the optic nerve (ganglion cells). The left and right optic nerves meet at a structure called the optic chiasma. Here, all the neurones that are carrying impulses from the half of the retina nearest to the nose cross over to the opposite optic nerve. As a result, the left optic nerve carries information from the right half of the field of vision and vice versa (see figure left). Beyond the optic chiasma, the neurones continue to the thalamus, where the information is processed. It is then carried to the visual cortex at the back of the brain, where further processing leads to formation of images.

Structure of the retina

Innate behaviour

The behaviour patterns of animals are classified into two types, innate and learned. Innate behaviour patterns develop independently of the environmental context. They are controlled by genes and are inherited from parents. They develop by natural selection, because they make members of a species better adapted to their environment and increase their chances of survival and reproduction. Innate behaviour is sometimes called species specific, because every member of an animal species usually shows it. *Innate behaviour is behaviour that normally occurs in all members of a species, despite variation in the environmental context.* In invertebrates most behaviour is innate and relatively simple. Taxes and kineses are behaviour patterns that increase the survival chances of many invertebrates.

TAXIS

Fly larvae (maggots) can perceive the direction of light and move away from it (top right). This is an example of a response called **taxis**. *A taxis is a movement towards or away from a directional stimulus.* The response of the fly larva is negative phototaxis.

The habitat of the fly larva is carrion – the carcass of a dead animal. Negative phototaxis ensures that the larva remains inside the carcass, where food is available and predators are less likely to catch it. Natural selection therefore favours larvae that show the response. A positive taxis is movement towards a stimulus.

KINESIS

Woodlice can perceive whether an area is humid or dry. They respond to humid areas by moving slowly and by changing direction after moving short distances. They respond to dry areas by moving more quickly and by moving further before changing direction. These responses are an example of **kinesis**. *Kinesis is response to a non-directional stimulus in which the rate of movement or the rate of turning depends on the level of the stimulus, but the direction of movement is not affected.* Woodlice are usually found in moist habitats. They use a tracheal system for gas exchange, which dehydrates rapidly in dry areas. Hygrokinesis ensures that woodlice spend more time in humid areas than dry areas if both are available. Natural selection therefore favours woodlice that show this response.

THE IMPORTANCE OF QUANTITATIVE DATA IN ANIMAL BEHAVIOUR STUDIES

Animal behaviour investigations often begin with careful observations. These help us to understand the natural history of a species. Observations often lead to the formulation of a hypothesis. To test the hypothesis, it is usually necessary to obtain quantitative data. Statistical tests can then be used to establish confidence levels for the data.

The figure (right) shows the results of a quantitative study of kinesis in *Porcellio scaber*, a woodlouse. The data can be used to calculate the number of turns per metre at each humidity.

Taxis response in a fly larva

direction

of light

light-proof barrier

Kinesis response in a woodlouse

dry area

humid area

Quantitative investigation of kinesis in a woodlouse

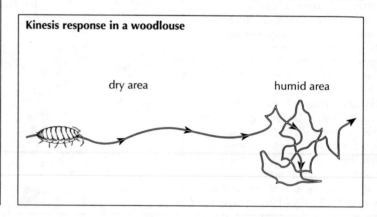

number of turnings

speed

% of time at rest

Number of turnings per hour

millimetres per minute

% of time at rest

Relative humidity

Reflexes in humans

SPINAL REFLEXES IN HUMANS

Spinal reflexes are examples of innate behaviour. For example, a spinal reflex action occurs when the hand touches a stinging plant. It is called the **pain withdrawal reflex**. Chemicals in the stings stimulate a pain receptor in the skin. The pain receptor passes a message to a sensory neurone, which carries it as a nerve impulse to the grey matter of the spinal cord. The message is passed via a linking neurone, called an **association neurone** in the grey matter to a motor neurone. The motor neurone carries the message to a muscle in the arm. The message stimulates the muscle to contract, pulling the hand away from the stinging plant. The muscle is called the **effector**. The series of neurones linking the receptor to the effector is called a reflex arc. Genes ensure that neurones in reflex arcs are connected up so that an appropriate response is made to a stimulus. The figure (below) shows a reflex arc. There are many spinal reflexes. If the skin on the sole of one foot receives a painful stimulus, a pain withdrawal reflex lifts this leg. Another reflex called the **crossed-extensor reflex** causes extensor muscles in the other leg to contract, so that it supports the body's weight.

CRANIAL REFLEXES IN HUMANS

The brain controls some reflexes. These are called cranial reflexes. The **pupil reflex** is an example. If a bright light shines into one eye, the pupils of both eyes constrict. Photoreceptor cells in the retina detect the light stimulus. Nerve impulses are sent in sensory neurones of the optic nerve to the brain. The brainstem processes the impulses and then sends impulses to circular muscle fibres in the iris of the eye. These muscle fibres contract, causing the pupil to constrict. The pupil reflex is sometimes tested in unconscious patients, to help to determine whether recovery is possible. If the pupil reflex and other brainstem reflexes have been lost, the patient has probably suffered brain death and will not recover. The conjunctival reflex is another example of a cranial reflex. If the conjunctiva is touched lightly, blinking occurs. The touch stimulus is passed to the brain along sensory neurones in the fifth cranial nerve. Messages are sent along motor neurones in the seventh cranial nerve to stimulate muscles in the upper and lower eyelids to contract and cause blinking. The figures (bottom) show the structure of the brain and some of its functions.

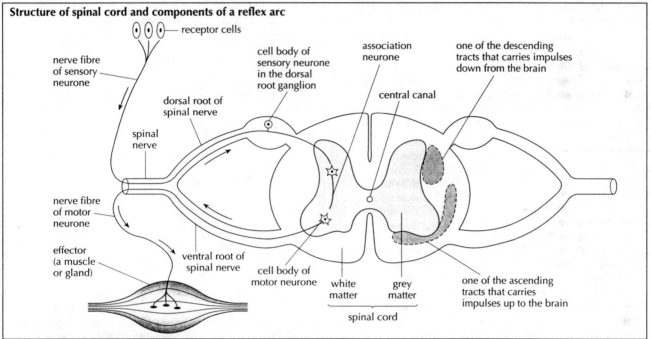

Structure of spinal cord and components of a reflex arc

receptor cells
nerve fibre of sensory neurone
cell body of sensory neurone in the dorsal root ganglion
association neurone
one of the descending tracts that carries impulses down from the brain
central canal
dorsal root of spinal nerve
spinal nerve
nerve fibre of motor neurone
effector (a muscle or gland)
ventral root of spinal nerve
cell body of motor neurone
white matter
grey matter
one of the ascending tracts that carries impulses up to the brain
spinal cord

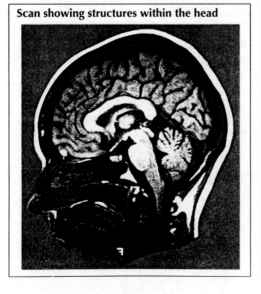

Scan showing structures within the head

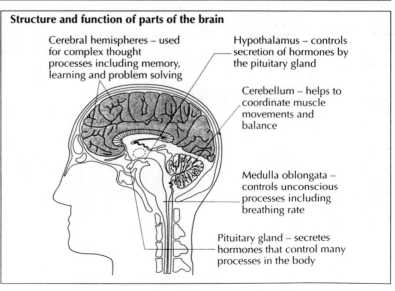

Structure and function of parts of the brain

Cerebral hemispheres – used for complex thought processes including memory, learning and problem solving

Hypothalamus – controls secretion of hormones by the pituitary gland

Cerebellum – helps to coordinate muscle movements and balance

Medulla oblongata – controls unconscious processes including breathing rate

Pituitary gland – secretes hormones that control many processes in the body

Learned behaviour

Learned behaviour patterns reflect the conditions that an animal experiences during development. Experiments in animal behaviour have revealed several different types of learning.

CLASSICAL CONDITIONING

Ivan Pavlov investigated the salivation reflex in dogs. He observed that dogs secreted saliva when they saw or tasted food. The sight or taste of meat is called the **unconditioned stimulus** and the secretion of saliva is called the **unconditioned response**. Pavlov then gave the dogs a neutral stimulus, such as the sound of a ringing bell or ticking metronome, before he gave the unconditioned stimulus – the sight or taste of food. He found that, after repeating this procedure for a few days, the dogs started to secrete saliva before they had received the unconditioned stimulus. The sound of the bell or the metronome is called the **conditioned stimulus** and the secretion of saliva before the unconditioned stimulus is the **conditioned response**. The dogs had learned to associate two external stimuli – the sound of a bell or metronome and the arrival of food. This is called classical conditioning. *Classical conditioning is an alteration in the behaviour of an animal as a result of the association of external stimuli.*

OPERANT CONDITIONING

Burrhus Frederic Skinner designed a piece of apparatus called a Skinner box to investigate learned behaviour in animals. When a rat or pigeon pressed a lever inside the box, a small pellet of food dropped into the box, which the rat could eat. The figure (below) shows the layout of a Skinner box. When a hungry rat is placed into the box it moves around, looking and sniffing at everything within the box. It eventually presses the lever by accident, but soon learns to associate pressing the lever with the reward of food. The food reward is called the **reinforcement**. Pressing the lever is called the **operant response**. This form of learning is called trial and error learning or operant conditioning. The more quickly the reinforcement is given, the more quickly the operant response develops. Surprisingly, Skinner found that if the reinforcement is not always given after the operant response, operant conditioning develops more strongly than if the reinforcement is always given. *Operant conditioning is behaviour that develops as a result of the association of reinforcement with a particular response, on a proportion of occasions.*

Rat receiving a reward in a Skinner box

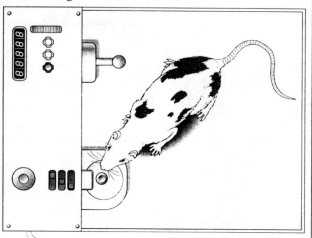

IMPRINTING

Konrad Lorenz investigated learning using greylag geese and other birds. In one experiment he removed half of the eggs that a female goose had laid and kept them in an incubator. Lorenz was with the goslings when they hatched out from these eggs, and he remained with them for a few hours. He was therefore the first moving object that they saw. The goslings did not show normal behaviour – they followed him around instead of their mother (below) and some of them even tried to mate with humans when they became adults. The goslings that hatched from the eggs left with the mother goose showed normal behaviour, following the mother around while they were young and eventually mating with other geese. Lorenz deduced that there is a sensitive period after the hatching of the birds during which they normally learn to identify and become attached to their mother. This is called **imprinting**. *Imprinting is learning a response to a stimulus during a sensitive period of development.* Further experiments showed that the newly hatched goslings did not become imprinted on any object – only moving objects. The moving object is an example of a **sign stimulus**. The birds have an innate ability to filter the stimuli that they are receiving and select the sign stimulus. This innate ability is called the **innate releasing mechanism**. The combination of sign stimulus and innate releasing mechanism allows an animal to learn to respond to significant stimuli in its environment and to ignore other stimuli. For each species of animal particular stimuli are significant, so sign stimuli and innate releasing mechanisms are an example of species specific behaviour.

Konrad Lorenz followed by birds that were imprinted on him

THE EFFECT OF LEARNING ON SURVIVAL CHANCES

There are many situations where survival chances can be increased as result of learning.
- Birds learn to avoid the evil-tasting black and orange caterpillars of the cinnabar moth by conditioning.
- Grizzly bears learn by operant conditioning how to catch salmon.
- Goslings learn who their mother is by imprinting and so avoid predators by remaining close to her.

Social behaviour

Animals that live in groups rather than alone are called social animals. When members of a group interact in terms of their behaviour, they are showing **social behaviour**. Some social animals form loose groups with limited interactions. Other social animals form groups that are highly organized with close interaction between the members of the group. Some mammals, such as naked mole rats and many insects including ants, termites and honey bees show this high degree of social behaviour.

SOCIAL ORGANIZATION OF HONEY BEES

Honey bees live in colonies consisting of up to 60 000 individuals. The colony acts like a super-organism that lives or dies together, and can reproduce to form extra colonies by swarming. There are three castes of honey bee, each of which has different tasks. The single queen bee is normally the only member of the colony to lay eggs. The worker bees do all the jobs that are needed to maintain the colony. The drones do nothing to help the colony to survive, but if they successfully mate with virgin queens they spread the genes of the colony to new colonies. Workers eject drones from the colony at the end of the season during which virgin queens are available. The table below summarises the tasks of the three castes.

Castes of honey bee

Caste	Gender	Tasks
Queen	Fertile female	Laying eggs Producing a pheromone to control the activities of workers
Drone	Fertile male	Mating with virgin females
Worker	Infertile female	Collecting nectar and pollen Converting pollen into honey Secreting wax and using it to build the comb Feeding and looking after larvae Guarding the hive

The figure (below) shows a worker bee performing a waggle dance to communicate the location of a good food source to the surrounding workers.

Waggle dance of honey bees

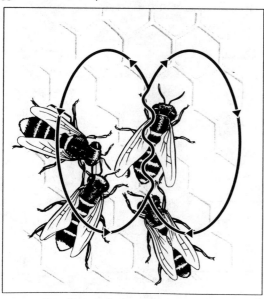

ALTRUISTIC BEHAVIOUR

If the behaviour of an animal is harmful to itself but helpful to another animal, it is altruistic. For example, a worker bee that dies while defending the colony against an attacking wasp is apparently behaving altruistically. A worker bee that feeds a larva is usually considered to be behaving altruistically, because it is not the parent of the larva. (Parental care is not considered to be altruism.)

Naked mole rats (below) are another example of altruistic behaviour. They live in burrow systems in parts of East Africa, in communities with different castes.

One dominant female mole rat acts like a queen bee. She is the only female in the community to reproduce. Three other castes of mole rat help her.

- 'Frequent workers' dig the tunnels and bring food.
- 'Infrequent workers' are larger and occasionally help with heavier tasks.
- 'Non-workers' live in the central nest, keeping the breeding female and her young offspring warm and defending the colony if it is attacked.

Naked mole rat non-worker

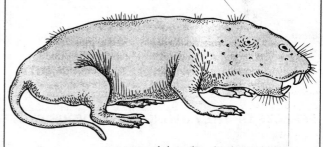

little hair as mole rats' burrows are maintained at a constant warm temperature

teeth bared to display aggression

As with honey bees, the mole rats in a colony are all genetically related, so although the workers are helping to rear offspring that are not their own, they are helping to ensure the survival of their own genes.

A stricter definition of altruism would include only acts that help an unrelated individual. One example is in vampire bats of South America. They live in groups and feed at night by sucking blood from larger animals. If one of the bats in the group fails to feed for more than two consecutive nights, it may die of starvation. However, bats that have fed successfully regurgitate blood for a bat that has failed to feed. Tests have shown that this is done whether the two bats are genetically related or not. There is an advantage for the whole group in this behaviour pattern. It is called reciprocal altruism, because the bat that donates food to a hungry bat may, in the future, receive blood when it is hungry.

Autonomic nervous systems

The autonomic nervous system (ANS) is the part of the peripheral nervous system that is used to control internal organs unconsciously. The heart, blood vessels, digestive system and smooth muscles of the bladder and other organs are controlled by the ANS. The figure (right) is a classification of the parts of the nervous system. It shows that there are two parts of the ANS – the sympathetic and the parasympathetic systems. They have opposite effects on many organs – they are antagonistic. The parasympathetic system keeps organs in a suitable state for non-threatening situations. The sympathetic system makes organs prepare for vigorous physical activity in response to threats or opportunities. It has effects similar to those of the hormone adrenalin. The table below shows the effects of each system on three organs.

Comparison of the effects of the two parts of the autonomic nervous system

Organ	Parasympathetic system	Sympathetic system
Heart	Heart rate is slowed as the body is relaxed and less blood flow is needed	Heart rate speeds up so that more blood can be pumped to the muscles
Salivary Glands	Saliva production is stimulated as food can be eaten in non-stressful situations	Saliva production is inhibited as feeding is not the main priority
Iris of the eye	Circular muscle fibres contract, so the pupil constricts to protect the retina	Radial muscles contract, dilating the pupil to give a better image of the threat

Classification of parts of the nervous system

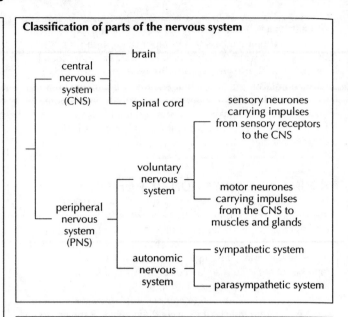

ACHIEVING CONSCIOUS CONTROL OVER AUTONOMIC REFLEXES

Although most autonomic reflexes are entirely unconscious, it is possible to achieve conscious control over some of them. This is an example of learned behaviour and involves the formation of new pathways in the brain by conditioning. Young children achieve conscious control over the sphincter muscles in their anus and bladder, so that they can release feces and urine voluntarily. Some adults have learned to achieve at least partial control over heart rate and other normally unconscious processes by meditation or yoga.

EFFECTS OF THE AUTONOMIC NERVOUS SYSTEM

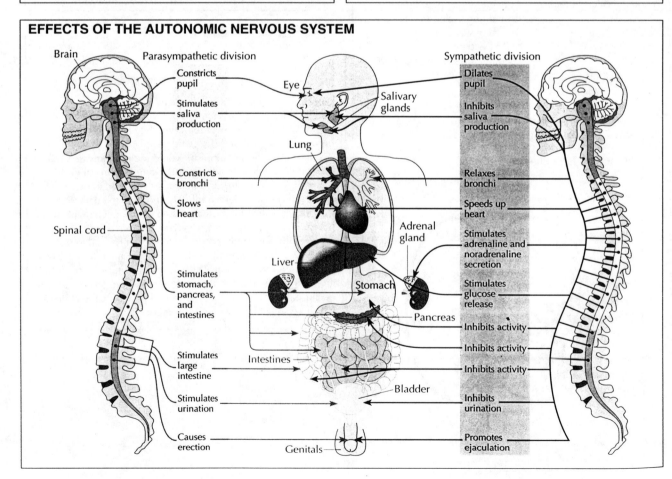

Neurotransmitters and synapses

NEUROTRANSMITTERS
There are many different neurotransmitters, but at each synapse only one is used. Synapses in the peripheral nervous system can be classified according to which neurotransmitter they use. Cholinergic synapses use acetylcholine. Most synapses in the parasympathetic nervous system are cholinergic. Synapses between motor neurones and muscle fibres, called neuromuscular junctions, are also cholinergic. Adrenergic synapses use noradrenaline. Most synapses of the sympathetic system are adrenergic.
A much wider range of neurotransmitters is used at synapses in the brain, including dopamine and enkephalins.

Electron micrograph showing adjacent neurones containing vesicles of different neurotransmitters

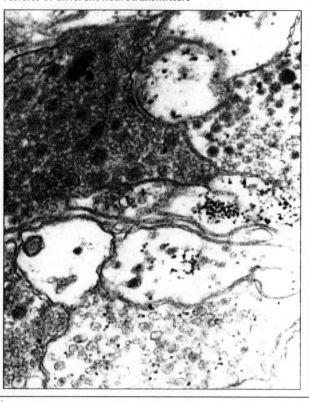

EXCITATORY AND INHIBITORY SYNAPSES
Neurotransmitters bind to receptors in the membrane of the post-synaptic neurone and cause a temporary change to the permeability of the membrane. Some neurotransmitters cause sodium ions or other positively charged ions to enter the post-synaptic neurone, helping to depolarize it and cause an action potential. Synapses where this type of neurotransmitter is used are called **excitatory synapses**.
Other neurotransmitters cause negatively charged chloride ions to move into the post-synaptic neurone, increasing its polarization. This effect, called **hyperpolarization**, makes it more difficult to depolarise a neurone sufficiently to cause an action potential. Synapses that cause hyperpolarization are therefore called **inhibitory synapses**. Most postsynaptic neurones have synapses with more than one pre-synaptic neurone, sometimes with hundreds (page 108). These may be a mixture of excitatory and inhibitory synapses. Whether an action potential is initiated in the post-synaptic neurone is decided by the summation of messages from all of these synapses. In this way, decisions can be made by the central nervous system.

PARKINSON'S DISEASE
Parkinson's disease was first described in 1817, but its cause was only recently discovered – death of neurones in a part of the brain called the substantia nigra. These neurones release a neurotransmitter called **dopamine** at synapses with neurones that help to control muscle contractions. The synapses at which dopamine is released are inhibitory. Muscle contraction cannot be controlled properly without dopamine and this causes the symptoms of Parkinson's disease. The early symptoms are feeling tired and shaky and loss of concentration. Eventually, the body becomes stiff because antagonistic muscles cannot relax. Uncontrollable shaking affects the hands and other parts and movements become very slow.

PAIN SENSATION
Pain receptors are located in the skin and other organs. They consist of free nerve endings, which perceive mechanical, thermal or chemical stimuli. Pain signals are sent from these nerve endings along nerve fibres to the spinal cord. The signals pass across synapses to neurones that carry them up in an ascending tract to the stem or thalamus of the brain. The signals may pass on in other neurones to sensory areas of the cerebral cortex, causing conscious pain sensation. Two types of nerve fibre carry impulses from the nerve endings to the brain – fast and slow. A painful stimulus therefore causes an initial sharp pain sensation, followed by slow, burning pain.

NATURAL PAINKILLERS
The sensation of pain is necessary to allow us to know when our body is being damaged, so that we can take avoiding action – pain withdrawal reflexes, for example. However, pain sometimes becomes excessive or stops us from concentrating on important activities. In these situations pain control systems in the brain and spinal cord can be used to reduce or prevent feelings of pain. Two natural painkilling substances are used – **enkephalins** and **endorphins**.
Pain control pathways in the brain lead to neurones that carry impulses down a descending tract of the spinal cord. These neurones cause the release of enkephalins at synapses where pain signals are passed to the neurones that carry them to the brain. Enkephalins block calcium channels in the membrane of the pre-synaptic neurones. They therefore block synaptic transmission, so the pain signals do not reach the brain. The pituitary gland releases endorphins to control pain. The endorphins are carried in the blood to the brain and other organs. They bind to receptors in the membranes of neurones that send pain signals and may block the release of a neurotransmitter that is used to transmit the pain signals to the brain. Endorphins are secreted during stressful times, after injuries and even during physical exercise such as running.

Psychoactive drugs

Drugs are chemical substances (see below) that are ingested, injected, inhaled or put into the body in some other way, to cause a change in the functioning of the body. Psychoactive drugs affect the functioning of the mind (below).

nicotine

cocaine

PSYCHOACTIVE DRUGS AND SYNAPSES

Most psychoactive drugs affect the mind by disrupting synaptic transmission in the brain. This can happen in various ways.

- Some psychoactive drugs have a chemical structure similar to a neurotransmitter and so bind to receptors for that neurotransmitter in post-synaptic membranes. They block the receptors, preventing the neurotransmitter from having its usual effect.
- Other psychoactive drugs with a chemical structure similar to a neurotransmitter have the same effect as the neurotransmitter. However, unlike the neurotransmitter, they are not broken down so when they bind to the receptor the effect is much longer lasting.
- Some psychoactive drugs interfere with the breakdown of neurotransmitters in synapses and so prolong the effect of neurotransmitters.

The overall effect of a psychoactive drug depends on whether synapse affected is excitatory or inhibitory.

INHIBITORY PSYCHOACTIVE DRUGS

Benzodiazepines, cannabis and alcohol are inhibitory psychoactive drugs because they decrease the activity of the nervous system. They act in different ways and so have different effects on behaviour.

Benzodiazepines bind to the same receptors as GABA, an inhibitory neurotransmitter, prolonging the inhibitory effects of GABA at synapses in the brain and spinal cord. This reduces anxiety and gives a feeling of relaxation. High doses cause drowsiness, slurred speech and loss of muscle co-ordination. Valium™ and Temazepan™ are examples of benzodiazepines that doctors prescribe as tranquillizers.

Cannabis contains a mixture of chemicals, but one of them called THC causes most of its psychoactive effects. THC binds to cannabinoid receptors in various parts of the brain, blocking synaptic transmission. Users claim that it increases the intensity of sensory perception, gives a feeling of emotional well-being and allows clear thinking about complex ideas. These claims are not backed up by any evidence of increases in mental performance. On the contrary, there is strong evidence that the ability to concentrate, control muscle contractions and judge times and distances is so much reduced that it is not safe to drive vehicles or operate machinery.

Alcohol acts as an inhibitor in at least two ways – by enhancing the effects of GABA, an inhibitory neurotransmitter, and decreasing the activity of glutamate, an excitatory neurotransmitter. In small quantities alcohol reduces inhibitions, making people more confident and talkative. It also impairs reaction times and fine muscle coordination, so it unsafe to drive vehicles. In larger quantities it causes loss of memory, slurred speech, loss of balance and poor muscle coordination. In some people alcohol causes violent behaviour.

EXCITATORY PSYCHOACTIVE DRUGS

Nicotine, cocaine and amphetamines are excitatory psychoactive drugs because they increase the activity of the nervous system. These drugs have different effects on behaviour.

Nicotine is the drug found in cigarettes and other forms of tobacco. It stimulates synaptic transmission at cholinergic synapses in many parts of the brain. Despite this, nicotine users claim that it has a calming effect. The reason is probably that nicotine is addictive and so an increase in the nicotine level of the blood reduces the craving for it.

Cocaine stimulates transmission at adrenergic synapses in the brain, causing increased energy, alertness, and talkativeness. It also gives an intense feeling of euphoria. Cocaine is usually absorbed through the skin inside the nostrils, where it causes constriction of blood vessels, delaying absorption. Crack is a form of cocaine that forms a vapour when it is heated. It can therefore be inhaled and absorbed very rapidly and gives very intense effects. These cause greater addiction and overdose problems than other forms of cocaine.

Amphetamines also stimulate transmission at adrenergic synapses, so have similar behavioural effects to cocaine. Whereas the effects of cocaine only last about 40 minutes, amphetamines continue to have an effect for between 2 and 4 hours. Amphetamines cause users to feel highly aroused and alert. They become hyperactive and sometimes aggressive. Ecstasy is a derivative of amphetamines that, like amphetamines, causes hyperactivity. This can lead to dangerous levels of overheating of the body. Ecstasy has some unusual behavioural effects – it causes feelings of empathy, openness and caring. It lowers feelings of aggression and increases sexual behaviour. However there is evidence that ecstasy causes long-term damage to neurones in the brain that it stimulates.

EXAM QUESTIONS ON OPTION E – NEUROBIOLOGY AND BEHAVIOUR

E1 Odorants are substances which can be detected by chemoreceptors in the nose. Many different odorants can be detected but each chemoreceptor cell is sensitive to only one type. The diagrams below show the mechanism used in the chemoreceptor.

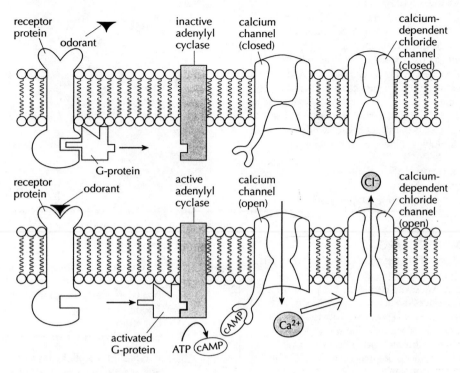

[Source of data: Gold et al, *Nature*, (1997), 385, page 677]

a) Deduce which part of the mechanism is different in chemoreceptor cells that are sensitive to different odorants. [2]

b) When the odorant binds to the receptor protein, the receptor protein starts activating G protein.
Using the data shown in the diagrams outline the effects of activated G protein. [3]

c) Predict the effect of entry of calcium ions and exit of chloride ions on the chemoreceptor cell. [1]

E2 a) State which type of receptor is found in the eye. [1]

b) Outline the neural pathway involved in the pupil reflex. [2]

c) State how this reflex can be used to find out the condition of the central nervous system. [1]

E3 a) State the name of the neurotransmitter substance normally produced at the effector in: [2]

(i) the sympathetic nervous system

(ii) the parasympathetic nervous system.

b) Atropine is a drug that depresses the activity of the parasympathetic neurons system. Predict the effect of atropine on the heart rate. [1]

c) Identify two other possible effects of atropine. [2]

Crop production

PLANTS AND PEOPLE

Plants are useful to humans in many ways:
- **Food** – plants produce much of the human diet. For example rice is the grain of rice plants.
- **Fuel** – wood and other parts of plants can be burned as fuel, or fuels can be made from plant material. For example, sugar cane crops can be grown and the sugar from them can be converted into alcohol, which is used as vehicle fuel.
- **Clothing** – fibres produced by plants can be spun and woven to produce material for making clothes. Cotton is an important example.
- **Building materials** – the wood that trees grow to provide support can also be used to construct floors, windows, doors and roofs of buildings. Pine is used in many parts of the world.
- **Esthetic value** – plants are grown indoors and outdoors because of their attractive appearance. Plants are also grown to produce cut flowers, for example, roses.

INTENSIVE MONOCULTURE

Monoculture is the system where the same crop is grown in an area every year. This system has some advantages. The most profitable crop can be grown every year. Also, by specializing in one crop the farmer reduces the amount of machinery needed. There are several potential problems with monoculture.
- The same nutrients are absorbed from the soil year after year by the crop. These nutrients tend to become depleted and large fertilizer applications are needed to prevent this.
- The pests, diseases and weeds that affect the crop tend to build up year after year, so large amounts of pesticide, fungicide and herbicide are needed to control them.

The aim of intensive systems is to produce the highest possible yield from the area of cultivation. This involves many actions.
- Land must be used efficiently with large field sizes and no areas left uncultivated.
- Fertilizers must be applied in large amounts, to prevent growth being checked by mineral deficiencies.
- Pesticides, fungicides and herbicides must be applied at the correct time, so that they are most effective at preventing damage to the crop.
- The sowing and harvesting of the crop must be done at the ideal times.

Planting rice

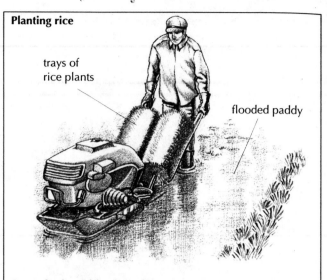

trays of rice plants

flooded paddy

Stages in the cultivation of rice crops

Month	Operations
April	Rice seed is sown in greenhouses or in fields under plastic sheeting, so that 3–4 week-old plants are ready at planting time Paddies are ploughed to kill weeds and bury trash NPK fertilizer is spread on the paddies
May	Paddies are flooded with irrigation water Soil is softened using a rotovator with a paddle wheel Rice plants from the nursery are planted using a planting machine.
June	Herbicides are sprayed on the crop to kill broad-leaved weeds Nitrogen fertilizer is applied to the growing crop
July	The crop is regularly weeded Insecticides are sprayed onto the crop if pests appear
August	The paddies are drained to allow the soil to start to dry out and the crop to ripen
September	The crop is harvested using a small combine harvester, which cuts the stems and threshes out the grain

RICE – AN EXAMPLE OF A CEREAL CROP

Rice is grown in many parts of the world in fields that can be flooded or drained, called paddies. Methods of cultivation vary. The table (above right) shows the sequence of operations that is followed on many farms in central Japan. Figures show planting (top right) and harvesting (right) of the crop.

Harvesting rice

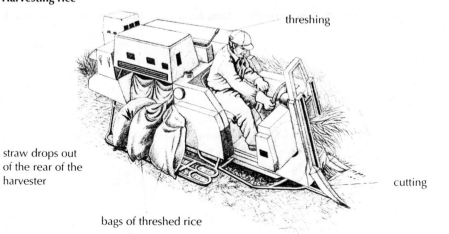

threshing

straw drops out of the rear of the harvester

cutting

bags of threshed rice

Plant productivity

FACTORS AFFECTING PLANT PRODUCTIVITY

Many factors affect plant productivity. The **genotype** of crop plants has a very significant effect. Crop plants have been selected for high yield, but varieties must be selected with a genotype that is adapted to the growing conditions. External factors also have significant effects on plant productivity. **Temperature, carbon dioxide concentration** and **light** all have an effect because they can limit the rate of photosynthesis. Water affects crop productivity because when water availability is low, plants close the stomata in their leaves. This slows down photosynthesis. **Availability of nutrients** also affects productivity because nutrients are needed for plant growth (see graph below). **Disease** affects the health of plants and therefore their growth rate. If **predators** consume leaves, they reduce photosynthesis and therefore growth. They also sometimes consume seeds, fruits or other parts of the crop that is harvested.

Effect of nitrogen fertilizer on yields of traditional (Peta) and new (IR20) varieties of wheat

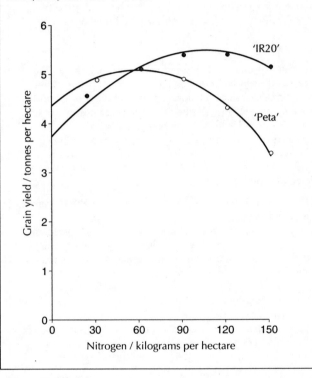

HYDROPONICS

Soil provides water, nutrients and anchorage for plants. If these requirements can be provided in a different way, soil is not necessary. Hydroponics involves growing plants in nutrient solutions. The crop is planted in sand, gravel, matting or some other material that provides an anchorage. A nutrient solution flows over the roots. In one system, fibreglass matting is placed on polythene sheeting on the floor of a greenhouse. The slopes are adjusted so that fluid drains through the matting from one end to the other. The nutrient content of the fluid that drains out is analysed by computer and nutrients are added to return it to an ideal composition before it is recirculated through the matting. This system is efficient because the polythene sheeting prevents leaching of water or nutrients and the roots of the crop can be continuously bathed in the ideal mixture of nutrients. Another advantage is that the matting can be sterilized to kill diseases much more easily than soil.

PLANT PRODUCTIVITY IN GREENHOUSES

The productivity of many crops can be increased by growing them in greenhouses. This is because many of the factors that affect plant productivity can be kept closer to the optimum level.

- Temperatures tend to be warmer inside greenhouses than outside. This is because short wave solar radiation passes through glass or plastic, but the longer wavelength radiation that is re-emitted cannot pass out, so the greenhouse warms up. If the temperature rises too high shading or ventilation can be used to reduce it.
- Carbon dioxide concentration can be increased from 0.035% to 0.1% or more by burning butane fuel in the greenhouse.
- Light levels can be increased using artificial lighting.
- Water availability can be maintained by irrigation. Humidity levels can also be kept high by reducing ventilation.

Greenhouse cultivation has other advantages. Birds and other animal predators can be excluded. Insect pests can be controlled using biological control methods more easily in greenhouses than in open fields. Damage due to wind, rain and hailstorms is prevented.

MEASURING CROP PRODUCTIVITY

Crop productivity is the rate at which plant matter is produced by a crop. The amount of plant matter can be measured either as fresh mass or dry mass. The rate of production can be measured in various ways:

- per gram of crop plant
- per unit area of cultivated land
- per unit area of crop plant leaf.

There are many measures of plant productivity.

Relative growth rate
Relative growth rate is the difference between the final mass and the initial mass, divided by the initial mass. The mass of one batch of plants is measured at the start of a defined period of time. The mass of an equivalent batch is measured at the end of the period.

Harvestable dry biomass
Harvestable dry biomass is the dry mass of the part of the crop that is harvested. For example, apple fruits are taken when ripe from an orchard, dried in an oven and weighed. The harvestable dry biomass is usually calculated per unit area of cultivated land.

Net assimilation rate
Net assimilation rate is the net increase in plant biomass per unit of leaf area and per unit of time. The mass of equivalent batches of plants is found at the start and end of a growth period. The leaf area of the crop is also found, by harvesting plants from a measured area, and removing and measuring the area of all of their leaves.

Livestock production

ANIMALS AND PEOPLE

Animals are useful to humans in many ways

- **Food** – foods rich in protein are produced for people using animals. For example, hens produce eggs and chickens produce meat.

- **Clothing** – leather for shoes and for garments is made from the skins of animals, for example cattle. Fur is made from the skin of mammals that have a thick coat of hair, for example mink.

- **Transport** – some animals are ridden or are used to provide traction for vehicles. Horses are an example of this.

- **Ploughing** – animals can be used to provide the traction for ploughing, for example, oxen.

- **Pets** – animals provide companionship and interest for people when they are kept as pets, for example cats.

CHICKENS – AN EXAMPLE OF ANIMALS REARED FOR MEAT

Chickens are reared in many parts of the world. Sometimes they are reared on a small scale, roaming freely and feeding on leftover scraps of food, but most chickens are now reared on a huge industrial scale in growing houses.

The table below shows the sequence of operations that is followed by many chicken production companies in California.

- The chickens are reared in 10 × 100 m houses, with a 100 mm depth of wood shavings or other absorbent material on the floor and food and water supplied automatically.
- The feed is usually a mixture of maize (corn), soybean meal, ground limestone and salt, with supplements of vitamins and other nutrients.
- Lighting, heating and ventilation are all artificially controlled.
- The aim is to rear at least five batches of chickens per year in each growing house.

Stages in the rearing of chickens

Week	Operations
1	Chicks hatch from eggs in incubators and after 48 hours are transferred to a growing house. A batch of 20 000–25 000 chicks is put in together. Supplies of food are unlimited (ad lib). The temperature is 32–35 °C
2 – 5	Ad lib feeding is continued. The temperature is reduced by 2.5 °C each week. Disease is controlled using vaccinations or antibiotics
6 – 7	When the chickens reach a mass of 1.8–2.2 kg they are slaughtered, by cutting veins in the neck. They are then de-feathered and have internal organs, feet, necks and heads removed. They are either frozen or refrigerated and sold
8 – 10	The growing house is emptied, cleaned, disinfected and left empty for 3 – 4 weeks before restocking, to avoid disease transmission between batches

VETERINARY TECHNIQUES AND LIVESTOCK PRODUCTION

Many modern veterinary techniques can help to increase livestock productivity, by improving health and fecundity. Fecundity is the fertility of the livestock – the number of offspring that they produce per year.

Artificial insemination (AI)

- This technique is most often used in cattle.
- Semen is obtained from bulls that have undergone progeny testing. Bulls that are found to pass on genetic defects or poor health are not used.
- The semen is diluted and placed in straws, which are stored in liquid nitrogen at −192 °C.
- When a farmer detects that a cow is in estrus, a straw of semen is thawed in warm water and placed in an insemination rod. This is inserted into the cervix, where the semen is expelled.
- The success rate of AI is very high, with few cows failing to start a gestation after one service.
- Semen from many different bulls can be used, preventing the health problems associated with inbreeding.

Vaccination

- When livestock are kept in crowded conditions, diseases can easily spread, reducing growth rates and sometimes causing death. Vaccination can prevent these problems. For example, chickens are often vaccinated against Newcastle disease, Marek's disease and infectious bronchitis.
- If cattle or sheep are vaccinated a few weeks before giving birth, they pass on antibodies to their offspring in the colostrum – the first milk that the udder secretes.

Nutrient supplementation

- Nutritional deficiencies can cause serious health problems in livestock.
- Cattle and sheep are sometimes affected by hypocalcemia (low blood calcium level) shortly before or after giving birth. This is because the onset of milk production causes a sudden increase in the demand for calcium. The animal dies very quickly if she is not given a calcium injection. Calcium supplements in the last weeks of gestation prevent hypocalcemia.
- Some animal feeds are deficient in particular nutrients. For example, soybeans are deficient in the amino acid methionine, so it is routinely added to chicken feed based on soybean meal.
- There are many other examples of the health of animals being greatly improved by the addition of small amounts of particular nutrients to the diet, for example retinol is often added to the diet of cattle and sheep housed in winter.

Controversies in agriculture

There are many different methods of farming and much disagreement about which methods are best. In the tables below biological arguments are indicated with the symbol B and ethical arguments with the symbol E.

BIOLOGICAL AND CHEMICAL PEST CONTROL METHODS

Arguments against biological pest control (*using a predator, parasite or pathogen to control the pest*)	**Arguments against chemical pest control** (*using a toxic chemical to control the pest*)
B The predator, parasite or pathogen may damage ecosystems by attacking organisms apart from the pest	B Pesticide residues in food sold for human consumption may cause health problems
B Biological controls often don't work outside greenhouses	B Pests almost always develop resistance to pesticides
B Biological controls cannot be introduced until there is a pest infestation, by which time there will have been some damage	B Non-target species, such as pollinating insects and natural predators of pests are sometimes killed by pesticides
E Some biological controls cause suffering to the pest before it is killed, for example the control of rabbits using a pathogen that causes the disease myxomatosis	B /E Pesticides tend to accumulate at higher and higher levels along food chains, reaching lethal levels in birds of prey and other top carnivores
E Biological controls involve disruption of the natural world when alien organisms are released into an ecosystem	B /E Workers may be forced to handle pesticides that cause long-term damage to their health

'ORGANIC' VS. 'NON-ORGANIC' FARMING METHODS

Arguments against organic farming (*farming without artificial fertilizer, insecticide, fungicide or herbicide*)	**Arguments against non-organic farming** (*farming using agrochemicals*)
E Yields are usually lower, and if world food production falls there is a risk of starvation in poorer countries	B Artificial fertilizers can leach into rivers and lakes, causing eutrophication problems
B Damage caused by pests causes blemishes, which makes the food less palatable	B/E Pesticide toxicity can cause problems – see table above. There is more wildlife on organic than non-organic farms
B Manure has to be applied before crops are sown, so nutrients released from it by decay may leach, causing water pollution	B Food may not taste as good if high levels of fertilizer cause rapid sappy growth
B Fungi cannot easily be controlled without fungicides. Some fungi produce toxins which are carcinogenic	B Without manure or other organic matter, soil structure deteriorates, increasing the danger of soil erosion

INTENSIVE LIVESTOCK REARING TECHNIQUES

Arguments for intensive-livestock rearing.	**Arguments against intensive livestock rearing.**
Yields of food per area of land and per worker are usually higher than with extensive techniques	Overcrowding increases the risk of outbreaks of disease or parasites, which can greatly reduce yields
E Large quantities of food can be produced cheaply, helping to solve the world food problem	E Overcrowding is cruel and causes stress for example when dominant animals attack weaker ones
E Less farmland is needed to produce sufficient food, so more can be left for wildlife	E Animals suffer because they are not able to follow their natural behaviour patterns
The health of the animals can be monitored more closely	E Pesticides have to be used, which damage ecosystems
Prices for meat from veal calves and other animals reared in small pens may be higher because the meat is more tender	Energy inputs are high. Yields per unit of energy used are lower than with extensive systems

ANTIBIOTICS AND GROWTH HORMONES IN LIVESTOCK PRODUCTION

Arguments for antibiotics and growth hormones	**Arguments against antibiotics and growth hormones**
Antibiotics increase growth rates by controlling infection, so increasing food availability	Antibiotics routinely included in feed rations increase the rate of evolution of antibiotic resistance in bacteria
Animals have as much right as humans to have the diseases that they suffer from controlled with antibiotics	Antibiotic residues in milk or other animal products can cause allergy problems in humans
Growth hormones boost production by increasing growth rates and milk secretion. More food is therefore available	Hormones are usually administered as an implant in the ear. After slaughter, cattle ears are often used in production of pet food, where the implants release their remaining hormones
Some foods obtained from plants contain larger amounts of estrogen than beef from cattle implanted with this hormone. Humans have eaten these foods without harmful effects for hundreds of years	Steroid hormones, which are sometimes used to promote growth, escape into the environment in animals' urine. If these hormones reach humans, for example through drinking water, they could alter development in young boys and girls or increase rates of breast cancer
Humans do not seem to be affected by consuming meat or milk containing growth hormones	Animals forced to grow or secrete milk at unnaturally high levels may suffer

Plant growth hormones

Plant growth hormones (sometimes also called plant growth regulators or plant growth substances) are chemical substances that act as messengers, controlling growth and development in plants. They are different from fertilizers such as nitrate and phosphate, which are needed for the synthesis of proteins and other essential substances.

AUXIN AND PHOTOTROPISM

The shoots of most plants grow towards the light. This response is called phototropism. Despite over a hundred years of research, the mechanism that controls it is still not fully understood. Shoot tips produce a hormone called auxin. Auxin acts as a growth promoter, possibly by causing secretion of hydrogen ions into cell walls, which loosens connections between cellulose fibres, allowing cell expansion.
According to a long-standing theory, auxin is redistributed in the shoot tip from the lighter side to the shadier side. It would then promote more growth on the shady side, causing bending towards the light. Recent experiments suggest that auxin may not move. Instead, an inhibitor of auxin may be produced on the lighter side of the shoot.

AUXIN AND APICAL DOMINANCE

If a shoot tip is cut off a plant, buds on the stem below start to grow, forming lateral shoots. While the shoot tip was present it exerted apical dominance, preventing the growth of the buds. Growth of lateral shoots can be prevented for a time after removal of some plants' shoot tips by applying auxin to the cut end of the stem. This suggests that high auxin concentrations in the shoot tip may be the cause of apical dominance. Growers of decorative plants sometimes remove shoot tips by pruning, to encourage lateral shoot growth and make the plant bushier.

MICROPROPAGATION

The use of artificially manufactured plant growth hormones has allowed the development of a new method of cloning, called micropropagation. The figure below shows the main stages in this method of cloning.

COMMERCIAL USES OF PLANT HORMONES

Plant hormones can be used in many ways to control or modify plant growth and development.

1. Promoting rooting of cuttings

Many plants can be propagated by taking cuttings. A cutting consists of a shoot, removed from the parent plant. The end of the cutting is inserted into a medium containing water, where it forms roots and starts to grow. Root formation can be encouraged in some plants by dipping the end of the cutting into a powder containing auxin.

2. Killing weeds

If auxin is sprayed onto a mixture of grasses and broadleaved plants at a carefully controlled rate, the broadleaved plants die and the grasses survive. The auxin causes excessive leafy growth in the broadleaved plants, which makes them so deformed that they die. Grasses either do not absorb so much of the auxin or they are not as sensitive to it. Auxin is used in this way to kill broadleaved weeds in lawns and in cereal crops.

3. Inducing fruit ripening

Ethene is a plant growth hormone that stimulates many fruits to ripen. Ripening fruits produce this gas, which can diffuse to other fruits, stimulating them to ripen. Chemicals that release ethene are sometimes sprayed onto fruit, either before or after it has been picked, to stimulate ripening to occur at the ideal time. For example, bananas are picked while they are green and unripe. They are often transported by ship. The bananas in the ship's holds are sprayed at a time that ensures that the fruits are just ripe when they reach the shops where they are sold.

4. Producing seedless fruit

Fruits do not normally develop unless seeds are developing inside them. If the flowers of fruit-producing plants are unpollinated, there is a danger that no fruit will be produced. With some types of fruit it is possible to stimulate fruit development even if pollination has failed, by spraying the flowers with a hormone. Auxin is used on tomatoes. Another plant growth hormone called gibberellin is used on pears, grapes and citrus fruits.
This procedure has the advantage that the fruits formed are seedless.

Stages in the micropropagation of plants

A small piece of tissue is removed from the plant that is being cloned. Often the tissue comes from a shoot tip. The tissue is sterilized. All apparatus and growth media must be sterilized to prevent infections. This is called aseptic technique.

The tissue is placed on sterile nutrient agar gel, containing a high auxin concentration. This stimulates cell growth and division.

An amorphous lump of tissue called callus grows, which can be cut up and made to grow more using the same type of nutrient agar containing auxin.

Eventually the callus is transferred to nutrient agar gel containing less auxin but high concentrations of cytokinin which stimulates plantlets with roots and shoots to develop. Gibberellin is sometimes added to increase shoot growth and prevent dormancy.

The plantlets are separated and transferred to soil, where they should grow strongly.

Plant and animal breeding

TECHNIQUES USED IN PLANT AND ANIMAL BREEDING

Plant and animal breeders use many different techniques in the quest for new improved varieties.

1. Inbreeding – *obtaining offspring by crossing closely related individuals together*
Example – pure-breeding strains of new rice and wheat varieties are developed using inbreeding. The new variety is grown each year in an area away from other varieties, so that no pollen from other varieties reaches it. Any plants that do not show the desired characteristics are removed. Over about 6 to 8 years the variety becomes homozygous for the desired allele of each gene, giving constant, predictable characteristics.

2. Outbreeding – *obtaining offspring by crossing unrelated individuals together.*
Example – most farmers use maize (corn) seed that is produced by crossing two unrelated varieties together. The plants that grow from this seed are F_1 hybrids and show much greater vigour than either of the parental varieties.
F_1 hybrid vigour – *the healthy growth, due to high levels of heterozygosity, that is seen in the offspring of crosses between unrelated parents.*

3. Interspecific hybridization – *obtaining offspring by crossing individuals from different species together.*
Example – cultivated wheat was developed by hybridization of three wild grass species (below)

4. Polyploidy – having more than two sets of chromosomes.
Example – the interspecific hybrids that gave rise to modern wheat were initially infertile. By doubling the chromosome number to produce polyploid cells, the plants became fertile (below).

BREEDING PROGRAMMES FOR RICE

Rice breeding programmes at the International Rice Research Institute in the Philippines have been continuing since 1959. The Institute initially collected many different rice varieties for assessment. Varieties that showed desirable characteristics were cross-bred by taking pollen from the anthers of one variety and brushing it onto the stigmas of another variety. The seed formed was sown and the plants that grew from it were assessed. Those that seemed promising were selected for further trials by saving seed from them.

One variety was bred by crossing a semi-dwarf variety from Taiwan with a vigorous fast-growing variety from Indonesia. The new variety produced was called IR8. It is short stemmed, with heavy ears of large grain (below). With large fertiliser and pesticide applications it can give yields twice or even three times as high as previous varieties. The breeding of IR8 and other new varieties was called the Green Revolution as it increased rice yields enormously in some countries, allowing India, for example, to become self-sufficient in rice.

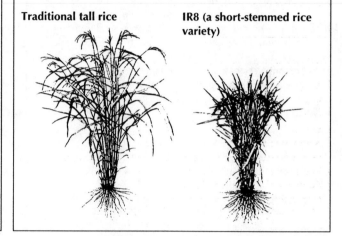

Traditional tall rice   **IR8 (a short-stemmed rice variety)**

Breeding of wheat

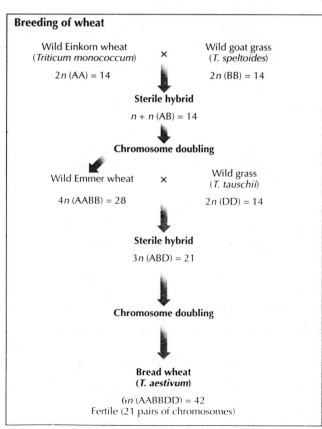

Wild Einkorn wheat
(*Triticum monococcum*)   ×   Wild goat grass
(*T. speltoides*)

$2n$ (AA) = 14 $2n$ (BB) = 14

Sterile hybrid
$n + n$ (AB) = 14

Chromosome doubling

Wild Emmer wheat   ×   Wild grass
(*T. tauschii*)

$4n$ (AABB) = 28 $2n$ (DD) = 14

Sterile hybrid
$3n$ (ABD) = 21

Chromosome doubling

Bread wheat
(**T. aestivum**)

$6n$ (AABBDD) = 42
Fertile (21 pairs of chromosomes)

BREEDING PROGRAMMES FOR DAIRY CATTLE

Cows produce milk for about ten months after calving. This period is called a lactation. Fifty years ago, in many parts of the world about 400 kg of milk was produced per lactation. As a result of breeding programmes, cows can now produce over 8000 kg per lactation. Most dairy farmers are continually breeding the cows in their herd to increase milk yields. The milk yield of each cow is recorded. The farmer can then identify the highest yielding cows in the herd. These chosen cows are crossed, usually by artificial insemination, with a bull whose progeny have been tested and shown to have high yields. Female calves from these crosses should give higher milk yields than the herd average, so they are reared and used to replace cows that have to be culled from the herd. Lower yielding cows in the herd are crossed with beef bulls and their offspring are fattened to produce meat.

THE NEED TO CONSERVE PLANT AND ANIMAL BIODIVERSITY

The wild plants and ancient breeds of livestock from which modern crop plants and breeds of livestock were developed still form a valuable reservoir of genetic diversity. Not only are many of these breeds often more suited to local conditions than the products of high-technology breeding, but they may also be the source of the alleles that are needed to change current varieties to meet future needs. The conservation of these breeds is therefore an important task.

Genetic engineering and agriculture

TRANSGENIC TECHNIQUES IN AGRICULTURE

Genetic engineering is the genetic manipulation of organisms to produce individuals with new genotypes. One type of genetic engineering is the transfer of genes from one species to another, to produce transgenic organisms. New types of transgenic crop plant and farm animal are produced every year. Two examples in chicken and maize are shown on page 29 and three others are described below.

1. Making crops glyphosate resistant

Almost all plants are killed by the herbicide glyphosate. A gene that causes resistance was discovered in a bacterium, *Salmonella typhimurium*. This gene has been transferred to maize (corn) and other crop plants. The transgenic crops can be sprayed with glyphosate to kill any weeds growing in them, without harming the crop plants.

2. Producing alpha-1-antitrypsin

Alpha-1-antitrypsin is a protein produced in humans that protects the lungs from damage during infections. The gene for making this protein has been transferred to sheep. The gene was pre-programmed so that it is expressed in mammary gland cells and the protein is secreted in the sheep's milk (right). The alpha-1-antitrypsin can be extracted and purified from the milk very easily and can then be used to treat patients with emphysema who cannot produce it themselves. Using farm animals to make products for pharmaceutical use is called pharming!

3. Using lectins as antifeedants

Lectins are proteins that have a glue-like effect on polypeptides and polysaccharides, causing them to agglutinate. A gene for lectin has been transferred from pea plants to potatoes. The leaves of the transgenic potato plants produce lectin, which protects them against insect pests. When insects feed on the leaves, the lectin disrupts their digestive systems so much that they stop feeding.

SENSE/ANTISENSE TECHNOLOGY

An ingenious form of genetic engineering has been developed to turn off particular genes in crop plants or farm animals. An antisense gene is constructed by reversing the orientation of the gene with respect to the promoter. The mRNA that is transcribed from this gene has the antisense base sequence, instead of the sense sequence. The antisense gene is transferred to the crop plant or animal. When this plant or animal tries to express the original gene, it transcribes both the original gene and the antisense version of it. The two types of mRNA produced are complementary, so they bind together and neither can be translated. The gene is therefore not expressed.

Sense/antisense technology has been used to control ripening in tomatoes. An enzyme called polygalacturonase makes tomato fruits ripen very quickly. Insertion of an antisense version of the gene for polygalacturonase delays ripening, so that the tomatoes stay firm for longer. They have a longer shelf-life in shops and also develop better flavour, hence the name of the genetically modified variety – Flavr-Savr™ .

Milking sheep to obtain proteins for pharmaceutical use

THE GMO CONTROVERSY

There has been a fierce debate about whether transgenic organisms should be used, which has even led to trade disputes between Europe and the United States. The table below shows some of the arguments that have been used for and against transgenic techniques.

Arguments against transgenic techniques	Arguments for transgenic techniques
It is impossible to be certain about the long-term effects on human health of the genes in GM foods that have been artificially inserted or the proteins made using these genes	The DNA and proteins of transgenic organisms are unlikely to cause any problems because they are digested in the human gut
The genes from transgenic organisms could escape and be transferred to wild plants or animals, possibly creating 'superweeds' or damaging natural ecosystems in other ways	Research suggests that transgenic organisms do not survive for long in natural ecosystems and if genes were transferred to wild species, natural selection would act against them
Transgenic techniques will allow the multinational companies that develop them to become too powerful	Transgenic techniques will reduce the cost of food production, which benefits both the producer and the consumer
Moving genes between species is an infringement of the right of every species to independent existence and could cause suffering in animals	Transfer of genes between species is a natural phenomenon. The human genome project has revealed that over two hundred genes were transferred to humans from bacteria

Sexual reproduction in plants

INSECT-POLLINATED AND WIND-POLLINATED FLOWERS

It is easy to distinguish between insect-pollinated and wind-pollinated flowers because of their different adaptations. A figure on page 110 shows the structure of an insect-pollinated flower. The figure (below) shows the structure of a wind-pollinated flower.

Structure of wind-pollinated grass flower

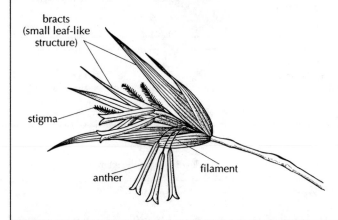

Insect-pollinated flowers	Wind-pollinated flowers
Large petals with colours that are visible to the pollinating insects, and often lines or other marks to guide the insect in	Small petals or equivalent structures with green or dull colouration only needed to protect the flower before it opens
Sturdy filaments hold the anthers in a precise position inside the flower where insects brush past and collect pollen	Long thin filaments hold the anthers loosely outside the rest of the flower, where pollen can be shaken off by the wind
Relatively small numbers of large, spiky pollen grains are produced, which stick firmly to the insects	Larger amounts of small, smooth pollen grains, with low density, easily carried by wind
Sturdy style holds the stigma in a precise position inside the flower where it can collect pollen as insects brush past	Large feathery stigmas protrude from the rest of the flower, to increase the chance of catching pollen from the wind
Petals often release scent and nectaries secrete nectar, to attract and reward insects	No scent or nectar

PHOTOPERIODIC CONTROL OF FLOWERING

Some plants only flower at the time of year when days are short and other plants only flower when the days are long. They are called short-day plants and long-day plants. Experiments have shown that it is not the length of day but the length of night that is significant. For example, chrysanthemums are short-day plants and only flower when they receive a long continuous period of darkness (below). They therefore naturally flower in the autumn (fall). Growers can produce pots of flowering chrysanthemums at all times of the year by keeping them in greenhouses with blinds. When the nights are not long enough to induce flowering, the blinds are closed to extend the nights artificially. In a similar way, petunias, which are long-day plants, can be induced to flower at times of the year when the days are short by being given extra light in greenhouses to reduce the length of the nights.

Response of chrysanthemums to different light/dark regimes

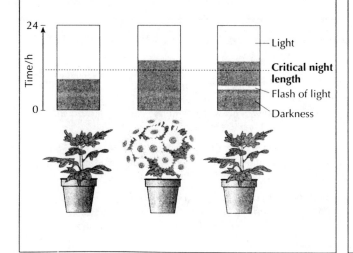

PHYTOCHROME AND PHOTOPERIODISM

Plants can measure the length of periods of dark to an accuracy of a few minutes. They do this using a pigment in their leaves called **phytochrome**, which exists in two interconvertible forms. One form is called P_r because it absorbs red light with a wavelength of 660 nm. P_r is the inactive form of phytochrome. When it absorbs red light it is rapidly converted into the active form, called P_{fr}. This form can absorb far red light with a wavelength of 730 nm and is then rapidly converted back to P_r. In normal daylight there is much more red light than far red light so phytochrome exists in the active P_{fr} form. In darkness P_{fr} reverts very slowly to P_r (above). This gradual reversion process is probably how the length of the dark period is timed. Enough P_{fr} remains in long-day plants at the end of short nights to stimulate flowering. In *Arabidopsis*, which is a long-day plant, a protein has been found to which P_{fr} binds. This protein probably acts as a transcription factor, causing genes involved in flowering to be switched on. In short day plants P_{fr} presumably acts as an inhibitor of flowering. At the end of long nights, enough P_{fr} has been converted to P_r to allow flowering to occur.

Interconversions of phytochrome

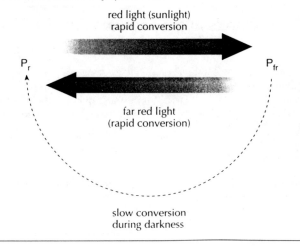

Asexual reproduction in plants

NATURAL METHODS OF ASEXUAL REPRODUCTION IN PLANTS

Many plants can reproduce asexually, using their roots, stems or leaves. Three methods of asexual reproduction are formation of tubers (see page 114), runners and bulbs (see right).

ARTIFICIAL METHODS OF ASEXUAL REPRODUCTION IN PLANTS

Many plants can be propagated asexually using a variety of artificial methods. Three of the most commonly used traditional methods are taking cuttings, grafting and layering (shown right and below). Micropropagation is a more recent method.

Bulbs

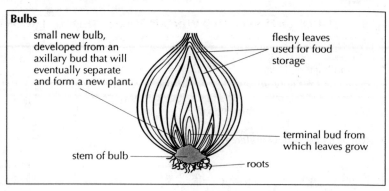

small new bulb, developed from an axillary bud that will eventually separate and form a new plant.

fleshy leaves used for food storage

terminal bud from which leaves grow

stem of bulb

roots

Runners

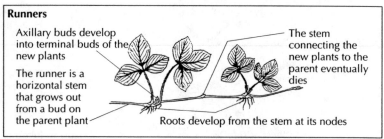

Axillary buds develop into terminal buds of the new plants

The runner is a horizontal stem that grows out from a bud on the parent plant

The stem connecting the new plants to the parent eventually dies

Roots develop from the stem at its nodes

Taking cuttings

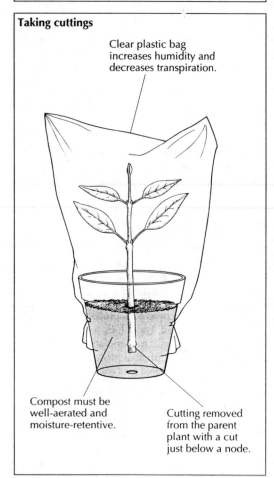

Clear plastic bag increases humidity and decreases transpiration.

Compost must be well-aerated and moisture-retentive.

Cutting removed from the parent plant with a cut just below a node.

Grafting

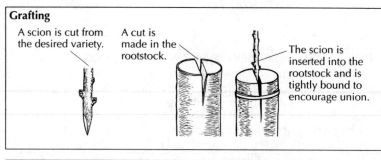

A scion is cut from the desired variety.

A cut is made in the rootstock.

The scion is inserted into the rootstock and is tightly bound to encourage union.

Layering

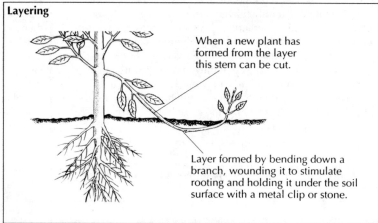

When a new plant has formed from the layer this stem can be cut.

Layer formed by bending down a branch, wounding it to stimulate rooting and holding it under the soil surface with a metal clip or stone.

ADVANTAGES AND DISADVANTAGES OF USING ASEXUAL METHODS OF PLANT PROPAGATION

Advantages

A plant with the combination of alleles that gives the ideal phenotype can be propagated to produce more plants of exactly the same type. For example, more plants with the combination of alleles of an F_1 hybrid could be produced

Many plants can be propagated asexually at any time of year, whereas most plants only reproduce sexually in one season

Many plants with the same characteristics can be propagated, which will need the same conditions, grow at the same rate and be ready for sale or harvest at the same time

Disadvantages

If only a few varieties are reproduced asexually, there will be a loss of diversity from the species and the gene pool will become smaller. This will reduce the capacity for evolution, or breeding of new varieties of the species

Plants propagated asexually from the same parent will all be vulnerable to the same diseases. A disease that infects one plant will spread quickly to other plants

Virus diseases are passed from parent to offspring in asexual methods of propagation (except some forms of micropropagation), whereas they are eliminated during sexual reproduction

EXAM QUESTIONS ON OPTION F – APPLIED PLANT AND ANIMAL SCIENCE

F1 The bar chart below shows the feed protein input and the protein yield for three production systems. In milk and intensive beef production systems, large amounts of cereals are fed to cattle. Beef cattle raised on rangeland feed mainly on fibrous plants such as grass and are free to wander over large areas.

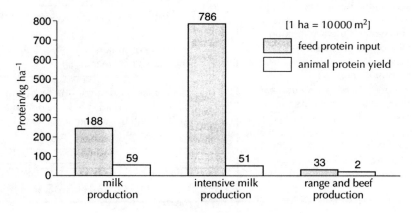

[Source: D & M Pimentel, Food, Energy and Society, (1979), Edward Arnold]

a) State which system produces the highest yield of protein per hectare (ha). [1]

b) (i) Compare the efficiency of conversion of feed protein into protein in milk or beef in these three production systems. [2]

(ii) Suggest a reason for the differences in efficiency. [1]

c) Suggest **two** advantages of the rangeland beef system compared with the two other systems. [2]

F2 a) Define harvestable dry biomass. [1]

b) List one measure of plant productivity other than harvestable dry biomass. [1]

c) Crops are often grown in intensive monoculture to maximise productivity and yield. State one disadvantage of intensive monoculture. [1]

F3 Flowering plants can be commercially produced by methods of vegetative propagation such as the rooting of cuttings.

a) Outline how plant growth substances can be used to promote the rooting of cuttings. [2]

b) Explain briefly the importance of vegetative propagation in production of flowering plants. [2]

Constructing pyramids of energy

MEASURING ENERGY FLOWS IN ECOSYSTEMS

Collecting the data needed to construct a pyramid of energy is very time-consuming and has only been done fully for a few ecosystems. Energy flow through each species in the ecosystem must be measured and then the species are classified into trophic levels.

The lowest bar of a pyramid of energy is the total amount of energy that flows through the producers in the ecosystem. This is also called the gross production.

Gross production is the total amount of organic matter produced by plants in an ecosystem. Gross production and all the other energy flows in a pyramid are measured in kilojoules of energy per square metre per year ($kJ\ m^{-2}\ year^{-1}$). Gross production does not have to be measured directly, as it can be calculated from net production and plant respiration.

Net production is the amount of gross production of an ecosystem remaining after subtracting the amount used by plants in respiration.

$$gross\ production = plant\ respiration + net\ production$$

Example – an old field community in Michigan, USA
net production = $20.79 \times 10^3\ kJ\ m^{-2}\ year^{-1}$
plant respiration = $3.68 \times 10^3\ kJ\ m^{-2}\ year^{-1}$
gross production = $(20.79 + 3.68) \times 10^3\ kJ\ m^{-2}\ year^{-1}$
= $24.47 \times 10^3\ kJ\ m^{-2}\ year^{-1}$

The upper bars of a pyramid of energy are the total amounts of energy that flow through the various groups of consumers. This is the amount of energy in the food that the consumers ingest.

CLASSIFYING ORGANISMS INTO TROPHIC LEVELS

When real food webs are constructed, many species are seen to exist partly in one trophic level and partly in another.

The following examples illustrate this.

- *Euglena*, a unicellular organism found in ponds, has chloroplasts and photosynthesizes, but it also feeds heterotrophically by endocytosis.
- Chimpanzees mainly feed on fruit and other plant matter, but they also sometimes eat termites and even larger animals such as monkeys, so they are both first and second consumers.
- Herring are second consumers when they feed on *Calanus* and other first consumers (right), but they are third consumers when they feed on sand eels and other second consumers.
- Oysters (*Ostrea* species) and many other filter feeders consume both ultraplanktonic producers and microplanktonic consumers, so they are first and second consumers. They also consume dead organic matter, so they are also detritivores.

It is difficult to decide into which trophic level these types of organism should be classified. The only practical solution is to classify each species according to its main food source.

USING DATA TO CONSTRUCT A PYRAMID OF ENERGY

The data below was obtained from an Arctic tundra ecosystem on Devon Island in northern Canada.

Trophic level	Energy flow ($kJ\ m^{-2}\ year^{-1}$)
Producers	4925
Primary consumers	24
Secondary consumers	4

This data can be used to construct a pyramid of energy. Each bar of the pyramid should be drawn to the same scale and labelled with the trophic level. Other examples are shown on page 39.

REASONS FOR SMALL NUMBERS AND LOW BIOMASS OF ORGANISMS IN HIGHER TROPHIC LEVELS

Pyramids of energy show that there are large losses of energy at each trophic level. The reasons for these losses of energy are explained on page 38. Losses of energy in ecosystems are accompanied by losses of biomass.

Biomass is the total dry mass of organic matter in organisms or ecosystems.

Respiration is an example of a process in which both energy and biomass are lost. When glucose or another respiratory substrate is oxidized in respiration, energy from the glucose is released for use in the cell and is then lost as heat. The mass of the glucose does not disappear – it passes into the carbon dioxide and water that are produced in respiration. When these waste products are excreted, biomass is lost. As a result of respiration and other processes, both energy and biomass are lost at each stage in a food chain. The energy content per gram of food does not decrease along a food chain. If anything, the food eaten by the higher trophic levels is richer in energy per gram than that eaten by lower trophic levels. However, the total biomass of food available to higher trophic levels is very small. It cannot support large numbers of organisms, especially as these organisms need to be large to overpower their prey. Higher trophic levels therefore usually contain very small numbers of large organisms, with a low total biomass per unit area of ecosystem.

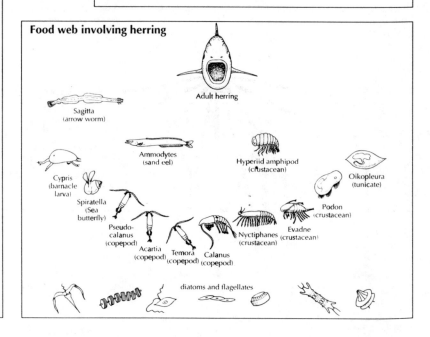

Food web involving herring

Adult herring

Sagitta (arrow worm)

Ammodytes (sand eel)

Hyperiid amphipod (crustacean)

Oikopleura (tunicate)

Cypris (barnacle larva)

Spiratella (Sea butterfly)

Podon (crustacean)

Pseudo-calanus (copepod)

Acartia (copepod)

Temora (copepod)

Calanus (copepod)

Nyctiphanes (crustacean)

Evadne (crustacean)

diatoms and flagellates

Distribution of plants and animals and succession

FACTORS AFFECTING THE DISTRIBUTION OF PLANT SPECIES

The distribution of a species is the range of places that it inhabits. The distribution of plants is closely linked to the levels of abiotic factors in the environment. The main abiotic factors are temperature, water, light, soil pH, salinity and mineral nutrients. *Avicennia germinans*, for example, is a tree found in mangrove swamps on the coast of Mexico. It grows where the climate is hot and the soils are waterlogged and anaerobic, with high levels of salinity, a pH close to neutral and high levels of mineral nutrients. Few plants can grow in these conditions, but *Avicennia germinans* thrives.

Sometimes the distribution of a plant species shows what conditions a plant prefers. The figure shows the distribution of *Asperula cynanchica* in Britain and Ireland. It is found in areas with alkaline soils formed from chalk or limestone rock. It is absent from colder northern areas even where the soils are alkaline.

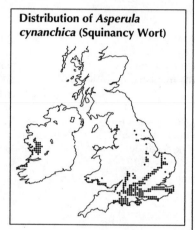

Distribution of *Asperula cynanchica* (Squinancy Wort)

FACTORS AFFECTING THE DISTRIBUTION OF ANIMAL SPECIES

The distribution of animal species is affected by both abiotic and biotic factors.

• **Temperature** – all animals are affected by external temperatures, especially those that do not maintain constant internal body temperatures. Extremes of temperature require special adaptations, so only some species can survive them.

• **Water** – animals vary in the amount of water that they require. Some animals are aquatic and must have water to live in and at the other extreme some animals including desert rats are adapted to survive in arid areas where they are unlikely ever to drink water.

• **Breeding sites** – all species of animals must breed at some stage in their life cycle. Many species need a special type of site and can only live in areas where these sites are available. For example, mosquitoes need stagnant water for egg laying.

• **Food supply** – many species of animal are adapted to feed on specific foods and can only live in areas where these foods are obtainable. For example, blue whales feed mainly on krill and so congregate in areas of the ocean where krill is abundant.

• **Territory** – some species of animal establish and defend territories, either for feeding or breeding. This tends to give the species an even rather than a clumped distribution. Pairs of tawny owls defend a single territory throughout their adult lives, for example.

ECOLOGICAL SUCCESSION

An ecological succession is a series of changes to an ecosystem, caused by complex interactions between the community of living organisms and the abiotic environment. The community causes the abiotic environment to change and as a result some species die out and others join the community. Although the community may continue to change in this way for hundreds of years during a succession, eventually a stable community develops, called the climax community.

The changes to the abiotic environment during ecological successions vary, but some often occur.

• The amount of organic matter in the soil increases as organic matter released by plants and other organisms accumulates.

• The soil becomes deeper as organic matter helps to bind mineral matter together.

• The soil structure improves as the organic matter content rises, increasing the amount of water that can be retained and the rate at which excess water drains through.

• Soil erosion is reduced by the binding action of the roots of larger plants.

• The amounts of mineral recycling increases, as the soil can hold larger amounts and more is held in the increasing biomass of the community.

AN EXAMPLE OF SUCCESSION

On the slopes of Volcan Osorno, in southern Chile, there are large areas of bare volcanic ash, released during recent eruptions of the volcano. Adjacent areas show the stages in an ecological succession.

• Mosses spread over the ash, eventually forming a complete cover.

• Small herbs join the mosses.

• Shrubs, including *Pernettya, Eucryphia* and *Embothrium*, enter the community and gradually replace the herbs and mosses (below).

• Trees, including *Nothofagus*, gradually spread to replace the shrubs with dense forest.

A stage in succession to forest on Volcan Osorno

Ecological niches

INTERACTIONS BETWEEN SPECIES

All living organisms are affected by the activities of other living organisms. A situation in which two species affect each other is called an interaction. The table below shows a classification of interactions.

Interaction	Terrestrial example	Marine example
Herbivory – a primary consumer feeding on a plant or other producer. The producer's growth affects food availability for the herbivore	The beetle *Epitrix atropae* feeds only on leaves of *Atropa belladonna*, often causing severe damage to them. To most other organisms the leaves are highly toxic	Algae growing on rocks in shallow seas are often heavily grazed. For example, a snail *Lacuna pallida* feeds on the brown seaweed *Fucus serratus* on rocky shores in Europe
Predation – a consumer feeding on another consumer. The numbers and behaviour of the prey affect the predator	The Canada lynx is a predator of the arctic hare. Changes in the numbers of hares (up or down) are followed by similar changes in lynx numbers	Bonitos feed on anchovetas in the Pacific Ocean west of Peru. When the anchoveta population crashed in the 1970s starving bonitos were found, with completely empty stomachs
Parasitism – a parasite is an organism that lives on or in a host and obtains food from it. The host is always harmed by the parasite	The tick *Ixodes scapularis* is a parasite of deer and of white-footed mice in northeast USA. The tick feeds by sucking blood from its hosts and therefore weakens them	Organisms that cause infectious diseases are all parasites. For example, *Sphingomonas* bacteria cause a disease in elliptical star corals on the Florida reef
Competition – two species using the same resource compete if the amount of the resource used by each species reduces the amount available to the other species	Douglas Fir and Western Hemlock grow together in mixed forests in Oregon and other states in northwest USA, competing with other each other for light, water and minerals	Species of coral compete with each other on coral reefs. *Pocillopora damicornis* competes with many other corals, including *Pavona varians*, which benefit when predators feed on *Pocillopora damicornis*
Mutualism – mutualists are members of different species that live together in a close relationship, from which both benefit	*Usnea subfloridana* and other lichens consist of a fungus and an alga growing mutualistically. The alga supplies foods made by photosynthesis and the fungus absorbs mineral ions	The cleaner wrasse is a small fish of warm tropical seas that cleans parasites from the gills and body of larger fish such as reticulate damselfishes. The cleaner benefits because the parasites that it removes are its food

THE NICHE CONCEPT

Studies of the distributions of organisms and of interactions between organisms show that there are many different ways of existing in an ecosystem. The mode of existence of a species in an ecosystem is its ecological niche. The niche includes:

• habitat – where the species lives in the ecosystem.

• nutrition – how the species obtains its food.

• relationships – the interactions with other species in the ecosystem.

If two species have a similar niche, they will compete in the overlapping parts of the niche, for example for breeding sites or for food. Because they do not compete in other ways they will usually be able to coexist.

However, if two species in an ecosystem have exactly the same niche they will compete in all aspects of their life and one of the two species will inevitably prove to be the superior competitor. This species will cause the disappearance of the other species from the ecosystem.

The principle that only one species can occupy a niche in an ecosystem is called the **competitive exclusion principle**.

Statistics in ecology

Ecologists often use statistics in their investigations. The use of the mean and standard deviation is described on page 36. The uses of two other statistics are described on this page.

THE *t*-TEST

The *t*-test can be used to find out whether there is a significant difference between the means of two populations. There are several versions of the *t*-test, with different formulae, for use in different situations. The formula shown below is used when the standard deviations of the two populations are not known, but are assumed to be equal, and both populations can be modelled on the normal distribution.

$$t = \frac{|\bar{x}_1 - \bar{x}_2|}{s\sqrt{\frac{1}{n_1} + \frac{1}{n_2}}}$$

$\bar{x}$ = mean.
s = estimate of common standard deviation.
n = number of entries in a set of data.
   degrees of freedom = $(n_1 + n_2 - 2)$.

It should be noted that *t* is nearly always now calculated using computer software and it is more important to be able to choose the correct formula than to learn one particular formula. When *t* has been calculated, it is compared with a table of critical values (below). The correct line of the table must be used, corresponding to the number of degrees of freedom. If the calculated value is greater than the relevant critical value, the difference between the means is considered to be significant.

Table of critical values of *t*

Level of significance (*P*)

Degrees of freedom	0.2	0.1	0.05	0.02	0.01	0.002
1	3.078	6.314	12.706	31.821	83.657	318.310
2	1.886	2.920	4.303	6.985	9.925	27.327
3	1.638	2.353	3.182	4.541	5.841	10.215
4	1.533	2.132	2.776	3.747	4.604	7.173
5	1.476	2.015	2.571	3.365	4.032	5.893
6	1.440	1.943	2.447	3.143	3.707	5.208
7	1.415	1.895	2.385	2.998	3.499	4.785
8	1.397	1.860	2.308	2.896	3.355	4.501
9	1.383	1.833	2.262	2.821	3.250	4.297
10	1.372	1.812	2.228	2.764	3.169	4.144
11	1.363	1.796	2.201	2.718	3.106	4.025
12	1.356	1.782	2.179	2.681	3.055	3.930
13	1.350	1.771	2.180	2.650	3.012	3.852
14	1.345	1.761	2.145	2.624	2.977	3.787
15	1.341	1.753	2.131	2.602	2.947	3.733
16	1.337	1.746	2.120	2.583	2.921	3.686
17	1.333	1.740	2.110	2.567	2.898	3.646
18	1.330	1.734	2.101	2.552	2.878	3.610
19	1.328	1.729	2.093	2.539	2.861	3.579
20	1.325	1.725	2.086	2.528	2.845	3.552

Example – sizes of two groups of lichens

To test whether there was any difference in the size of lichens growing on the top and side of a stone wall, some data were collected. The diameters of a random sample of ten lichens on the top and ten on the side were measured and a *t*-test was used to find out if there was a significant difference.

Surface	Diameter of lichen (mm)									
Top	22	10	24	45	9	26	5	34	10	13
Side	22	12	23	13	7	13	5	24	3	10

Mean diameters: top = 19.8 mm   side =13.2 mm

t = 1.406 (calculated using the equation shown above). When this value is compared with critical values (for 18

THE SIMPSON DIVERSITY INDEX

It is sometimes useful to have an overall measure of species richness in an ecosystem. The Simpson index is one of the most commonly used.

Method
1. Use a random sampling technique to search for organisms in the ecosystem.
2. Identify each of the organisms found.
3. Count the total number of individuals of each species.
4. Calculate the index (*D*).

$$D = \frac{N(N-1)}{\sum n(n-1)}$$

N = total number of organisms.
n = number of individuals of each species.

Example – species richness in a river in Sweden

Organisms were found and identified in the River Enningdalselva in a part of Sweden where some lakes and rivers have been affected by acid rain. Six sites in the river were chosen randomly and kick sampling was used at each site along a 10 m transect. Nets with a 25 cm × 25 cm opening and 0.5 mm mesh were used. The results are shown below.

Group	Species	Name	Count
Ephemerida	*Dixa* sp.	Mayfly larva	8
Odonata	*Tipula* sp.	Dragonfly larva	5
Trichoptera	Sp. unidentified	Caddis fly larva	4
Plecoptera	*Nemoura variegata*	Stonefly larva	4
Hemiptera	*Gerris* sp.	Pond skater	3
Isopoda	*Asellus aquaticus*	Water louse	2
Acari	*Arrhenurus* sp.	Water mite	1
Platyhelminth	*Dendocoelum lact.*	Flatworm	4
Platyhelminth	*Dugesia* sp.	Flatworm	3
Hirundinea	Sp. unidentified	Leech	1
Oligochaeta	*Lumbriculides*	Annelid worm	2
Gastropoda	*Lymnaea* sp.	Snail	4
Bivalvia	*Margaritifer*	Pearl mussel	1

$$D = \frac{42\,(42-1)}{140} = 12.3$$

The high diversity index suggests that the river has not been damaged by acid rain, or any other disturbance. This fits in with observations of a thriving salmon population in the river.

USES OF THE SIMPSON DIVERSITY INDEX

Sampling and calculation of the diversity index each year, using exactly the same methods, allows monitoring of the health of the ecosystems like the River Enningdalselva. Sampling of a group of similar ecosystems, for example a group of rivers in an area, allows comparison of their health. Identification of high-diversity ecosystems allows informed decisions to be made about priorities in wildlife conservation.

degrees of freedom and *P* = 0.05) we see that 1.406 is less than 2.101, so there is no significant difference between the mean diameters on the two surfaces. If a larger sample size was used, for example one hundred lichens on each surface, the difference might be shown to be significant.

Biodiversity

The word biodiversity was only invented in 1986. It is an abbreviation of 'biological diversity' and encompasses the diversity of ecosystems on Earth, the diversity of species within them, and the genetic diversity of each species.

REASONS FOR THE CONSERVATION OF BIODIVERSITY

Economic reasons
- New commodities, for example medicines or materials may be found in organisms growing in the wild.
- New crop plants or farm animals could be developed from wild species or existing varieties could be improved using genes from wild species.
- Ecotourism could provide considerable income.

Ecological reasons
- Native species are adapted to local conditions, whereas alien species that might replace them are unlikely to be so well adapted.
- Species in natural communities are interdependent, so if one species becomes extinct the rest of the community is threatened.
- Damage to ecosystems can have widespread effects including soil erosion, silting up of rivers, flooding and even changes to weather patterns.

Ethical reasons
- Every species has a right to life, regardless of whether it is useful to humans or not.
- The wildlife of each area has cultural importance to the local human population and it is therefore wrong to destroy it.
- It would be wrong to deprive humans in the future of the rich experiences that the Earth's biodiversity provide to us.

Esthetic reasons
- Natural ecosystems and species in the wild are beautiful and give us great enjoyment.
- Painters, writers and composers have been inspired by nature around them.

EXTINCTION OF SPECIES

When the last members of a species die, the species becomes extinct. The rate of species extinctions is probably at an all-time high at the moment, as a result of human activities. There are unfortunately many extinct species from which to select examples for study, including the passenger pigeon and the dodo. Two less famous examples are described here.

1. Conuropsis carolinensis – the Carolina Parakeet

These brightly coloured parrots were once common in forests to the east of the Mississippi, from New York to Florida, feeding on seeds of trees and herbs. Clearance of forests reduced their habitat and they started to feed on crops. Farmers killed many of them. Others were caught to obtain feathers, which were used to make fashionable women's clothing. They were also trapped and kept as pets. By 1900 there were no Carolina Parakeets in the wild and the last specimen died in Cincinnati Zoo in 1918.

MONITORING ENVIRONMENTAL CHANGE

Problems in natural ecosystems can be detected quickly if there is frequent environmental monitoring. Abiotic factors can be measured directly, but another technique is the use of living organisms to detect changes. Indicator species are very useful, as they need particular environmental conditions and therefore show what the conditions in an ecosystem are. Lichens are valuable indicator species because their tolerance of sulphur dioxide varies considerably from the most tolerant to the least tolerant species. Indicator species are also often used to assess pollution levels in aquatic ecosystems. Stonefly, mayfly and caddis fly larvae (below) require unpolluted, well-oxygenated water. Other aquatic species, including chironomid midge larvae, rat tailed maggot larvae and tubifex worms indicate low oxygen levels and excessive levels of suspended organic matter, from untreated sewage for example. The indicator species found in the River Enningdalselva (page 157) show how unpolluted it was.

Indicator species is aquatic ecosystems

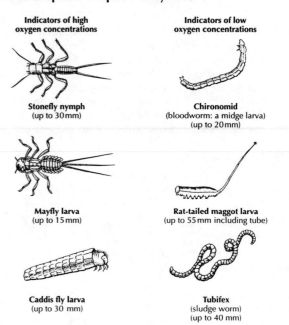

To obtain an overall environmental assessment of a river or other ecosystems, a biotic index can be calculated. There are various methods, which usually involve multiplying the number of individuals of each indicator species by its pollution tolerance rating. An abundance of tolerant species gives a low overall score and an abundance of intolerant species gives a high score.

2. Calochortus indecorus – the Sexton Mountain mariposa lily

Calochortus indecorus was discovered growing over a small area on Sexton Mountain in Oregon. The building of an interstate highway destroyed the habitat and with it this rare plant disappeared. No other sites are known where *Calochortus indecorus* grows, so it is almost certainly extinct

Conservation

NATURE RESERVES AND *IN SITU* CONSERVATION
The best place to conserve a species is in its own habitat. This is called *in situ* conservation. Many terrestrial and marine nature reserves have been established for this purpose. *In situ* conservation has several advantages.

- Species remain adapted to their habitats.
- Greater genetic diversity can be conserved.
- Animals maintain natural behaviour patterns.
- Species interact with each other, helping to conserve the whole ecosystem.

Despite these advantages, *in situ* conservation is not always enough to ensure the survival of a species.
- Some species become so rare that it is not safe to leave them unprotected in the wild.
- Sometimes destruction of a natural habitat makes it essential to remove threatened species from it.

In these situations *ex situ* measures are needed.

MANAGEMENT OF NATURE RESERVES
Nature reserves often need active intervention – this is called management.
- Alien species must be eliminated, especially alien species of predator and invasive plants.
- Areas that have been degraded by human activities must be restored.
- Special measures may be needed to help encourage threatened species, supplementary feeding or clearing vegetation, for example.
- Exploitation by humans must be controlled, for example the hunting of animals for bushmeat.

INTERNATIONAL ORGANIZATIONS AND CONSERVATION
Wildlife does not recognize frontiers between countries, so international co-operation over conservation is vital. Both voluntary and governmental agencies have important roles.

WWF – an example of a voluntary organization
The World Wildlife Fund is the largest privately supported conservation organization in the world, with millions of members and over 10 000 conservation projects so far undertaken. WWF is involved in political lobbying, monitoring of endangered species and establishing nature reserves. It also tries to involve local populations in conservation projects.

CITES – an example of a governmental organization
The Convention on International Trade in Endangered Species is the largest conservation convention, with over 100 member states. It regulates trade in threatened wild plant and animal species. Every 2 years there is a review of the species that are listed in two appendices to the convention. Trade in species listed in Appendix I is banned. Trade in species listed in Appendix II is only allowed with a licensing system that allows the trade to be monitored. The African elephant and the Alerce tree of South America are examples of species listed in Appendix I. Listing of the African elephant in 1989 stopped the rapid fall in numbers caused by poaching.

EX SITU CONSERVATION
1. **Captive breeding** – some or all members of a species are caught and moved to a zoo, where they are encouraged to breed. When numbers are high enough, some are returned to the wild to re-establish a natural population. An example of a species helped by captive breeding is the Hawaiian kestrel.
2. **Botanic gardens** – sites where many different species of plants are cultivated, either in greenhouses or in the open. One of the largest, Royal Botanic Gardens of Kew has more than 50 000 of the world's 250 000 known species in its collection.
3. **Seed banks** – seeds are kept in cold storage at –10 °C to –20 °C. Seeds of most species remain viable for more than a hundred years in these conditions. Other species that are not as long lasting can be germinated and grown to produce replacement seed before viability is lost. The Kew Millennium Seed Bank will eventually hold seed of 25 000 endangered species.

CONSERVATION OF FISH
Wild populations of fish are an important food source for many human populations. If a population is overexploited and the numbers of adult fish fall below a critical level, spawning fails. The disastrous collapse in the Peruvian anchoveta fishery is an example of this. Industrial scale exploitation of the anchoveta began in 1940 and grew at a rapid rate until 1973, when the annual catch dropped from 12 million tonnes to zero. The fall in anchoveta egg production in the years preceding the population crash is shown below. An El Niño event was partly responsible, but over-fishing was also a major factor. The anchoveta is a key species in its ecosystem. Many predators rely on it for food, including bonitos, cormorants, gannets and pelicans and populations of all of these species were greatly reduced.

Graph showing a collapse in anchoveta egg production

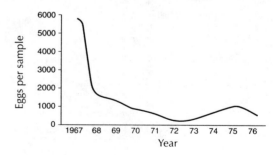

International measures are needed to promote fish conservation because most fish live in international waters, where ships from any country can catch fish. Various measures would help.
- Monitoring of stocks and of reproduction rates.
- Quotas for catches of species with low stocks.
- Moratoria on catching endangered species.
- Minimum net sizes, so that immature fish are not caught.
- Banning of drift nets, which catch many different species of fish indiscriminately.

Some of these measures have been used already in parts of the world, with limited success. Enforcement is very difficult and relies on a level of international trust and co-operation that is not always seen.

Ⓗ Nitrogen cycle

BIOGEOCHEMICAL CYCLES

All chemical elements in living organisms can be reused endlessly, as a result of biogeochemical cycles. These cycles involve living organisms, the atmosphere, water and land. There are two alternating sections in most biogeochemical cycles.

> Inorganic forms of the element are absorbed from the environment by autotrophic living organisms and are converted into complex organic forms

> The organic forms of the element are passed along food chains but are ultimately converted back into an inorganic form and released into the environment

The carbon cycle is described on page 43.
The nitrogen cycle is shown below.

THE ROLE OF BACTERIA IN THE NITROGEN CYCLE

Bacteria have many essential roles in the nitrogen cycle.

1. Nitrogen fixation
Free-living *Azotobacter* and *Rhizobium* living mutualistically in root nodules both fix nitrogen. Nitrogen fixation is conversion of nitrogen from the atmosphere into ammonia, using energy from ATP.

2. Nitrification
The conversion of ammonia to nitrate (nitrification) involves two types of soil bacteria. *Nitrosomonas* converts ammonia to nitrite and *Nitrobacter* converts nitrite to nitrate. Nitrification happens very rapidly, as long as soils are well aerated with abundant supplies of oxygen.

3. Denitrification
Nitrate is sometimes converted into nitrogen in a type of anaerobic respiration. As this process reduces nitrate levels in soil it is called denitrification. *Pseudomonas denitrificans* is an example of a bacterium that carries out denitrification. Nitrate is broken down when it is used as a terminal electron acceptor in respiration instead of oxygen. Anaerobic soils therefore encourage denitrification. Bad drainage and waterlogging are a frequent cause of anaerobic conditions in soils.

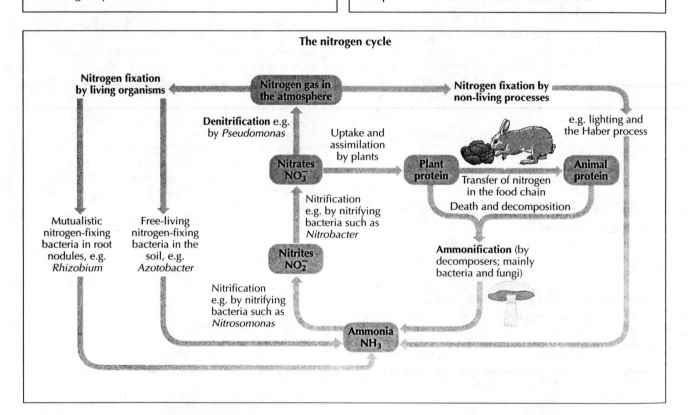

The nitrogen cycle

CHEMOAUTOTROPHY AND THE NITROGEN CYCLE

A chemoautotroph is an organism that obtains energy by oxidizing inorganic compounds and uses it to synthesize ATP. Some of the ATP is used to make organic compounds from carbon dioxide and other inorganic compounds. Chemoautotrophy is only found in bacteria. There are many different types of chemoautotrophic bacteria, including nitrifying bacteria. Nitrifying bacteria obtain the energy for synthesizing ATP by oxidizing ammonia or nitrite. They fix carbon and synthesize sugars and other carbon compounds using the Calvin cycle, as in plants.

THE NITROGEN CYCLE AND SOIL FERTILITY

Farmers and gardeners can increase the growth of plants by applying nitrogen fertilizers such as ammonium nitrate. Promoting the natural processes of the nitrogen cycle can produce the same effect. Ploughing or digging increases the aeration of the soils and therefore encourages nitrification and discourages denitrification. Crop rotation can increase the nitrogen content of soils if a legume is included in the rotation. Legumes develop root nodules in which nitrogen-fixing *Rhizobium* bacteria thrive. When the remains of the crop after harvest are ploughed or dug into the soil, nitrogen fixed by *Rhizobium* is released.

Ozone depletion and acid rain

OZONE AND ULTRA-VIOLET RADIATION

At low altitudes in the atmosphere, the concentration of ozone is usually about 0.01 ppm, but at 20 – 50 km above the Earth's surface, in the stratosphere, ozone is much more concentrated – about 1–10 ppm. This is called the ozone layer. Ozone absorbs short wave radiation, especially ultra-violet. The amount of ultra violet radiation reaching the Earth's surface is greatly reduced by the ozone layer. Ultra-violet radiation has very damaging effects on living organisms.

- It increases mutation rates, by causing damage to DNA.
- It can cause cancers, especially of the skin.
- It causes severe sunburn and cataracts of the eye.
- It reduces photosynthesis rates in plants and algae and so affects food chains.

OZONE DEPLETION

Measurements of ozone concentrations in the stratosphere have shown that there has been depletion throughout the world. Since the 1980s an ozone 'hole' has appeared over the Antarctic every year between September and October, which persists for several months.

CHLORINE AND OZONE DEPLETION

CFCs are the main cause of ozone depletion. They are chemical compounds manufactured by humans and released into the atmosphere. Ultra-violet light causes CFCs to dissociate and release atoms of chlorine. These chlorine atoms are highly reactive and cause complex reactions in which ozone is converted to oxygen. The reactions form a cycle, with the chlorine atoms being released again, so that they can go on to cause the destruction of more ozone. One chlorine atom can potentially cause the destruction of hundreds of thousands of ozone molecules.

FIGHTING OZONE DEPLETION

CFCs were used very widely in the 1970s and 1980s:
- in refrigerators as the refrigerant
- in aerosol cans as the propellant
- in gas-blown plastics used for fast-food packaging.

In 1987, after research had shown that CFCs damage the ozone layer, an international treaty called the Montreal Protocol was signed. This treaty set targets for the replacement of CFCs with other chemicals that do not damage the ozone layer. Another measure that has been introduced widely is the collection of CFCs from obsolete refrigerators, to prevent them escaping into the atmosphere. Although levels of CFCs are continuing to rise, they should start to fall by 2010. CFCs are stable chemicals and so levels will only fall slowly, but forecasts made using computers suggest that by 2050 ozone holes over the poles will no longer form.

ACID PRECIPITATION

Carbon dioxide dissolves in droplets of water in clouds and makes the precipitation that falls from the clouds slightly acidic. Sulfur dioxide and nitrogen oxides have the same effect, but can make the precipitation much more acidic – as low as pH3. Although there are some natural sources of these gases, human activities are the main source. The figure below shows the origins of these acid pollutants and the processes involved in the formation of acid precipitation.

Sulfur dioxide emissions have been reduced in many countries, but acidification continues to be a problem where levels of nitrogen oxides are still increasing.

THE BIOLOGICAL CONSEQUENCES OF ACID PRECIPITATION

1. Aluminium becomes water soluble in acidified soils and leaches into streams and lakes. Aluminium ions are toxic to fish and in many acidified lakes and rivers all fish have been killed.
2. When soils become acidified, potassium (K^+), magnesium (Mg^{2+}) and calcium ions (Ca^{2+}) leach out, making the soil less fertile and reducing plant growth.
3. Trees affected by acid precipitation show premature leaf fall and dieback of branches. Conifers seem to be particularly vulnerable, perhaps because acid mist condenses on their needles in winter.

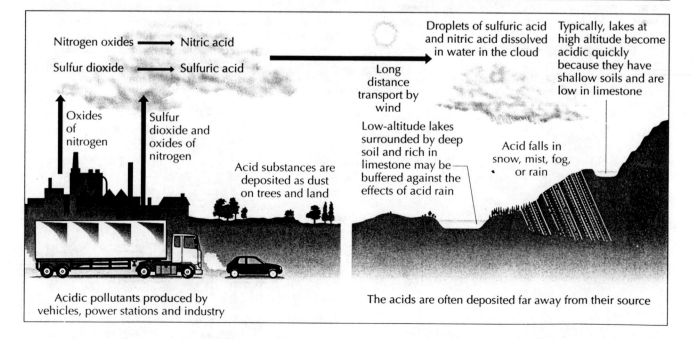

Nitrogen oxides → Nitric acid

Sulfur dioxide → Sulfuric acid

Oxides of nitrogen

Sulfur dioxide and oxides of nitrogen

Acid substances are deposited as dust on trees and land

Long distance transport by wind

Droplets of sulfuric acid and nitric acid dissolved in water in the cloud

Typically, lakes at high altitude become acidic quickly because they have shallow soils and are low in limestone

Low-altitude lakes surrounded by deep soil and rich in limestone may be buffered against the effects of acid rain

Acid falls in snow, mist, fog, or rain

Acidic pollutants produced by vehicles, power stations and industry

The acids are often deposited far away from their source

EFFECTS OF RAW SEWAGE ON RIVERS

Raw sewage and other forms of pollution that are rich in organic matter have profound effects on river ecosystems. The effects can be studied by measuring biotic and abiotic variables at different distances below the sewage input (right).
1. Bacteria consume the organic matter and proliferate.
2. The bacteria use oxygen in aerobic cell respiration. As the numbers of bacteria rise, they consume more and more oxygen, so the **biochemical oxygen demand** of the river water increases. The aeration of the water does not increase, so the water becomes **deoxygenated**. Fish and other animals that depend on dissolved oxygen are sometimes killed.
3. Digestion of organic matter by bacteria causes release of ammonia and phosphate. The ammonia is converted to nitrate by nitrifying bacteria. The increase in levels of nitrate, phosphate and other mineral nutrients is called **eutrophication**.
4. Eutrophication causes proliferation of algae and photosynthetic bacteria (formerly known as blue-green algae). If the numbers of algae are high enough they cause a discolouration of the water, called an **algal bloom**. The algae absorb nutrients from the water, reducing concentrations in the water.
5. The algae release oxygen by photosynthesis, so the water becomes reoxygenated.
6. Primary consumers feed on the algae, reducing their numbers. Conditions in the river are then similar to those before the sewage input – the river has recovered.

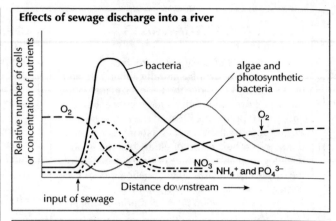

Effects of sewage discharge into a river

EFFECTS OF NITRATE FERTILIZER ON RIVERS

Nitrates can also have profound effects on rivers.
1. Nitrate ions are very soluble and are leached from soils very easily if excessive amounts are applied to crops. If phosphate and other minerals also reach a high concentration, a river becomes eutrophic.
2. As with eutrophication caused by sewage pollution, algae proliferate and **algal blooms** develop. Nitrate from fertilisers sometimes cause such excessive growth of algae that some of the algae are deprived of light and die.
3. Bacteria decompose the dead algae. The bacteria create an increased **biochemical oxygen demand** and **deoxygenation** of the water.
4. Low oxygen levels kill fish and other aquatic animals.

BIOLOGICAL FUELS

Biomass already provides large amounts of fuel, in the form of wood, crop residues and dried manure. Methods now exist for converting biomass into fuels that are more convenient to use, such as ethanol and methane.

METHANE GENERATION

Methane is sometimes called marsh gas, because it is naturally produced in anaerobic conditions by methanogenic bacteria. These conditions are recreated in bioreactors used for methane generation (below). Any organic waste can be the raw material, but sewage and manure or slurry from farms are most commonly used. The raw material is loaded into the bioreactor where anaerobic conditions encourage the growth of three groups of naturally occurring bacteria.
The first groups convert organic matter into organic acids and alcohol.
The second group convert organic acids and alcohol into carbon dioxide, hydrogen and acetate.

The third group of bacteria are the methanogens –they produce methane from carbon dioxide, hydrogen and acetate.

$$\text{Carbon dioxide} + \text{hydrogen} \rightarrow \text{methane} + \text{water}$$
$$CO_2 + 4H_2 \rightarrow CH_4 + 2H_2O$$

$$\text{Acetate} \rightarrow \text{methane} + \text{carbon dioxide}$$
$$CH_3COOH \rightarrow CH_4 + CO_2$$

Methanobacterium and *Methanococcus* are examples of methanogens.
The gas that is produced in bioreactors is sometimes called biogas and is 40–70% methane. It can be used as a fuel for cooking or lighting. If it is compressed in cylinders it can be used as a fuel in vehicles. It is renewable and non-polluting. Production of it helps to dispose of potentially polluting wastes. The organic matter left over at the end of methane generation can be used as an organic fertilizer, rich in mineral nutrients.

A bioreactor

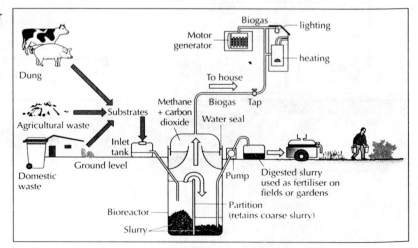

OPTION G – ECOLOGY AND CONSERVATION

G1 Food chains are difficult to study in natural ecosystems, so a group of ecologists set up communities in culture vessels. They used them to investigate the effects of varying nutrient concentrations. In all of the vessels an aquatic bacterium, *Serratia marcescens,* was present. Three concentrations of the nutrients on which *S. marcescens* feeds were used. In some of the cultures *Colpidium striatum,* a predator of *S. marcescens,* was added. In some of these cultures *Didinium nasutum,* a predator of *C. striatum* was added. The cultures therefore each had one, two or three trophic levels. The population density of *S. marcescens* at the end of the experiment is shown in the bar charts below.

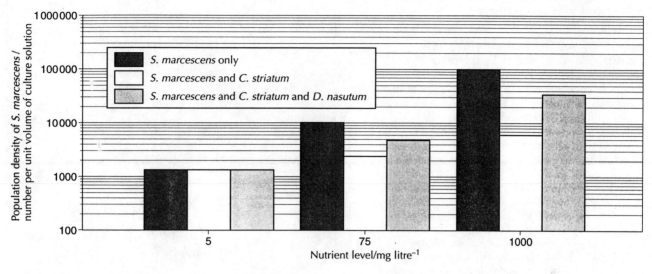

[Source: Kaunzinger, *Nature* (1998), 395, pages 495–496)

a) (i) Explain the effect of the nutrient concentration on the population density of *S. marcescens*. [1]

   (ii) Explain the effect of the presence of C. *striatum* on the population density of *S. marcescens*. [1]

   (iii) Explain the effect of the presence of *D. nasutum* on the population density of *S. marcescens*. [2]

b) In the culture with the lowest nutrient level *D. nasutum* eventually died out but *C. striatum* survived. Explain the reasons for *D. nasutum* dying out. [2]

c) Using the results of this investigation, predict a relationship between nutrient levels and length of food chain in natural ecosystems. [1]

G2 a) Explain how indicator species may be used. [2]

   b) Outline two ex situ methods of conservation of endangered species. [2]

G3 a) (i) Outline the role of *Nitrobacter* in the nitrogen cycle. [1]

   (ii) State the environmental factor that favours the action of this bacterium. [1]

   b) Outline the possible harmful effects of overuse of nitrogen fertilizer in agriculture. [3]

Hormonal control

Hormones are chemical messengers. They are secreted directly into the blood by endocrine glands and are carried to target cells where they elicit a response. A wide range of chemical substances is used as hormones in humans.
- Steroids, e.g. estrogen, progesterone, testosterone.
- Peptides, e.g. ADH (vasopressin), TRH, insulin.
- Tyrosine derivatives, e.g. thyroxin.

MODE OF ACTION OF HORMONES

Hormones do not all work in the same way. There are two main types of mechanism.

1. Steroid hormones enter target cells by passing through the plasma membrane. They bind to receptor proteins in the cytoplasm of target cells, to form a hormone–receptor complex. This complex acts as a regulator of gene transcription, by binding to specific genes. Transcription of some genes is promoted and other genes are inhibited. In this way steroid hormones control whether or not particular enzymes or other proteins are synthesized. They therefore can help to control the activity and development of target cells.
2. Peptide hormones do not enter cells. Instead they bind to receptors in the plasma membrane of target cells. The binding of the hormone causes the release of a secondary messenger inside the cell. The secondary messenger causes a change to the activities of the cell, usually by activating or inhibiting an enzyme.

THE HYPOTHALAMUS AND PITUITARY GLAND

The figure (above right) is a diagram of the hypothalamus and pituitary gland. The hypothalamus is a small part of the brain that links the nervous and endocrine systems. It controls hormone secretion by the pituitary gland located below it. The anterior and posterior lobes of the pituitary gland are controlled in a different way by the hypothalamus. The control of thyroxin secretion by the anterior lobe is shown in the figure (right). The control of ADH secretion by the posterior lobe is described below. The hormones released by the pituitary gland control hormone secretion by other endocrine glands.

CONTROL OF ADH SECRETION

Neurosecretory cells in the supra-optic nucleus of the hypothalamus synthesize ADH, transport it down their axons and store it in nerve endings in the posterior pituitary gland. Osmoreceptor cells in the hypothalamus monitor the concentration of the blood plasma. If the plasma becomes too concentrated, impulses are passed to the ADH-secreting neurosecretory cells, which convey the impulses to their nerve endings in the posterior pituitary. The impulses stimulate release of ADH into the blood from the stores in the nerve endings. ADH causes a reduction in the concentration of the blood plasma, by stimulating the kidney to produce hypertonic urine (see page 106). If the osmoreceptor cells detect that the concentration of blood plasma is too low, the neurosecretory cells are not stimulated to release ADH and the blood ADH level rapidly drops.

Structures of the hypothalamus and pituitary gland

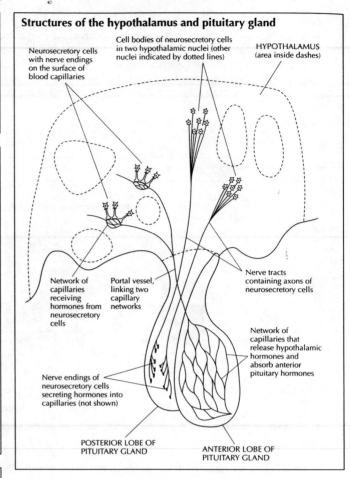

Negative feedback and thyroxin secretion

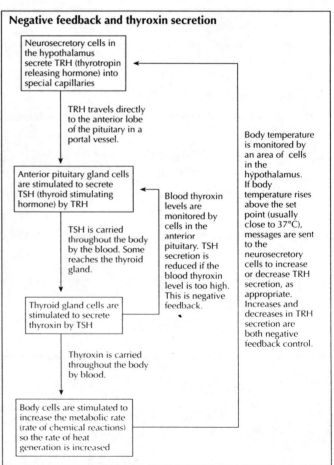

Secretion of digestive juices

Food is digested as it passes along the alimentary canal (see page 47). Longitudinal and circular muscle fibres in the wall contract and relax, squeezing the food and breaking up large solid lumps. Digestive juices, containing enzymes, are mixed with the food. The enzymes digest proteins, nucleic acids, starch and other macromolecules. Digestive juices are secreted by the salivary glands, by glands in the wall of the stomach and by the pancreas.

EXOCRINE GLANDS

Digestive juices are secreted by **exocrine glands**.
The secretory cells in an exocrine gland are in a layer that is only one cell thick. The total area of the layer of secretory cells can be very large because of invagination and branching. The digestive juice is released from the cells by exocytosis. It is then discharged from the gland by travelling along ducts. One group of secretory cells, clustered around the end of a duct is called an **acinus**.
The ducts and acini in part of the pancreas that secretes pancreatic juice are shown below.

Structure of exocrine gland tissue in the pancreas

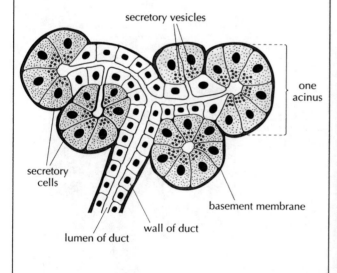

secretory vesicles

one acinus

secretory cells

basement membrane

wall of duct

lumen of duct

EXOCRINE GLAND CELLS

Exocrine gland cells have distinctive features.
- One or two prominent nucleoli inside the nucleus, for production of ribosomes.
- An extensive area of rough endoplasmic reticulum, for protein synthesis.
- Golgi apparatuses for processing proteins.
- Many large vesicles, sometimes called secretory granules, for storage of the substances being secreted and transport of them to the plasma membrane. The vesicles are usually densely stained because of the concentration of proteins.
- Mitochondria, to provide ATP for protein synthesis and other cell activities.
The figure (above right) is an electron micrograph of a pancreas cell and shows these distinctive features.

CONTROL OF SECRETION OF DIGESTIVE JUICE

Both nerves and hormones are involved in the control of digestive juice secretion. The control of gastric juice secretion is described here as an example.
Before food reaches the stomach, gastric juice is already being secreted, as a result of a reflex action. The sight or smell of food stimulates the brain to send nerve impulses to exocrine gland cells in the wall of the stomach. The gland cells start to secrete gastric juice in response.
Much more gastric juice is secreted when food enters the stomach. The food is detected by touch receptors and chemoreceptors in the lining of the stomach and by stretch receptors in the stomach wall. Impulses are sent from these receptors to the brain, which sends more nerve impulses to the exocrine gland cells.
When food is in the stomach, impulses are also sent to endocrine gland cells in the stomach lining that secrete a hormone called gastrin. Gastrin is carried to the exocrine gland cells in the stomach wall, where it stimulates them to increase the secretion of hydrochloric acid. This causes the pH of the food that has entered the stomach to fall to about pH 3.0.

Electron micrograph of an exocrine gland cell in the pancreas (× 6000). The duct into which the cell secreted fluid is visible at the bottom of the micrograph.

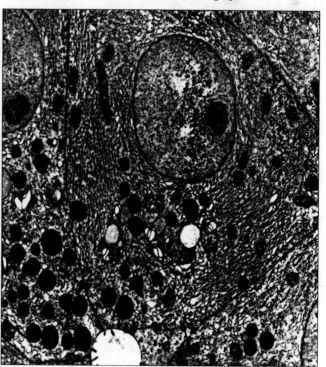

Digestive enzymes

SOURCES OF DIGESTIVE ENZYMES

Food contains many different types of substance that have to be digested before they can be absorbed. Digestion therefore involves many different enzymes, secreted by exocrine glands. The table below lists the main digestive enzymes and other substances in saliva, gastric juice and pancreatic juice.

Digestive juice	Source	Content
Saliva	Salivary glands	– Salivary amylase – Mucus
Gastric juice	Glands in stomach wall	– Pepsinogen – Hydrochloric acid – Mucus
Pancreatic juice	Pancreas	– Pancreatic amylase – Pancreatic lipase – Phospholipase – Trypsinogen – Carboxypeptidase – HCO_3^- ions

The wall of the small intestine also produces many different enzymes that complete the process of digestion. These enzymes are not secreted, but instead they remain in the plasma membranes of cells on the surface of the villi (epithelium cells). The active sites of the enzymes are exposed to the food in the small intestine. They can digest their substrates and the products of digestion can then immediately be absorbed. Epithelium cells tend to be lost from the tips of villi by abrasion, but the membrane-bound enzymes continue to work as they become mixed into the food in the small intestine.

DIGESTION OF LIPIDS

The digestion of lipids poses special problems, because they are insoluble in water. Foods and the digestive juices added to them are mainly composed of water. In the alimentary canal, lipids in foods melt and form liquid droplets. Because of their insolubility, these droplets tend to coalesce to form larger droplets. Lipase is water soluble so it does not enter the lipid droplets, but its active site is hydrophobic (see figure on page 68) and hydrolyses lipids on the surface of droplets. The droplets gradually decrease in size as the lipids on their surface are digested. However food does not remain in the alimentary canal long enough for large droplets to be digested completely. Bile helps to overcome this problem. It contains substances called **bile salts**, which are natural detergents. Bile salt molecules have a hydrophobic and a hydrophilic end. They are therefore attracted to both water and lipids. They coat lipid droplets, causing them to break up into smaller droplets. This process is called **emulsification**. Bile is secreted by the liver and stored in the gall bladder. When it is discharged into the small intestine it emulsifies lipids, which speeds up their digestion, because many small droplets have a larger total surface area than one large droplet of the same volume. With the help of bile, lipids can be completely digested in the small intestine.

DIGESTION OF PROTEINS

A series of enzymes is used in the alimentary canal to hydrolyse large polypeptides into single amino acids. Pepsin and trypsin are involved in the early stages. Both of these enzymes are **endopeptidases** – they hydrolyse peptide linkages in polypeptide chains, breaking large polypeptides into smaller ones.

Pepsin and trypsin are potentially very harmful to the exocrine gland cells that secrete them. They are therefore secreted as inactive precursors, called **pepsinogen** and **trypsinogen**. Pepsinogen is activated by hydrochloric acid, which converts it into pepsin. Different cells in the wall of the stomach secrete pepsinogen and hydrochloric acid (below). Pepsinogen is therefore only activated after it has been secreted.

An enzyme, enterokinase, which is secreted by the lining of the small intestine, activates trypsinogen. Activation therefore only happens when trypsinogen enters the small intestine.

Digestion of proteins is completed by **exopeptidases** – proteases that hydrolyse peptide linkages at the ends of peptide chains, releasing single amino acids. Different exopeptidases are needed to remove amino acids from the two ends of a polypeptide – the amino and the carboxyl terminals.

Structure of the stomach wall

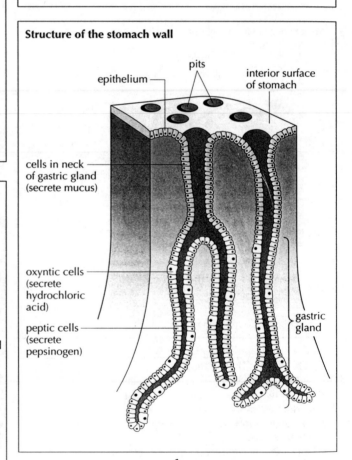

CELLULOSE

Fruit, vegetables and other foods of plant origin contain large amounts of cellulose. The digestion of cellulose involves the enzyme cellulase, which cannot be produced by humans. Cellulose therefore, remains undigested in the alimentary canal and is egested in feces. Feces contains other substances that are not digested or absorbed, including lignin and bile pigments. Feces also contains many bacteria and some intestinal cells that have become detached.

Absorption of digested foods

Digested foods are absorbed in the small intestine, mainly in latter part, called the ileum. The structure of the wall of the ileum allows it to absorb digested foods efficiently. The figure (below left) is a light micrograph of the ileum in longitudinal section and below right is a drawing of the ileum in transverse section.

Micrograph of ileum in longitudinal section (× 40)

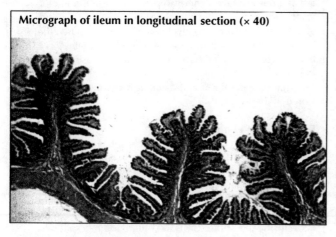

Transverse section of ileum

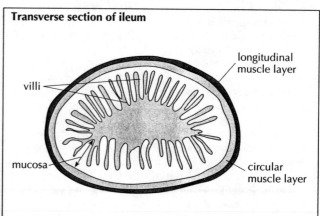

STRUCTURE AND FUNCTION OF VILLUS EPITHELIUM CELLS

Digested foods are absorbed by villi in the ileum. The structure of a villus is shown on page 47. The outer layer of cells where absorption occurs is the epithelium. The figure (right) is an electron micrograph of epithelium cells, showing the structural features that are typical of this cell type. The plasma membranes of adjacent cells are firmly linked together near the free surface by structures called **tight junctions**. These structures prevent molecules from leaking between the epithelium cells. To be absorbed, digested foods have to pass through the plasma membrane of the epithelium cells, and absorption can therefore be carefully controlled. The table below describes the mechanisms used to absorb foods and the structural features used in these mechanisms.

Micrograph of villus epithelium cells (× 2500)

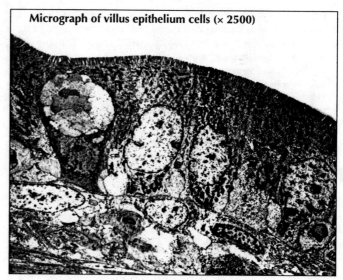

Relationships between structure and function in villus epithelium cells

Structural feature	Function
Microvilli – protrusions of the free surface of the plasma membrane into the lumen of the ileum; about 1 μm long and 0.1 μm wide	Microvilli greatly increase the surface area of plasma membrane exposed to the digested food in the ileum. This increases the rate of absorption of foods by diffusion. Lipids, and other foods that can pass easily through the hydrophobic centre of the plasma membrane of the epithelium cells are absorbed by **simple diffusion** Fructose and some other hydrophilic food substances that are at a low concentration inside body cells are absorbed by **facilitated diffusion**. There is a steep enough concentration gradient for absorption of these substances by diffusion, but they need assistance to pass through the plasma membrane. Channel proteins help them to cross the hydrophobic centre of the membrane
Mitochondria – there are many mitochondria scattered through the cytoplasm	Mitochondria produce the ATP that is needed for absorption of substances by **active transport**. Pump proteins in the plasma membrane of the microvilli carry out the active transport. Glucose, amino acids and mineral ions including sodium, calcium and iron are absorbed in this way
Pinocytotic vesicles – there are many small vesicles, especially near the microvilli	Pinocytic vesicles are formed by **endocytosis**. Each vesicle contains a small droplet of fluid from the lumen of the ileum. The membranes of these vesicles are formed from the plasma membrane and so contain channels for facilitated diffusion and pumps for active transport. Digested foods can be absorbed from the vesicles into the cytoplasm

Liver

The liver is the largest organ in the human abdomen. It contains huge numbers of cells called hepatocytes, which carry out many vital processes.

BLOOD FLOW THROUGH THE LIVER

The liver is supplied with blood by two vessels – the **hepatic portal vein** and the **hepatic artery**. One vessel, the **hepatic vein** carries blood away.

The blood brought by the hepatic portal vein is deoxygenated, because it has already flowed through the wall of the stomach or the intestines. The level of nutrients in this blood varies considerably, depending on the amount of digested food that is being absorbed. One of the main functions of the liver is to regulate levels of nutrients before the blood flows on to the rest of the body. Excessively high levels of glucose and other nutrients would cause damage to the organs of the body, especially the brain.

Inside the liver the hepatic portal vein divides up into vessels called **sinusoids**. These vessels are wider than normal capillaries and have more porous walls, consisting of a single layer of very thin cells, with many pores or gaps between the cells and no basement membrane. Blood flowing along the sinusoids is therefore in close contact with the surrounding hepatocytes. The sinusoids drain into wider vessels that are branches of the hepatic vein. Blood from the liver is carried by the hepatic vein to the right side of the heart via the inferior vena cava.

The hepatic artery supplies the liver with oxygenated blood from the left side of the heart via the aorta. Branches of the hepatic artery join the sinusoids at various points along their length, providing the hepatocytes with the oxygen that they need for aerobic cell respiration.

The figure (below) shows the relationships between blood vessels in liver tissue.

Structure of a sinusoid in the liver

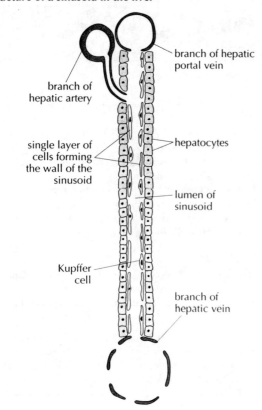

FUNCTIONS OF THE LIVER

Storage of nutrients

When certain nutrients are in excess in the blood, hepatocytes absorb and store them, releasing them when they are at too low a level. For example, when the blood glucose level is too high, insulin stimulates hepatocytes to absorb glucose and convert it to glycogen for storage. When the blood glucose level is too low, glucagon stimulates hepatocytes to break down glycogen and release glucose into the blood.

Iron, retinol (vitamin A) and calciferol (vitamin D) are also stored in the liver.

Breakdown of erythrocytes

Erythrocytes, also called red blood cells, have a fairly short lifespan of about 120 days. The plasma membrane becomes fragile and eventually ruptures releasing the hemoglobin into the blood plasma. The hemoglobin is absorbed by **phagocytosis**, chiefly in the liver. Some of the cells in the walls of the sinusoids are phagocytic. They are called **Kupffer cells**. Inside the Kupffer cells the hemoglobin is split into heme groups and globins. The globins are hydrolysed to amino acids, which are released into the blood. Iron is removed from the heme groups, to leave a yellow-coloured substance called bile pigment or bilirubin. The iron and the bile pigment are released into the blood. Much of the iron is carried to bone marrow, where it is used in the production of hemoglobin in new red blood cells. The bile pigment is absorbed by hepatocytes and forms part of the bile.

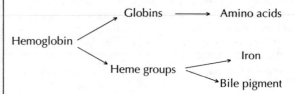

Bile secretion

Hepatocytes secrete bile into narrow tubes that permeate liver tissues, called **canaliculi**. The bile contains hydrogen carbonate ions (HCO_3^-), bile salts, bile pigment and cholesterol. It drains through the canaliculi to the **gall bladder**, where it is concentrated by reabsorption of water and eventually discharged into the small intestine via the **bile duct**.

Synthesis of plasma proteins

The rough endoplasmic reticulum of hepatocytes produces 90% of the proteins in blood plasma. There are three main types of plasma protein: albumin, globulin and fibrinogen. All of the albumin and fibrinogen is produced in hepatocytes and much of the globulin. Some globulins are antibodies and these are made by lymphocytes.

Synthesis of cholesterol

Although some cholesterol is absorbed from food in the intestine, a larger quantity is synthesized each day by hepatocytes. Some of this cholesterol is used in the liver, for example in the production of bile. The rest is transported by the blood for use elsewhere in the body.

Cardiac cycle

The sequence of actions occurring repeatedly in a beating heart is called the cardiac cycle. The figure below shows pressure and volume changes in the left atrium, left ventricle and aorta during two cycles. It also shows electric currents (electrocardiogram) and sounds (phonocardiogram) generated by the beating heart.

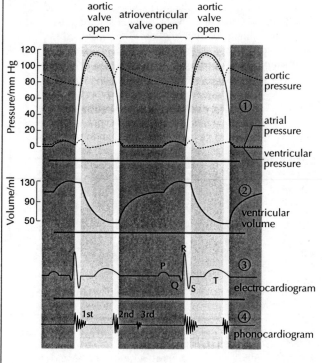

EVENTS OF THE CARDIAC CYCLE

The cardiac cycle is described briefly on page 48
Contraction of the chambers of the heart is called **systole** and relaxation is called **diastole**.

1. Atrial systole

The cardiac cycle begins with the contraction of the wall of the atrium. This happens when the ventricle is already 70% full. The contraction of the atrium pumps move blood into the ventricle, filling it to its maximum capacity before the start of ventricular systole.

2. Ventricular systole

Contraction of the ventricle wall causes a rapid increase in pressure inside the ventricle. This causes the closure of the atrio-ventricular valve, with resulting vibrations in the valve and adjacent walls of the heart. These vibrations are the first heart sound. The pressure in the ventricle rapidly rises above the pressure in the aorta, causing the aortic (semi-lunar) valve to open. Blood can then be pumped from the ventricle into the aorta, raising the aortic blood pressure and decreasing the volume of blood in the ventricle to a minimum. While the ventricle is contracting, the atrium is relaxing and blood enters it from the pulmonary veins.

3. Ventricular diastole

Relaxation of the ventricle wall causes pressure in the ventricle to fall below the pressure in the aorta. The semi-lunar valve therefore closes, with resulting vibrations that are the cause of the second heart sound. When the pressure in the ventricle falls below the pressure in the atrium, the atrio-ventricular valve opens and blood that has accumulated in the atrium flows into the ventricle causing a rapid rise in ventricular volume. With both the atrium and ventricle relaxed, blood continues to drain from the pulmonary veins through the atrium into the ventricle until by the end of the cycle it is about 70% full.

CONTROL OF THE HEART BEAT

Heart muscle cells are stimulated to contract by electrical impulses. Interconnections between adjacent cells (see page 1) allow impulses to spread through the wall of the heart, stimulating it to contract. A small region in the wall of the right atrium initiates each impulse (right). This region is called the SA node (sinoatrial node) and acts as the pacemaker of the heart. Impulses initiated by the SA node spread out in all directions through the walls of the atria, but are prevented from spreading directly into the walls of the ventricles by a layer of fibrous tissue. Instead, impulses have to travel to the ventricles via a second node, called the AV node (atrioventricular node). This node is positioned in the wall of the right atrium, close to the junction between the atria and ventricles. Impulses reach the AV node 0.03 seconds after being emitted from the SA node. There is a delay of 0.09 seconds before impulses pass on from the AV node, which gives the atria time to pump blood into the ventricles before the ventricles contract. Impulses are sent from the AV node along two bundles of conducting fibres that pass through the septum between the left and right ventricles, to the base of heart. Narrower conducting fibres called Purkinje fibres branch out from these bundles and carry impulses to all parts of the walls of the ventricles, causing almost simultaneous contraction throughout the ventricles.

The effects of nerves and hormones on the heart beat rate are described on page 48.

Structures involved in the control of the heart beat

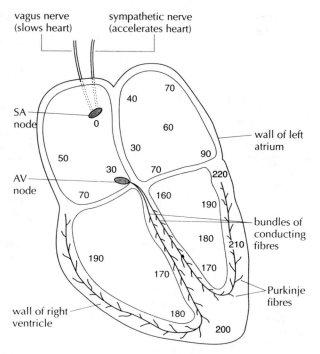

Numbers represent the time taken for impulses from the pacemaker to reach different parts of the heart wall

Lymph, lipoproteins and CHD

FORMATION OF TISSUE FLUID AND LYMPH

Tissue fluid is found in the spaces between cells. It is formed from blood plasma in blood capillaries (below) and seeps through the spaces between cells, supplying nutrients and oxygen and receiving waste products. Most of it is eventually re-absorbed into the capillaries. The tissue fluid that is not re-absorbed by capillaries drains into blind-ended tubes called lymph capillaries. This fluid is called **lymph** and drains through the lymphatic system eventually entering the blood system in the veins near the heart.

The lymphatic system is involved in the transport of lipids, including triglycerides and cholesterol. Lymph capillaries in the villi of the small intestine absorb droplets of lipid that were taken in from the digested food in the small intestine.

Interrelationships between blood and lymphatic systems

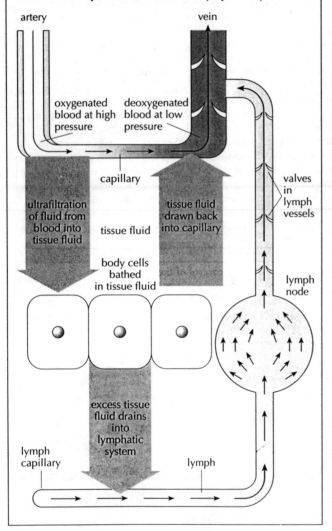

LIPID TRANSPORT IN THE BLOOD

The droplets of lipid that are carried in lymph and in blood plasma are associated with protein. When these droplets of lipoprotein enter the blood from the lymphatic system they have a very low density (VLDL). Adipose tissue and other parts of the body gradually absorb triglycerides from VLDL. The remaining lipoprotein has a high concentration of cholesterol and is called low density lipoprotein (LDL). Another form of lipoprotein, high density lipoprotein (HDL), is used to remove cholesterol from tissues and transport it to the liver for excretion in bile.

ATHEROSCLEROSIS

Atherosclerosis is a degenerative disease of large and medium sized arteries. Phagocytes are attracted to sites of damage to the inner lining of the arteries. The phagocytes release growth factors that stimulate the muscle and fibrous tissues in the artery wall to thicken. LDL may penetrate the damaged areas and release cholesterol, which can build up to form large deposits. The growth of wall tissue and accumulation of cholesterol cause the artery wall to bulge inwards, reducing or even preventing the flow of blood. The thickened wall loses its elasticity and calcium salts are sometimes deposited in it, making it hard. The figure (below) shows a coronary artery that shows signs of atherosclerosis.

Structure of an artery showing atherosclerosis

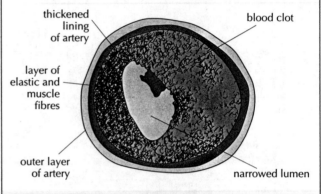

CORONARY THROMBOSIS

The rough inner surface of atherosclerotic arteries tends to cause blood clots to form. The formation of clots is called thrombosis. The wall of the heart is supplied with blood by the coronary arteries. If a blood clot blocks one of these arteries, part of the wall of the heart is deprived of its supply of oxygen. The cells in this part of the wall are unable to respire and so stop contracting. This is either called myocardial infarction or a heart attack.

RISK FACTORS AND CORONARY HEART DISEASE

Atherosclerosis and coronary thrombosis are together known as coronary heart disease (CHD). The rates of CHD vary widely between countries and within populations. Much research has been done to try to identify factors that increase the risk of CHD. Increasing age, being male rather than female and having a family history of CHD are risk factors that cannot be avoided. Obesity, physical inactivity, high blood pressure and tobacco smoke are risk factors that can usually be avoided. The effect of diet is more equivocal. High levels of LDL in the blood are associated with an *increased* risk of CHD. High levels of triglycerides in the blood are also associated with an *increased* risk of CHD, but this may be because they are correlated with high LDL levels. High levels of HDL are associated with a *reduced* risk of CHD. However, it is not certain that there are causal links between HDL and LDL levels and CHD. Also, lipoprotein levels in the blood are not due solely to diet – genetic factors are also important. This may explain why some populations that consume large quantities of cholesterol and saturated triglycerides have extremely low CHD rates – the Maasai of Kenya, for example.

Oxygen transport

Oxygen is transported from the lungs to respiring tissues by hemoglobin in red blood cells. Hemoglobin is a protein that is highly adapted to its function.

OXYGEN DISSOCIATION CURVES

If air with the normal oxygen content is bubbled through a sample of blood, oxygen binds to the hemoglobin until almost all of the hemoglobin molecules have four oxygen molecules bound. The hemoglobin is nearly 100% saturated. If air with a low oxygen content is then bubbled through, some of the oxygen dissociates from the hemoglobin, reducing its percentage saturation. The oxygen content of the air is measured as a **partial pressure**. *Partial pressures are the pressures exerted by each of the gases in a mixture of gases.* The percentage saturation of hemoglobin with oxygen at each partial pressure of oxygen is shown on an oxygen dissociation curve. The figure (below) shows the oxygen dissociation curves of hemoglobin and myoglobin.

Myoglobin is a protein consisting of one globin and one heme group that is used to store oxygen in muscles. The oxygen curve for myoglobin is to the left of the curve for adult hemoglobin because myoglobin has a higher affinity for oxygen. At moderate partial pressures of oxygen, adult hemoglobin releases oxygen and myoglobin binds it. Myoglobin only releases its oxygen when the partial pressure of oxygen in the muscle is very low. The release of oxygen from myoglobin delays the onset of anaerobic respiration in muscles during vigorous exercise.

The dissociation curves for myoglobin and hemoglobin have different shapes. The curve for hemoglobin is S-shaped and that for myoglobin is not. Myoglobin consists of one heme group attached to a globin, whereas hemoglobin has four heme groups, each attached to different globins that interact with each other. As oxygen molecules dissociate from hemoglobin, conformational changes occur, which make it easier for other oxygen molecules to dissociate. Blood containing adult hemoglobin therefore releases large amounts of oxygen over a narrow range of oxygen partial pressures, corresponding to the conditions normally found in respiring tissues.

Oxygen dissociation curves of hemoglobin and myoglobin

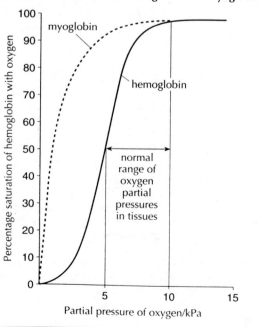

THE BOHR SHIFT

The release of oxygen by hemoglobin in respiring tissues is promoted by an effect called the Bohr shift. Hemoglobin's affinity for oxygen is reduced as the partial pressure of carbon dioxide increases (below). Respiring tissues have high partial pressures of carbon dioxide, so oxygen tends to dissociate. The lungs have lower partial pressures of carbon dioxide, so oxygen tends to bind to hemoglobin.

Effect of CO_2 on the oxygen dissociation curve of hemoglobin

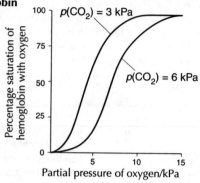

FETAL HEMOGLOBIN

The hemoglobin in the red blood cells of a fetus is slightly different in amino acid sequence from adult hemoglobin. The figure (below) shows that it has greater affinity for oxygen and so, in the placenta, the oxygen that dissociates from adult hemoglobin binds to fetal hemoglobin, which only releases it once it enters the tissues of the fetus.

Oxygen dissociation curves of adult and fetal hemoglobin

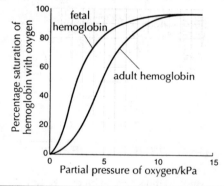

GAS EXCHANGE AT HIGH ALTITUDE

The partial pressure of oxygen at high altitude is lower than at sea level. Hemoglobin may not become fully saturated as it passes through the lungs, so tissues of the body may not be adequately supplied with oxygen. A condition called mountain sickness can develop, with muscular weakness, rapid pulse, nausea and headaches. This can be avoided by ascending gradually to allow the body to acclimatize to high altitude. During acclimatization the ventilation rate increases. Extra red blood cells are produced, increasing the hemoglobin content of the blood. Muscles produce more myoglobin and develop a denser capillary network. These changes help to supply the body with enough oxygen. Some people who are native to high altitude show other adaptations, including a high lung capacity with a large surface area for gas exchange, larger tidal volumes and hemoglobin with an increased affinity for oxygen.

Carbon dioxide transport

Carbon dioxide is produced by aerobic respiration in cells and then either diffuses directly into capillaries or into tissue fluid that is drawn into capillaries. Carbon dioxide is carried by the blood to the lungs in three different ways. A small amount (7%) is carried dissolved in the plasma. The remainder is either converted to hydrogen carbonate ions or binds to hemoglobin.

CONVERSION TO HYDROGEN CARBONATE IONS

Carbon dioxide can be converted into hydrogen carbonate ions within a fraction of a second of entering the blood. About 70% of carbon dioxide is carried in this way. After diffusing into red blood cells, carbon dioxide combines with water to form carbonic acid. This reaction is catalysed by carbonic anhydrase. Carbonic acid rapidly dissociates into hydrogen carbonate and hydrogen ions. The hydrogen carbonate ions move out of the red blood cells by facilitated diffusion. A carrier protein is used that simultaneously moves a chloride ion into the red blood cell. This is called the **chloride shift** and prevents the balance of charges across the membrane from being altered. The figure (below) shows the reactions that produce hydrogen carbonate ions and the chloride shift.
The hydrogen ions that dissociate from carbonic acid bind to hemoglobin in the red blood cells, preventing an excessive change in pH. This is called pH buffering. Plasma proteins also act as pH buffers in blood.

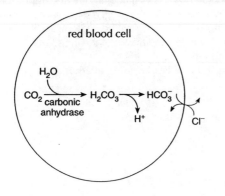

BINDING OF CARBON DIOXIDE TO HEMOGLOBIN

In respiring tissues, carbon dioxide binds reversibly to hemoglobin, to form carbaminohemoglobin.
In the lungs, **carbaminohemoglobin** dissociates and the carbon dioxide is released. Between 15% and 25% of carbon dioxide is carried in this way. The binding of carbon dioxide and hydrogen ions to hemoglobin lowers its affinity for oxygen. This causes the Bohr shift (page 171).

THE EFFECT OF EXERCISE ON VENTILATION

During vigorous exercise, the energy demands of the body can increase by over ten times. The rate of aerobic respiration in muscles rises so there is an increase in the amount of CO_2 entering the blood and the concentration rises. This reduces the pH of the blood and is rapidly detected by cells in the walls of arteries, which monitor blood pH and concentrations of oxygen and carbon dioxide in the blood. These cells are called **chemosensors**. The chemosensors send nerve impulses to the parts of the medulla of the brain that control the ventilation rate, called the **breathing centres**. The breathing centres also monitor blood pH and carbon dioxide concentration. If the concentration of carbon dioxide in the blood rises and the blood pH falls below its normal level of pH 7.4, the breathing centres increase the rate of inspiration and expiration. This is done by sending nerve impulses to the diaphragm and intercostal muscles, causing them to increase the rate at which they contract and relax. The increase in the ventilation rate helps to remove from the body the CO_2 produced in aerobic cell respiration. It also helps to increase the rate of oxygen uptake, which allows aerobic cell respiration to continue in the muscles and it helps to repay the oxygen debt after anaerobic cell respiration.
After exercise, the level of CO_2 in the blood falls, the pH of the blood rises and the breathing centres cause the ventilation rate to decrease.
The figure (below) shows the relationship between blood pH, partial pressure of carbon dioxide in blood and ventilation rate.

Effect of varying blood pH and CO_2 level on the ventilation rate

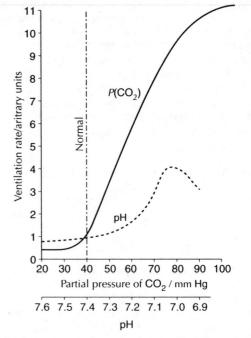

LUNG CANCER

Inhaling carcinogens can cause cancer of the lung. Most cases are caused by cigarette smoking. Tobacco smoke contains many different carcinogens. Atmospheric pollution can also cause lung cancer. Workers in certain industries suffer an increased risk, especially mining and metal refining where the metal ores contain radioactive substances.
Lung cancer is treated by removing part or the entire affected lung, with obvious consequences for gas exchange.

ASTHMA

During asthma attacks the muscles in the wall of the bronchi contract excessively, narrowing the bronchi. Ventilation is a struggle and gas exchange is reduced.
Asthma is an allergic reaction, often to house dust mites, but sometimes also to pollen, pets and some fungi. According to a recent theory, living in very clean homes increases the risk. Without enough pathogens to fight, the immune system starts to react against harmless substances, causing allergies to develop.

EXAM QUESTIONS ON OPTION H – FURTHER HUMAN PHYSIOLOGY

H1 The causes of lung cancer were investigated by researchers in Italy using lichens. Lichens grow on the trunks of trees and vary in their ability to grow in polluted air. As air becomes more polluted fewer species of lichen are able to survive, so lichen biodiversity falls. Mortality due to lung cancer in men was found in each municipality in the Veneto region of north east Italy, using medical records. The biodiversity of lichens growing in each municipality was then measured. The results are shown in the scattergram below.

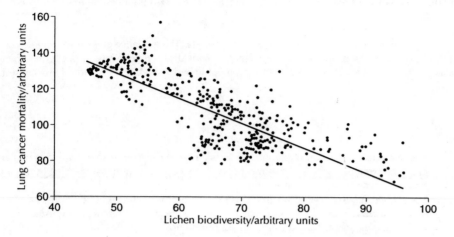

[Source: Cislaghi and Nimis, Nature (1997), 387 page 463]

a) (i) Using only the data in the scattergram, identify the relationship between mortality due to lung cancer and lichen biodiversity in the Veneto region. [1]

   (ii) Explain the relationship between mortality due to lung cancer and lichen biodiversity. [2]

b) Only men who had lived in one municipality for their entire life were included in the investigation. Suggest **one** reason for excluding other men. [1]

c) Suggest **two** reasons for the points on the scattergram not all lying on the line of best fit. [2]

H2 a) Outline how the atria of the heart are stimulated to contract. [2]

   b) Explain the origin of the heart sounds. [2]

H3 State how the supply of oxygen to respiring tissues is helped by: [1]

   a) the diaphragm

   b) myoglobin. [1]

   c) the Bohr shift. [1]

During the IB Biology course, your teacher will help you to improve your skills in planning and performing investigations. When your teacher thinks that you are ready, your skills will be assessed. This will be done during lessons at your school, so it is called internal assessment (IA). Exams that are sent off to an examiner are called external assessment. In IB Biology, 24% of the marks are for IA.

Although your teacher will help you as much as possible, you cannot be given higher marks than you deserve because samples of work from your school are checked to see whether they have been marked to the right standard. This is called external moderation. You must therefore demonstrate high levels of skill to gain high marks in IA.

Eight skills are assessed in IA – these are called the IA criteria. For each criterion, you can gain a mark of 3, 2, 1 or 0. You will be assessed at least twice for each criterion and two of these scores will be selected. The maximum possible mark for IA is therefore 48 (3 for each of eight criteria twice).

The table below explains how to improve your level of skill in each of the eight criteria. The guidance given encourages you to use standard procedures for carrying out an investigation. Research scientists do not always follow these procedures – many great discoveries have resulted from doing things differently. However, for IA it is probably best to stick to the standard procedures!

IA criterion	Guidance
Planning (a) (In this part of planning you outline the main principles of an investigation)	• The first stage in this part of planning is to ask a question. Your teacher will give you an area of biology to investigate and you must decide on a question about something in this area. You must state it clearly in your write-up. • The second stage is to propose an answer to the question – this is either a prediction or a hypothesis. You must base your prediction or hypothesis on scientific knowledge and understanding and explain this fully in your write-up. You can now start to design an experiment to test your prediction or hypothesis. In most experiments there is one factor that you vary deliberately. You choose what levels of this factor to use. This is called the independent variable. You then measure the level of another variable that is affected by the independent variable. This is called the dependent variable. There will be other variables that could affect the dependent variable. You must keep these constant during the experiment and they are therefore called the controlled variables. • The third stage of Planning (a) is to state clearly what the independent variable, the dependent variable and the controlled variables will be in your experiment.
Planning (b) (In this part of planning you describe in detail your method for an experiment)	In this part of planning there are three things that you must decide and then describe in your write-up. • The apparatus and materials that you plan to use – obviously they must be suitable. You can simply list them, but you must be precise, for example specifying sizes of glassware and concentrations of solutions. You can also show the apparatus that you intend to use using labelled or annotated diagrams. • How you intend to control each of the variables – the independent variable and the controlled variables. • How you intend to collect sufficient data – this involves the method that you will use to measure the dependent variable, the range of levels of the independent variable and the number of repeats that you will do. Repeats are two or more measurements of the dependent variable, with the same level of the independent variable. Repeats help you to assess whether your results are reliable or not.
Data collection	Usually the data that you collect will be quantitative – measurements with S.I. units. You should record these measurements as accurately as possible during your experiment. You could do this by data logging using a computer or calculator, or manually using a results table on paper. If working with a partner, do not rely on them to record the results – do it yourself. After the experiment, you should draw up a clear and accurate results table. Show every result that you obtained, not just the mean results. Repeat results should be numbered. The column headings on results table should show both the quantity being measured and the S.I. units. Sometimes your data will be qualitative – drawings of structures, colour changes or other observations. Drawings should be large, with sharp lines and labels or annotation to interpret the structures shown. Measure the size of the specimen and the drawing and calculate the magnification. Remember to include a title for both drawings and results tables.

Data processing and presentation	The results that are collected during an experiment are called raw data. It is usually necessary to process this raw data in some way. This might involve calculating mean results, or performing a statistical test on the data. It might involve drawing a graph or displaying the data in some other way.
	If you are drawing a graph, remember to put the independent variable on the *x*-axis and the dependent variable on the *y*-axis. Join the points with a curve or straight lines, depending on whether you know where intervening points would have been or not. Check that you have labelled both of the axes with the quantity and the units, for example, mass (grams).
	If your raw data consisted of drawings, you can process them by constructing a diagram to show significant features of the structure.
Conclusion and evaluation	In this stage, you review the results of an investigation to see what can be learned from it. The most important task is to draw conclusions. You can use the following questions to guide you towards valid conclusions based on the results.
	• What trends are shown by the data? You might have to decide whether there is a significant difference between two means, or deduce a relationship between an independent and a dependent variable, shown by a graph.
	• What is the explanation for the observed differences or relationships?
	• How does the data compare with data from similar experiments in textbooks or scientific journals?
	• What conclusions can be drawn from the investigation? (if any!)
	Another task is to evaluate all aspects of the investigation. You can use the following questions to guide you.
	• Are there any results that did not fit in with the rest? These are called anomalous results.
	• Were there any errors made during the experiment that explain the anomalous results?
	• How successfully did the method used in the experiment generate reliable results? You can often decide whether results are reliable or not by how close repeats are to each other.
	• What were the main weaknesses in the investigation?
	• What could be done to make genuine improvements to the investigation, if it was done again?
Manipulative skills	If you have done enough practical work in Biology, your manipulative skills should be excellent. You probably will not need the following reminders!
	• Study instructions carefully before starting work so that you know what you are doing.
	• Be sensible about asking for help from your teacher. Try to work out what to do yourself. Use your own initiative to decide how to modify a procedure yourself when necessary. But, if you have not been given full enough instructions or are worried about the safety of the procedure, ask for help.
	• Make sure that you know about any potential risks in the procedure that you are following.
	• Work in a careful and systematic way – arrange your apparatus tidily and do not waste time, but work without rushing.
Personal skills (a) (In this part of personal skills you show that you can work effectively with others)	Scientists often work in teams, so the ability to cooperate with others is important. Ask yourself these questions about the way that you work.
	• Do you find it easy to work cooperatively with others?
	• Do you recognize the needs of others?
	• Are you only interested in your own views or do ask for the views of others?
	• Do you exchange ideas with others and help to combine them so that the team completes a task more effectively than any one individual could?
Personal skills (b) (In this part of personal skills you show that the qualities of your work as an individual)	Scientists also sometimes work by themselves and need to take personal responsibility for all that they do. Ask yourself these questions about the way that you work.
	• Are you persistent enough to continue working on a task until you finish it, even if there is a problem?
	• Do you always make sure that the data that you obtain is accurate and you avoid the temptation to 'improve' your data to fit a hypothesis?
	• Do you ensure that living organisms do not suffer during your experiments?
	• Do you ensure that your experiments do not damage the environment?

Many IB students choose a biological research question for their extended essay. There are unlimited opportunities for novel and interesting work because of the diversity of life. Many excellent Biology extended essays are written each year. Every essay is an individual effort and there is no formula for writing the perfect essay. The steps shown below are intended to help you to avoid some common faults, without preventing you from writing the essay that you want to write.

While you are working on the essay, your most important resource will be the teacher who is supervising you. If you need help at any stage, fix a time to talk things over. You should make sure that your supervisor always knows what you are doing – discuss how things are going as frequently as possible. If you don't, you could waste a lot of time on an unproductive approach to the work. Remember two important maxims: 'things take time' and 'if something can go wrong it will'. Assume from the start that you'll have to do a second run of any experiments or observations and allow time for this. Start work as soon as possible and then you will have time to produce the finest essay that you can. You can also earn extra points towards your diploma.

Planning and data collection

Choose a suitable topic	Pick the field of biology that interests you most and gradually narrow down to one small section of it. You must choose a truly biological topic – one that involves living organisms and interactions between them. It must be a topic in which you can have a personal input – this isn't easy with some topics, for example diagnosis and treatment of diseases, so these are best avoided.
Choose an approach	There are two main types of approach. 1. Doing experiments/making careful observations. 2. Searching in books, journals or on the internet for relevant data. Most of the best essays use both approaches. If you cannot design and do experiments in your chosen topic, reconsider your choice of topic!
Do some preliminary work	Try out some experiments – this should allow you to find out whether your approach is likely to be successful. Avoid experiments that cause unnecessary risks, suffering to animals or environmental damage. Do some background reading and take careful notes of important relevant information. Preliminary work should get you thinking and asking questions about your topic.
Formulate a research question	This should be a question worth asking – not one with an obvious answer. It should be narrow enough to be fully answered in a 4000-word essay, based on 40 hours work. It can either be phrased as a question or a hypothesis – a prediction that you are going to test. It is absolutely vital at this stage to talk to your main helper – your teacher.
Plan your methods	If you are following advice given earlier, you will be designing experiments or planning how to make careful observations. Although you may use some standard protocols, there must be a personal input to the experimental design, even if you are working in a research lab. If your methods are too complex for a personal input, they are too complex! There are no extra marks for complexity.
Collect the data that you need	Remember the things that you have been taught when planning experiments for Internal Assessment – variables must be controlled and repeats are needed to allow you to assess the reliability of your data. If you aren't doing your own experiments, you must obtain the published results of experiments, not just the conclusions that were drawn from these experimental results.

Writing up your essay

Write an introduction	This can be quite brief. There is no need to include large amounts of background material, especially if it is straightforward biology. Instead, say why the topic is worth writing about and give the background information that is needed to understand the essay. State the research question precisely and then outline the approach that you are using to investigate it.
Describe your methods	This section shouldn't be very long – if it is then your methods were probably too complex. Explain clearly and fully what you did and why. Make clear how the experiments tested your hypothesis or gave the evidence needed to answer the research question. Remember to explain how variables were controlled and why you did repeats. Take credit for the things that you designed yourself.
Display your results	Use clear tables or other formats to display the data that you have obtained – the results of your experiments. You only need to put raw data into an appendix if you have huge amounts of it. Use graphs or other charts to display the most significant features of the data, for example mean results. If you are using data from other scientists, you must display and manipulate it in an original way.
Analyse your data	This should be a long and detailed section of the essay. You should discuss whether the data is reliable – were the repeats close? do the results show a consistent trend? – what confidence level do statistical tests show? Then use all the understanding of your topic that you have built up to discuss possible explanations for any trends, with reasons for rejecting or accepting them.
Draw your conclusions	Only a short section is needed here. It should not include new information or views different from those expressed in earlier sections of the essay. Instead you should sum up what answer you have found to your research question or whether your hypothesis is supported or undermined. Your conclusions should be based only on the data that you have obtained and analysed.
Write an abstract	You must summarize your whole essay in 300 words. You must state your research question clearly, how you investigated it and what conclusion you reached. The usual purpose of an abstract is to give the reader a quick impression of the contents of a long article so that he or she can decide whether it is worth reading or not! Obviously, your essay will be well worth reading!
Add the finishing touches	You now need to write a contents page and a bibliography. The contents page lists the sections of the essay with the number of the page on which each section begins. The bibliography is a numbered list of the published sources that you used. In the text of your essay you should put a reference, in the form of a superscript number, wherever you have used information from these sources.

If you want to do well in your final exams, you must prepare for them very carefully in the weeks beforehand.
The most important task is to memorize all the facts that you have been taught. For a high grade, you will need a comprehensive knowledge of them. You will need to spend many hours on revision and find tactics that work for you.
You should also practise answering exam questions. You can use the questions at the end of topics in this book, after you have revised each topic. Your teacher should also give you some past exam papers to try.

There are three styles of question in IB Biology exams.
- **Multiple choice questions –** These are questions where you choose one of four possible answers. Read all of them before choosing one. If you cannot decide on one answer, try to eliminate answers that are obviously wrong, to narrow down the possibilities. Leave difficult questions until you have answered the straightforward ones. Give an answer to every question – marks are not deducted for wrong answers.
- **Structured questions –** These questions are broken up into small sections, each of which you answer in the space or on the lines provided. If you run out of space, continue your answer elsewhere on the paper – it will be marked as long as you indicate clearly what you have done. The number of marks for each section is indicated and this helps show you how detailed your answer needs to be. Some structured questions involve data analysis. Look through the data questions in this book to see some of the ways in which data can be presented. You should always study the data very carefully before answering the questions, for example the scales and labelling on the axes of graphs. If there are calculations, remember to show your working and give units with your answer, for example grams or millimetres.
- **Free response questions –** These questions require long and detailed answers on lined paper. You can decide what style of answer to give. Usually continuous prose is best, but sometimes ideas can be shown on a table or on a carefully annotated diagram. There may be a choice of free response question. Read the whole of each question before making your choice. There may be marks for the quality of construction of your answer. If the question is divided up into sections – (a), (b) and so on, you must answer it in these sections. Try to express your ideas clearly so that the examiner understands what you mean. Plan out your answer on scrap paper, so that you arrange your ideas in a logical sequence and do not include irrelevant material. As with all questions, you must write legibly or the examiner may not be able to mark your work. This may mean that you have to write more slowly than normal.

If you do revise carefully and build up a comprehensive knowledge of the facts on the syllabus you should find many of the questions straightforward. This is because in IB Biology exams, 50% of the marks are for simple factual recall. These questions will start with words like *list, state, outline* or *describe*. The other 50% of the marks involve more than simple factual recall – they involve expressing ideas that are more complex or involve using your knowledge to develop an answer that you probably haven't been taught.
The word at the start of each question tells you what to do. These words are therefore called action verbs.

Explain – Sometimes this involves giving the mechanism behind something – often a logical chain of events, each one causing the next. This is a 'how' sort of explanation. A key word is often 'therefore'.
Sometimes it involves giving the reasons or causes for something. This is a 'why' sort of explanation. A key word is often 'because'.

Discuss – There won't be a simple straightforward answer to these questions. Sometimes your answer should include arguments for and against something. Try to give a balanced account. Sometimes your answer should consist of a series of alternative hypotheses – you could indicate how likely each one is but you don't need to make a make a final choice.

Suggest – Don't expect to have been taught the answer to these questions. Use your overall biological understanding to find answers – as long as they are possible, they will get a mark.

Compare – This type of question involves assessing how similar or different two or more things are. You cannot do this by describing the things separately. Every point that you make should be a similarity or a difference. There may be more similarities or more differences – all of them are relevant.

Distinguish – This is similar to a compare question, except that only differences need to be included in your answer. The key word in this type of question is often 'whereas'.
In both compare and distinguish questions a table is often the best way to arrange your answer. Use the columns of the table for the things that you are comparing and the rows for the individual similarities or differences.

Evaluate – This usually involves assessing the value, importance or effects of something. You might have to assess how useful a technique is, or how useful a model is in helping to explain something. You might have to assess the expected impacts of something on the environment. Whatever it is that you are evaluating, you will probably have to use your judgement in composing your answer.

There are other action verbs that are used in questions, but they are more straightforward and you are unlikely to have difficulty in understanding what sort of answer to give.

TOPIC 1 CELLS

1 (a) X = rough endoplasmic reticulum Y = mitochondria (b) nuclear membrane is the curved structure on the left hand side (c) proteins because there is rough endoplasmic reticulum with ribosomes which make protein; ATP because there are many mitochondria which make ATP.

2 (a)(i) phospholipid (ii) head is hydrophilic and tails are hydrophobic (b)(i) II is integral (ii) any two of: III is a pump protein; transfers specific substances; uses energy from ATP to move substance against the concentration gradient.

3 (a) mitosis is division of the nucleus; cytokinesis is division of the cell/cytoplasm (b)(i) animal cells use centrioles in mitosis but plant cells do not (ii) plants construct new cell walls/cell plates across the equator whereas the plasma membrane divides the cytoplasm in animal cells (c) any four of: DNA replication; transcription of genes; protein synthesis; growth/increase in volume of cytoplasm; construction of more membrane/thickening of cell wall; division of mitochondria/chloroplasts.

TOPIC 2 THE CHEMISTRY OF LIFE

1 (i) DNA (ii) DNA (iii) RNA (b) experimental error (c)(i) DNA is double stranded; A pairs with T and C pairs with G; one base in each pair is therefore A or G, so A + G = 50%; (ii) any two of A = T; C = G; C + G = 50% ; A + G / C + G = 1.00 (d)(i) influenza virus (ii) RNA contains uracil instead of thymine; single stranded so amounts of G and C not equal.

2 (a) ring of five carbon atoms and one oxygen; H and OH groups correctly shown; (b) any three of: OH groups are hydrophilic; form hydrogen bonds with water; size of molecule is small (c) nitrogen/sulfur.

3 (a) radical/variable portion of the amino acid (b) C – N bond; O= linked to C and H- linked to N (c) 70S ribosomes in prokaryotes vs. 80S ribosomes in eukaryotes; free ribosomes in prokaryotes vs. ribosomes sometimes linked to rough endoplasmic reticulum in eukaryotes.

TOPIC 3 GENETICS

1 (a) O group individual must be genotype ii because it is due to a recessive allele; B group individual in generation 2 must be $I^B i$ because the parent that was blood group A could not have passed on I^B; B group individual in generation 3 must have been $I^B i$ because the O group parent must have passed on i

(b) parents could have been group O; parents could have been group A with genotype $I^A i$; parents could have been group B with genotype $I^B i$ (c) blood transfusion.

2 (a) $C^r C^r$ $C^W C^W$ and $C^r C^W$ (b) The allele for red flowers is dominant in peas but codominant in *Mirabilis* (c) gametes C^r and C^W; genotypes $C^r C^r$ $C^r C^W$ $C^W C^r$ and $C^W C^W$; phenotypes red pink pink and white, respectively.

3 (a) A group of organisms with identical genotypes; nucleus removed from a cell in an adult animal; nucleus removed from an egg cell and replaced with nucleus from adult (c) any two of: possible identity/psychological problems; risk of abnormalities/miscarriage; risk of premature aging; wrong to interfere with nature/religious objections.

TOPIC 4 ECOLOGY AND EVOLUTION

1 (a) sigmoid/S-shaped (b)(i) line reaching a plateau by year 8 (ii) any two of: food supply; predation; breeding sites; disease (c)(i) population would have reached carrying capacity more quickly (ii) carrying capacity would have been the same.

2 (a) I = secondary consumers II = primary consumers III = producers (b) chemical energy (c) arrow from the sun to box III (d) any two of: arrows represent energy losses; heat produced because energy transformations are never 100% efficient; energy not passed along the food chain to another organism.

3 (a) methane causes an increase in the Earth's temperature by the greenhouse effect; temperature only increases as a result of an increase in atmospheric methane; methane emissions to the atmosphere must be greater than losses (b) methane emission is a natural process, e.g. swamps and marshes; humans cause methane emission, e.g. coal burning/cattle and sheep/rice paddies; most emissions are caused by humans/humans have increased emissions considerably (c) any three of: drain swamps and marshes; reduce cattle and sheep farming; stop growing rice in paddies; control releases of natural gas; reduce burning of coal; prevent forest fires/burning of biomass.

TOPIC 5 HUMAN HEALTH AND PHYSIOLOGY

1 (a) T_b is higher than T_a; T_b is constant whereas T_a is decreasing (b) heat absorbed from the environment; heat generated by cell respiration (c) active during darkness because it maintains constant high body temperature as a result of cell respiration.

2 (a)(i) ingestion of pathogens (ii) in blood; in body tissues (b) to allow the production of any different types of antibody; to fight many different diseases.

3 (a) I = trachea II = bronchioles (b) maintains concentration gradients of oxygen and CO_2 between air in alveoli and blood; ensures rapid diffusion/gaseous exchange (c) asthma/lung cancer/emphysema/bronchitis.

TOPIC 6 NUCLEIC ACIDS AND PROTEINS

1 (a)(i) higher than 40 °C; initial rate was faster; then reaction stopped due to denaturation (ii) lower temperature than 40 °C because the rate is slower; 30 °C because the rate is half that at 40 °C (b)(i) curve drawn above the curve W (similar to curve Y) (ii) curve drawn above the curve W; gradient of curve decreasing markedly with time showing increasing inhibition as the substrate concentration falls.

2 (a) synthesis of DNA/cDNA; from RNA (b)(i) retroviruses (ii) HIV (c) any three of: mRNA can be obtained quite easily; genes can be hard to find; gene consisting of DNA can be made from RNA no introns in the gene using reverse transcriptase; gene can then be inserted into another organism; cDNA/probes can be used to locate other genes.

3 (a) globular (b) number and sequence of amino acids (c)(i) X is a beta pleated sheet and Y is alpha helix (ii) hydrogen bonding (d) any two of: tertiary structure determines the enzyme's shape; determines the active site's shape; makes the enzyme substrate specific; shape ensures that when the substrate binds it is distorted/induced fit.

TOPIC 7 CELL RESPIRATION AND PHOTOSYNTHESIS

1 (a) Any two of: double membrane; cristae/infoldings of inner membrane; ovoid shape; (b) double outer membrane shown; inner membrane shown folded in to form a crista (c)(i) label indicating the matrix (ii) label indicating the inner membrane/cristae (iii) label indicating the cytoplasm outside the mitochondria.

2 (a) peaks in the red and blue sections of the spectrum; minimum in the green section at about half of maximal rate (b) action and absorption spectra are closely correlated; because pigments absorb the light energy used in photosynthesis; the more light absorbed at a wavelength the more photosynthesis.

3 Any two of: proton gradient is a store of energy; energy given out as protons flow across the membrane; used to make ATP; used to couple electron flow to ATP synthesis (b) any two of: gradient is generated by the inner mitochondrial membrane; cristae/infoldings give a large area of inner mitochondrial membrane; protons stored between inner and outer membrane; small volume between membranes so large proton concentration develops more easily.

TOPICS 8 AND 9 GENETICS AND HUMAN REPRODUCTION

1 (a) polygenic (b) AaBb; blue-flowered (c) all gametes shown with one allele of each gene only; four homozygous genotypes shown AABB AAbb aaBB and aabb; four double heterozygous genotypes shown AaBb; eight other genotypes shown AABb AAbB aaBb aabB AaBB aABB Aabb and aAbb; all sixteen phenotypes indicated (d) 9 blue 3 red and 4 white (e) gene A converts white to red and gene B converts red to blue.

2 reassortment of genes into different combinations from those of the parents (b) black body long wing; grey body vestigial wing (c) genes are linked/found on the same chromosome; parental combinations are kept together; unless there is a cross-over between the genes (d) any two of: find which chromosome a gene is located on; identify all of the genes in a linkage group; estimate how far apart the loci of genes on a chromosome are.

3 (a) Any four of: both contain a haploid nucleus; both have a plasma membrane; the sperm has a tail but the egg does not; the egg has much more cytoplasm; mitochondria in sperm are helical but in the egg are ovoid; the egg has cortical granules but the sperm does not; the sperm has an acrosome but the egg does not; (b) stimulates gametogenesis in both men and women; promotes development of follicles in women and primary spermatocytes in men; stimulates estrogen secretion in women but not testosterone secretion in men; (c) both stimulate the development of the corpus luteum; both stimulate the secretion of progesterone; before fertilization by LH and after fertilization by HCG.

TOPICS 10 AND 11 DEFENCE AGAINST INFECTIOUS DISEASE AND NERVES MUSCLES AND MOVEMENT

1 (a)(i) immunity developed after chicken pox infection/other similar example (ii) immunity due to antibodies made by the mother and received by a fetus via the placenta/received by a baby from colostrum; (iii) immunity to tetanus developed after vaccination/other similar example (b) vaccine injected; vaccine contains weak or killed form of the bacterium or virus; lymphocytes stimulated to produce antibodies (c) clone of plasma cells is already present/memory cells have already been formed.

2 (a)(i) actin (ii) regions II and III (b)(i) II would increase in length (ii) I and III would increase in length.

3 (a) Any two of: more energy needed for surface swimming; penguin and sea otter both use more energy for surface swimming (b) Any two of: humans are not streamlined; legs adapted to walk not swim; humans swim on the surface which takes more energy than underwater; lack of buoyancy so energy used to remain at surface.

TOPICS 12 AND 13 EXCRETION AND PLANT SCIENCE

1 Any two of: waste products of metabolism must be removed; toxins must be removed; substances in excess must be removed (b) Any two of: blood in artery more oxygenated than blood in vein; blood in artery contains more urea than blood in vein; blood in artery contains less carbon dioxide than blood in vein; salt/water content of artery more variable than vein (c) loop of Henle makes the medulla hypertonic by raising the sodium/ mineral ion concentration; allows the production of hypertonic urine (d) basement membrane of glomerulus.

2 (a) One mark for at least two and two marks for at least four of: Bowman's capsule; proximal convoluted tubule; loop of Henle; distal convoluted tubule; collecting duct (b)(i) proximal convoluted tubule (ii) root hairs/phloem sieve tubes in leaves/guard cells.

3 (a)(i) 6 pm to 6 am/sunset to sunrise (ii) 6 am to 4.30 pm (b) CAM plant is the xerophyte because it opens its stomata at night; less water loss during cooler conditions in the night (c)(i) partial closure between 11 am and 12 am; followed by reopening (ii) plant needs to limit transpiration during the hottest part of the day.

OPTION A – DIET AND HUMAN NUTRITION

1 (a)(i) 10.0 MJ day^{-1} (ii) 10.3 MJ day^{-1} (b)(i) energy requirement increases as PAL increases (ii) energy requirement increases as body mass increases (c)(i) energy needed for muscle contraction (ii) more energy needed to move a heavier body (d)(i) girl has a lower energy requirement (due to better insulation and lower heat loss) (ii) adult has lower energy requirement (because the adult does not need energy for growth).

2 (a) eggs/meat/fish/milk/cheese/nuts/mycoprotein (b) replacing hair/skin/blood cells/damaged cells/growth of new tissues (c) any two of: deamination of amino acids; remaining part used as source of energy in cell respiration; nitrogen/urea from amino acids is excreted.

3 (a) milk/butter/cheese/yoghurt (b) cyanocobalamin not found in plant based foods so there is a danger of deficiency; but it can be obtained from any two of: vitamin supplement pills/yeast extract/soya milk/margarine/foods to which it is added artificially.

OPTION B – PHYSIOLOGY OF EXERCISE

1 (a) humans store less oxygen per kg of body tissue (b) Any three of: both store most in blood; seal stores a higher proportion in blood than human; smallest proportion stored in lung of seal but muscle of human; human stores higher proportion in lung than seal; seal stores higher proportion in muscle than human (c) any three of: size of muscle; ratio of fast and slow fibres; concentration of myoglobin in muscle; amount of blood in muscle.

2 (a) central nervous system (b) Any three of: released when an impulse arrives at the end of the neurone before the synapse; carries a message across the synapse; causes an electrical impulse in the neurone beyond the synapse.

3 (a) Any three of: monitor how stretched the muscle is; senses when a muscle is become more stretched; send messages to the brain/cerebral cortex; helps the brain to decide if the muscle is contracted enough (b) motor areas of the cerebral cortex.

OPTION C – CELLS AND ENERGY

1 (a) CO_2 concentration (b)(i) temperature; rate of photosynthesis rises as temperature rises; (ii) temperature controls the rate of photosynthesis between 35 and 40 °C; but is not the factor nearest to its minimum level/is supra-optimal
(c) light is the limiting factor at low light intensity; temperature therefore does not affect the rate of photosynthesis.

2 (a) (i) malonate inhibits succinate dehydrogenase/other example (ii) copper/mercury/silver ions/other example; (b) similarity: both types of inhibitor reduce the rate of catalysis; difference between competitive and non-competitive: inhibitor structure similar to substrate v.s not similar/inhibitor binds to active site v.s binds elsewhere.

3 (a) CO_2 (b) ATP is an energy source; NADH + H$^+$; used as an hydrogen/ energy source in oxidative phosphorylation.

OPTION D – EVOLUTION

1 (a)(i) positive correlation (ii) any two of: primate brains are larger in relation to body mass; but there is much variation; largest primates have relatively small brains (iii) any two of: scattergram shows that human brain has the largest size; primates with a larger body mass have a smaller brain; human brain mass is furthest above the line of best fit (b) easier to climb trees/speed of movement/less food needed.

2 (a) endosymbiotic theory; chloroplasts and mitochondria originated as symbiotic bacteria inside host cells (b) Any two of: double membrane of chloroplast/mitochondrion suggests enclosure of a cell in a vesicle; chloroplasts/mitochondria have their own DNA; chloroplasts/mitochondria have 70S ribosomes like bacteria; chloroplasts/mitochondria can divide to produce daughters.

3 (a) $p^2 + 2pq + q^2 = 1$ and q^2 is the frequency of homozygous recessives; frequency = 0.23/23% (b) 35% (c) carriers have increased resistance to malaria; selective advantage over homozygous dominants so the sickle cell allele survives.

OPTION E – NEUROBIOLOGY AND BEHAVIOUR

1 (a) receptor protein; each receptor protein's shape is complementary to a specific odorant; (b) Any three of: G protein activates the enzyme adenylyl cyclase; enzyme converts ATP to CAMP; CAMP causes calcium channel to open; calcium causes chloride channel to open (c) membrane of chemoreceptor cell depolarizes/action potential created/chemoreceptor cell passes an impulse to a sensory neurone.

2 (a) photoreceptors (b) in sensory neurones from the retina to the brain; in motor neurones from the brain to the circular muscle fibres in the iris (c) no response when a light is shone into eye of unconscious patient indicates damage to the brainstem.

3 (a) (i) noradrenaline (ii) acetylcholine (b) increases the heart rate (c) prevents contraction of circular muscle/dilates pupil; less saliva produced/other valid answer.

OPTION F – APPLIED PLANT AND ANIMAL SCIENCE

1 (a) milk production (b)(i) efficiency of protein conversion in milk is much higher than in beef systems; intensive and rangeland beef systems have similar efficiencies (ii) some protein in beef cattle is used to grow skin/hoof/inedible parts, whereas all of the milk can be used (c) any two of: less labour needed/cheaper feed/less spread of disease/parasites/less cost of housing/lower energy inputs/cattle behaviour is more natural.

2 (a) dry biomass of the plant of the plant that is of commercial value (b) net assimilation rate/other measure (c) disease or pest can destroy the entire crop/soil becomes depleted of nutrients/large fertilizer/pesticide applications needed.

3 (a) dip base of cutting in hormone rooting powder; auxin (b) any two of: allows a combination of genes to be conserved; allows more of a desirable genotype to be produced; uniformity of growth requirements/growth rate/characteristics.

OPTION G – ECOLOGY AND CONSERVATION

1 (a)(i) S. marcescens feeds on the nutrients so more grow at high nutrient levels (ii) C. striatum reduce the numbers by predation (iii) D. nasutum increases the numbers because it feeds on C. striatum; which reduces the predation of S. marcescens (b) low population of S. marcescens at low nutrient levels; therefore very low levels of C. striatum on which D. nastum feeds
(c) longer food chain with higher nutrient levels.

2 (a) indicator species need particular environmental conditions; can be used to give a measure of an pollution levels/levels of an environmental variable (b) any two of: captive breeding and release of endangered species; growth of endangered plants in botanic gardens; storage of frozen seeds of endangered species in seed banks.

3 (a) converts nitrite to nitrate (b) oxygen in the soil (c) any three of: excessive fertilizer run-off into rivers/lakes; causes rapid increase in algal growth; as light cannot penetrate organisms in deeper water die; bacteria that decompose the dead organisms cause oxygen depletion/high BOD.

OPTION H – FURTHER HUMAN PHYSIOLOGY

1 (a)(i) negative correlation (ii) lichen biodiversity gives a measure of air pollution; air pollution is a cause of lung cancer
(b) different air pollution experienced before moving to the area (c) any two of: random variation in lung cancer mortality; air pollution not the only risk factor; levels of other risk factors may vary between municipalities; heredity/genetic factors;

2 (a) SAN/pacemaker sends out a signal; signal spreads out through the walls of the atria (b) any two of: lub dup sounds made when valves close; closing valve causes vibration of blood in ventricle; rushing sounds due to flow of blood.

3 (a) helps inflate the lungs when it contracts/increases thorax volume (b) stores oxygen in the muscles (c) causes more oxygen to be released in tissues with high CO_2 level.

A.S.	Page
1.1.1	4
1.1.2	4
1.1.3	5
1.1.4	2
1.1.5	2
1.1.6	1
1.1.7	3
1.1.8	3
1.1.9	3
1.1.10	1
1.1.11	1
1.1.12	1
1.2.1	5
1.2.2	5
1.2.3	5
1.3.1	6
1.3.2	6, 8
1.3.3	6
1.3.4	6
1.3.5	6
1.4.1	7
1.4.2	7
1.4.3	7
1.4.4	8
1.4.5	8
1.4.6	8
1.4.7	8
1.4.8	8
1.5.1.	9
1.5.2	9
1.5.3	9
1.5.4	9
1.5.5	9
1.5.6	9
1.5.7	9
2.1.1	12
2.1.2	12
2.1.3	12
2.1.4	12
2.1.5	11
2.1.6	11
2.2.1	12
2.2.2	12
2.2.3	12
2.2.4	12
2.2.5	13
2.2.6	13
2.2.7	13
2.2.8	13
2.2.9	13
2.2.10	13
2.3.1	14
2.3.2	14
2.3.3	14, 15
2.3.4	14
2.3.5	15
2.4.1	18
2.4.2	18
2.4.3	18
2.4.4	18
2.4.5	18
2.5.1	18
2.5.2	18
2.5.3	18
2.6.1	19
2.6.2	19
2.6.3	19
2.6.4	19
2.6.5	19
2.6.6	19
2.7.1	16
2.7.2	16
2.7.3	16
2.7.4	16
2.8.1	17
2.8.2	17
2.8.3	17
2.8.4	17
2.8.5	17
2.8.6	17

A.S.	Page
2.8.7	17
2.8.8	17
3.1.1	21
3.1.2	21
3.1.3	21
3.1.4	21
3.1.5	28
3.1.6	28
3.2.1	21
3.2.2	21
3.2.3	22
3.2.4	22
3.2.5	22
3.2.6	23
3.2.7	23
3.3.1	21,25,27
3.3.2	23
3.3.3	24
3.3.4	24
3.3.5	24
3.3.6	25
3.3.7	25
3.3.8	25
3.3.9	25
3.3.10	25
3.3.11	25
3.3.12	21,23,24,25
3.3.13	26
3.4.1	31
3.4.2	31
3.4.3	31
3.4.4	31
3.4.5	27
3.4.6	27
3.4.7	30
3.4.8	30
3.4.9	29
3.4.10	29
3.4.11	29
3.4.12	29
3.4.13	28
3.4.14	30
3.4.15	30
3.4.16	30
4.1.1	33,35,40,43,45
4.1.2	44
4.1.3	41
4.1.4	40
4.1.5	42
4.1.6	40
4.1.7	40
4.1.8	42
4.1.9	41
4.1.10	41
4.1.11	42
4.1.12	42
4.1.13	43
4.1.14	43
4.1.15	43
4.2.1	37
4.2.2	37
4.2.3	37
4.2.4	37
4.2.5	37
4.2.6	35
4.2.7	35
4.2.8	35
4.2.9	35
4.2.10	36
4.2.11	36
4.3.1	38
4.3.2	38
4.3.3	38
4.3.4	36
4.3.5	39
4.3.6	38
4.3.7	39
4.3.8	39

A.S.	Page
4.4.1	33
4.4.2	33
4.4.3	33
4.4.4	33
4.4.5	33
4.4.6	34
4.5.1	44,45
4.5.2	44,45
4.5.3	44,45
5.1.1	47
5.1.2	47
5.1.3	47
5.1.4	47
5.1.5	47
5.1.6	47
5.1.7	47
5.2.1	48
5.2.2	48
5.2.3	48
5.2.4	48
5.2.5	48
5.2.6	48
5.3.1	49
5.3.2	49
5.3.3	49
5.3.4	49
5.3.5	49
5.3.6	50
5.4.1	50
5.4.2	50
5.4.3	50
5.4.4	50
5.4.5	50
5.5.1	51
5.5.2	51
5.5.3	51
5.5.4	51
5.5.5	51
5.6.1	52
5.6.2	53
5.6.3	52
5.6.4	52
5.6.5	53
5.6.6	52
5.6.7	53
5.6.8	52
5.6.9	52
5.7.1	54
5.7.2	54,56
5.7.3	54
5.7.4	55
5.7.5	55
5.7.6	55
5.7.7	55
5.7.8	55
5.7.9	57
5.7.10	57
5.7.11	55
5.7.12	57
5.7.13	57
6.1.1	60
6.1.2	60
6.1.3	59
6.2.1	59
6.2.2	59
6.2.3	60
6.3.1	62
6.3.2	60
6.3.3	61,62
6.3.4	61,62
6.3.5	60
6.3.6	62
6.3.7	62
6.4.1	63
6.4.2	63
6.4.3	65
6.4.4	64
6.4.5	64,65
6.4.6	60
6.5.1	66,67

A.S.	Page
6.5.2	68
6.5.3	68
6.5.4	68
6.6.1	71
6.6.2	69
6.6.3	69
6.6.4	70
6.6.5	71
7.1.1	73
7.1.2	73
7.1.3	76
7.1.4	74,75
7.1.5	75
7.1.6	76
7.1.7	74
7.2.1	80
7.2.2	87
7.2.3	78,81
7.2.4	78
7.2.5	79
7.2.6	80
7.2.7	77
7.2.8	77
7.2.9	81
8.1.1	89
8.1.2	88
8.1.3	88,89
8.1.4	87
8.1.5	83
8.1.6	83
8.2.1	84
8.2.2	87
8.2.3	86
8.3.1	84
8.3.2	88
8.3.3	88
8.3.4	88
8.3.5	88
8.4.1	85
8.4.2	85
9.1.1	90
9.1.2	90
9.1.3	92
9.1.4	91
9.1.5	91
9.1.6	92
9.1.7	92
9.1.8	92
9.2.1	93
9.2.2	94
9.2.3	94
10.1.1	96
10.1.2	97
10.1.3	96
10.1.4	97
10.1.5	96
10.1.6	98
10.1.7	98
10.1.8	98
11.1.1	52,100
11.1.2	100
11.1.3	99
11.1.4	99
11.1.5	100
11.2.1	102
11.2.2	102
11.2.3	102
11.2.4	102
11.2.5	101
11.2.6	101
12.1.1	104
12.1.2	104
12.1.3	104
12.2.1	105
12.2.2	105
12.2.3	105
12.2.4	106
12.2.5	106
12.2.6	106
12.2.7	105

A.S.	Page
12.2.8	104
13.1.1	107
13.1.2	107
13.1.3	108,109
13.1.4	108
13.1.5	107
13.1.6	107
13.2.1	109
13.2.2	109
13.2.3	109
13.2.4	109
13.2.5	108
13.2.6	108
13.2.7	108
13.2.8	108
13.2.9	108
13.2.10	108
13.3.1	110
13.3.2	110
13.3.3	109
13.3.4	110
13.3.5	110
13.3.6	110
A1.1	112
A1.2	113
A1.3	113
A1.4	113
A1.5	112
A1.6	112
A1.7	112
A2.1	113
A2.2	113
A2.3	113
A2.4	113
A2.5	112
A2.6	112
A2.7	113
A2.8	113
A2.9	113
A2.10	113
A2.11	113
A2.12	114
A3.1	115
A3.2	115
A3.3	115
A3.4	115
A3.5	115
A3.6	116
A3.7	116
A3.8	116
A3.9	114
A3.10	114
A3.11	114
A3.12	115
A3.13	116
A3.14	116
A3.15	116
B1.1	118
B1.2	118
B1.3	102,118
B1.4	102
B1.5	118
B1.6	118
B1.7	101,118
B1.8	101,118
B1.9	120
B2.1	52
B2.2	100,119
B2.3	119
B2.4	119
B2.5	119
B3.1	120
B3.2	120
B3.3	120
B3.4	120
B3.5	120
B3.6	120
B3.7	120
B4.1	121
B4.2	121

A.S.	Page
B4.3	121
B4.4	121
B4.5	121
B5.1	121
B5.2	121
B5.3	121
C1.1	66,67
C1.2	68
C1.3	68
C1.4	68
C2.1	71
C2.2	69
C2.3	69
C2.4	70
C2.5	71
C3.1	73
C3.2	73
C3.3	76
C3.4	47,75
C3.5	75
C3.6	76
C3.7	74
C3.8	75
C4.1	80
C4.2	77
C4.3	78,81
C4.4	78
C4.5	79
C4.6	80
C4.7	77
C4.8	77
C4.9	81
C4.10	77,78,79,81
D1.1	124
D1.2	124
D1.3	124
D1.4	124
D1.5	124
D1.6	124
D2.1	125
D2.2	125
D2.3	125
D2.4	124
D2.5	124
D3.1	126
D3.2	127
D3.3	127
D3.4	127
D3.5	127
D3.6	126
D3.7	126
D3.8	127
D3.9	127
D3.10	126
D3.11	125
D4.1	128
D4.2	128
D4.3	128
D4.4	129
D4.5	129
D4.6	129
D4.7	128
D4.8	129
D4.9	129
D5.1	131
D5.2	131
D5.3	39
D5.4	130
D5.5	130
D5.6	132
D5.7	132
D5.8	132
D5.9	130
D6.1	130
D6.2	130
D6.3	130
D6.4	130
D6.5	130
D6.6	130,131

A.S.	Page
E1.1	136,138
E1.2	136,138
E1.3	136
E1.4	134
E1.5	136
E2.1	135
E2.2	135
E2.3	135
E2.4	135
E2.5	135
E2.6	135
E2.7	135
E3.1	136
E3.2	137
E3.3	137
E3.4	137
E3.5	137
E3.6	137
E3.7	137
E3.8	136
E3.9	136
E3.10	136
E4.1	138
E4.2	138
E4.3	138
E4.4	138
E4.5	138
E4.6	138
E4.7	138
E5.1	139
E5.2	139
E5.3	139
E6.1	140
E6.2	140
E6.3	140
E6.4	140
E6.5	140
E7.1	141
E7.2	141
E7.3	141
E7.4	141
E7.5	142
E7.6	142
E7.7	142
F1.1	144
F1.2	144
F1.3	145
F1.4	145
F1.5	145
F1.6	145
F1.7	145
F1.8	144
F1.9	144
F1.10	144
F1.11	147
F1.12	147
F2.1	146
F2.2	146
F2.3	147
F2.4	146
F2.5	147
F3.1	148
F3.2	148
F3.3	148
F3.4	148
F3.5	148
F4.1	149
F4.2	149
F4.3	149
F4.4	149
F4.5	149
F5.1	150
F5.2	150
F5.3	150
F6.1	151
F6.2	151
F6.3	152
F6.4	152
F6.5	152
F6.6	151
F6.7	151

A.S.	Page
G1.1	155
G1.2	155
G1.3	157
G1.4	156
G1.5	156
G2.1	156
G2.2	154
G2.3	154
G2.4	154
G2.5	154
G2.6	154
G2.7	155
G2.8	155
G3.1	158
G3.2	158
G3.3	157,158
G3.4	158
G3.5	159
G3.6	159
G3.7	159
G3.8	159
G3.9	159
G3.10	159
G4.1	160
G4.2	160
G4.3	160
G4.4	160
G4.5	160
G4.6	160
G4.7	160
G4.8	160
G5.1	161
G5.2	161
G5.3	161
G5.4	161
G5.5	162
G5.6	161
G5.7	162
G5.8	162
H1.1	164
H1.2	164
H1.3	164
H1.4	164
H1.5	164
H1.6	164
H2.1	165
H2.2	165
H2.3	165
H2.4	166
H2.5	165
H2.6	166
H2.7	166
H2.8	166
H2.9	166
H2.10	166
H3.1	167
H3.2	167
H3.3	167
H3.4	166
H4.1	168
H4.2	168
H4.3	168
H4.4	168
H4.5	168
H4.6	168
H5.1	169
H5.2	169
H5.3	169
H5.4	170
H5.5	170
H5.6	170
H5.7	170
H6.1	171
H6.2	171
H6.3	172
H6.4	171
H6.5	172
H6.6	172
H6.7	171

Index

A

ABO blood groups 24
absorption of food 167
absorption spectrum of pigments 77
acetyl coenzyme A 74
acetylcholine 100
acid rain 161
acrosome reaction 93
actin 101
action potential 99
action spectrum of photosynthesis 77
activation energy 69
active sites 14, 69, 70
active transport 8
adaptation 38
adenosine triphosphate 16
ADH 106, 164
adrenalin 48, 120
aerobic respiration 16, 75
AI (artificial insemination) 146
AIDS 50
albinism 26
alcohol 142
algal blooms 162
alien species 45
alleles 21
allostery 71
alpha helix 66
altitude (effect on gas exchange) 171
altruism 139
alveoli 51
amino acids 12, 66, 68, 113
amniocentesis 55
amniotic sac 55
amphetamines 142
anaerobic respiration 16, 120
anemia 115
angiospermophytes 107
animal behaviour (examples) 134
animal breeding 149
animal species distribution 155
antagonistic muscles 102, 119
antibiotic resistance 39, 147
antibiotics 49, 147
antibodies 50, 96, 97
anti-diuretic hormone 164
antigens 50, 97
apical dominance 148
arteries 48
arthropods 102
artificial insemination 146
asexual reproduction 9, 152
assimliation (of food) 47
asthma 172
atherosclerosis 115, 170
atoms 12
ATP production 16, 73, 75, 76, 78
autonomic nervous system 140
autosomal linkage 87
autotrophs 41
auxin 148

B

bacteria 5, 29, 49, 160
balanced diets 112

balanced polymorphism 131
base pairing 18, 59
base substitution mutations 28
B-cells 97
behaviour (examples) 134
benzodiazepines 142
beta pleated sheets 66
bile 168
bile salts 166
binomial system 33
biochemical oxygen demand 162
biodiversity 149, 157, 158
biogeochemical cycles 43, 160
biological fuels 162
biological pest control 147
biomass 154
biosphere 44
biotechnology 15, 29, 30, 150
bipedalism 128
birth 55
blood 48
blood clotting 96
blood groups 24
BOD 162
body temperature 53
Bohr shift 171, 172
bones 118
botanic gardems 159
brain size 128
breeding 149
bryophytes 107
bulbs 152

C

calcium 12
Calvin cycle 79
CAM plants 107
cancer 9, 172
cannabis 142
capillaries 48
captive breeding (in zoos) 159
capture mark release method 35
carbohydrates 13, 113
carbon cycle 43
carbon dioxide transport 172
carbon fixation 17, 79
cardiac cycle 169
carriers (of recessive alleles) 25
carrying capacity 37
cattle breeding 149
cell division 9
cell respiration 16, 51, 73-76
cell theory 4
cell walls 5, 6
cells 1, 4-6
cellulose 6, 114, 166
centrioles 9
centromere 21
cerebrum (of brain) 119, 137
channels 8
CHD 115, 170
chemical pest control 147
chemoautotrophs 160
chi squared test 86
chiasmata 88

chicken farming 146
childbirth 55
chlorophyll 17
chloroplasts 6, 80
cholesterol 115, 168, 170
chromatids 21
chromosomes 21
CITES 159
classical conditioning 138
classification 33
clones and cloning 30, 31
cloning of plants 148, 152
clotting of blood 96
cocaine 142
codominance 24, 84
codons 19, 64
colour blindness 25
combustion 43, 162
communication 134
communities 40
competition 156
competitive exclusion 156
competitive inhibition (enzymes) 70
condensation reactions 13
conditioning 138
condom 57
conifers 107
conjugated proteins 67
conservation 158, 159
consumers 40, 41
continuous variation 85
copulation 55
coronary heart disease 115, 170
coronary thrombosis 170
corpus luteum 56
cortical reaction 93
courtship 134
cranial reflex 137
creation of life 124
crop production 144, 145
crossing over 88
cultural evolution 129
cuttings 148, 152
cyclic photophosphorylation 81
cystic fibrosis 131
cytokinesis 9
cytoplasm 1, 5, 9
cytotoxic T-cells 96

D

Darwin 38
Darwin's finches 125
de-amination 113
deficiency diseases 114, 116
denaturation 14
denitrification 160
deoxyribonucleic acid 18
depolarisation 99
detrivores 41
dialysis 104
diaphragm 51
dietary fibre 114
diets 112
differentiation 1
diffusion 8

digestion 47, 165, 166
dihybrid crosses 83, 84
diploid 21
disaccharides 13
disease 28, 49, 50, 114, 115, 170
disease transmission 49
diversity index 157
DNA 18, 59
DNA fingerprinting 31
DNA profiling 31
DNA replication 18, 59
DNA transcription 19, 62
Dolly the sheep 30, 31
dominant alleles 23
dopamine 141
Down's syndrome 21, 22, 55
drugs 142
drugs in sport 121

E

earthworms 102
ecological efficiency 42
ecological niches 156
ecological succession 155
ecology 43
ecosystems 43, 154, 155
egestion 47
eggs (human) 92
elbow 102
electron microscopes 2
electron transport chain 76
elements 12
embryos (human) 55
end product inhibition 71
endergonic reactions 69
endocrine system 52
endocytosis 8
endoplasmic reticulum 6,8
endorphins 141
endosymbiotic theory 124
energy 16
energy efficiency 42
energy flow 41, 154
energy loss in food chains 41, 42, 154
energy pyramids 42, 154
energy requirements (human) 112
energy storage compounds 13
environmental monitoring 158
enzyme inhibition 70, 71
enzyme specificity 14
enzymes 14, 15, 69, 70
epistasis 84
epithelium 167
ER 6
erythrocytes 48
estrogen 54, 56, 94
ethanol 16, 142
ethics in agriculture 147
eukaryotes 6, 60
eutrophication 162
evidence for evolution 39, 125, 126
evolution 38, 39, 125-132
evolution of humans 129
excretion 52, 104, 105
exercise (and ventilation) 120, 172
exergonic reactions 69
exocrine glands 165
exocytosis 8
extinction 158
eye 135

F

F1 hybrids 23
family planning 57
fast and slow muscle 120
fatty acids 12
ferns 107
fertilisation 55, 93
fetus 55
fibre (dietary) 114
fibrous proteins 68
filicinophytes 107
fish locomotion 102
fish conservation 159
fitness 121
Flavr-Savr tomatoes 150
flight in birds 102
flowering of plants 151
flowering plants 107
flower structure 110, 151
fluid mosaic model 7
food additives 116
food chains 40
food handling 116
food labelling 113, 116
food storage 108
food webs 42, 154
fossil fuels 43
fossilisation 127
fossils 126,127
fruit ripening 148
FSH 56,92
fungi 33, 40, 49

G

gametogenesis 90, 91, 93
gas exchange 51, 171
gel electrophoresis 31
gender 25
gene expression 60, 61
gene interaction 84
gene linkage 87, 88
gene locus 21
gene mutation 28
gene pool 130
gene regulation 60
gene therapy 28
gene transfer 150
genes 21
genetic code 19, 63
genetic diseases 27, 28
genetic engineering 29, 150
genetic screening 27
genetic variation 22, 88
genetically modified organisms 29, 150
genome 21
genotype 23
genus 33
geogrphical distribution 126
germination 110
gibberellin 110
glands 164, 165
global warming 44
globular proteins 68
glucagon 53
glycerides 13
glycerol 12
glycogen 53, 120, 168
glycolysis 73
glycoproteins 7
GMO 29, 150

goitre 114
golgi apparatus 6, 8
gradualism 130
grafting of plants 152
greenhouse crops 145
greenhouse effect 44
grooming 134
growth hormones in livestock 147
growth rates of crops 145

H

habitat 45
haemoglobin 67
haemophilia 25
half-life 127
haploid 21
Hardy-Weinberg Principle 130
HCG 94
heart 48
heart 169
heart action 169
heart attacks 170
heart beat 169
helper T-cells 97
hemoglobin 67, 127, 168, 171, 172
hemophilia 25
herbicides 148
herbivory 156
heterotrophs 41
heterozygous 23
HIV 50, 62
homeostasis 52, 53
hominids 129
homology 126
homologous chromosomes 21
homozygous 23
honey bees 139
hormones 7
hormones in humans 52-57, 164, 165
hormones in plants 148
human ancestors 128
human classification 128
human diets 112
human evolution 129
human genome project 30
human impacts 44, 45
human origins 128
human reproduction 54, 90-94
hybridisation 149
hydrogen bonds 11, 18, 59, 66
hydrolysis reactions 13
hydrophytes 107
hydroponics 145
hygeine 116
hypothalamus 164

I

IDD 114
identification of organisms 34
ileum 167
immunisation 96, 98
immunity 50, 96, 97
imprinting 138
in vitro fertilisation 57
inbreeding 149
independent assortment 83
induced fit hypothesis 69
infections 50
inheritance of acquired characteristics 125
inhibititors (enzyme) 70, 71

injuries in sport 121
innate behaviour 136
inoculation 96, 98
inorganic compounds 12
insect pollination 110
instinct 136
insulin 29, 53
intensive agriculture 147
interphase 9
intestines 47
intramolecular bonding 66
introns 60-62
iodine deficiency 114
ions 12
iron 12
IVF 57

J
joints 102, 118

K
karyotypes 21
keys (dichotomous) 34
kidney machines 104
kidneys 52, 105
kinesis 136
Krebs cycle 74

L
Lac operon 60
lactate (lactic acid) 16, 120
ladybugs (ladybirds) 125
Lamarck 125
layering of plants 152
learned behaviour 138
leaves 3, 108
leukocytes 48
LH 56, 92
light microscopes 2
light-dependent reactions 78
light-independent reactions 79
limiting factors (photosynthesis) 81
link reaction 74
linkage (genes) 87
lipase 68
lipid digestion 166
lipids 13, 113, 115, 170
lipoproteins 170
liver 168
livestock production 146, 147
local impacts 45
locomotion 102
loop of Henle 106
Lorenz (Konrad) 138
lung cancer 172
lungs 51
lymph 170
lysis (of cells) 1
lysosomes 6

M
macroevolution 130
magnification 2, 3
malnutrition 114
management of wildlife reserves 159
mate selection 134
mean 35, 36
meiosis 22, 87-89

membrane proteins 7, 68
membranes 7
Mendel 2, 83
mesosomes 5
metabolic pathways 71
metabolism 5, 52, 71
metal tolerance 39
methane generation 162
microevolution 130
micropropagation 148
microscopes 2
microvilli 167
migration 134
Miller and Urey 124
mineral absorption (plants) 109
mineral elements 12
minerals in food 113
mitochondria 6, 76
mitosis 9
monitoring environments 158
monoclonal antibodies 98
monoculture 144
monohybrid crosses 23
monosaccharides 13
motor neurones 100
multicellular organisms 1
multiple alleles 24
muscle 101, 102, 118-120
muscle contraction 101, 118
muscle fatigue 120
muscular dystrophy 26
mutation 28, 131
mutualism 156
myofibrils 101, 118
myoglobin 120, 171
myosin 101, 118

N
natural selection 38, 39, 125, 130, 131,
nature reserves 159
negative feedback 53, 56, 164
neoteny 128
nephrons 105, 106
nerve impulses 99
nervous system 52, 140
neurones 100, 119, 135
neurotransmitters 100, 119, 141
niches 156
nicotine 142
nitrates in water 162
nitrification 160
nitrogen 12
nitrogen cycle 160
nitrogen fertilizer 145
nitrogen fixation 160
nomenclature 33
non-competitive inhibition 70
non-disjunction 22
normal distribution 36
nucleosomes 60
nucleotides 18
nucleus 1
nutrient cycles 43, 160
nutrient supplementation 146
nutrition (human) 112, 113

O
oestrogen (estrogen) 54, 56, 94
Okazaki fragments 59
oogenesis 91-93

operant conditioning 138
operons 60
organelles 1
organic compounds 12
organic farming 147
organs 1
origin of cells 5
origin of life 124
osmoregulation 106
osmosis 8, 104, 106, 109, 170
osteoporosis 115
outbreeding 149
ovary (human) 54, 91
oxidation reactions 73
oxidative phosphorylation 75
oxygen dissociation 171
oxygen transport 171
ozone 161

P
pacemaker 48
pain 141
painkillers 141
palaeontology 126
pancreas 47, 53, 165
panspermia 124
parasitism 156
parasympathetic system 140
Parkinson's disease 141
partial pressure 171
passive transport 8
pathogens 49
Pavlov (Ivan) 138
PCR 31
pectinase 15
pedigree charts 24, 26
penis 54
peptide linkage 13
pesticide use 147
phagocytes 50
phenotype 23
phenylketonuria 131
phloem 108
phospholipids 7
phosphorus 12
photolysis 17, 78
photoperiodism 151
photophosphorylation 78, 81
photoreceptors 135
photosynthesis 17, 77-81
phototropism 148
phylogeny 127
phytochrome 151
pigments 77
pituitary gland 164
PKU 131
placenta 55, 94
plant breeding 149
plant classification 33, 107
plant distribution 155
plant growth hormones 148
plant productivity 145
plant propagation 152
plant reproduction 110, 151, 152
plant uses 144
plasma membrane 1, 5
plasma proteins 168
plasmids 29
platelets 48, 96
polarity of molecules 11
pollination 110, 151

polygenic inheritance 85
polymerase chain reaction 31
polymorphism 130,131
polypeptides 13, 19, 64
polyploidy 149
polysomes 64
population growth curves 37
populations 35-37
positive feedback 55, 56
pre-biotic Earth 124
predation 156
pregnancy 55, 94
primates 128
producers 40
progesterone 56, 94
prokaryotes 5, 33, 64, 124
proprioceptors 119
prosthetic groups 67
protease 15, 166
protein digestion 166
protein functions 7, 68
protein structure 66, 67
protein synthesis 64
proteins in food 113
protoctista 33
psychoactive drugs 142
puberty 54
pumps (membrane) 7
pupil 137, 140
pyramids of energy 42, 154
pyruvate 16, 73, 74

Q
quadrats 35

R
random sampling 35
receptors 135
recessive alleles 23
recombination 87
recycling 43
reduction reactions 73
reflexes 137
renal dialysis 104
replication of DNA 18, 59
reproductive systems (human) 54
resolution 2
respiration 16, 51, 73-76
resting potential 99
restriction enzymes 29, 31
retina 135
retroviruses 62
reverse transcriptase 62
ribonucleic acid 19
ribose 12
ribosomes 5, 6, 63, 64
rice breeding 149
rice farming 144
rickets 115
RNA 19
roots 109
rough endoplasmic reticulum 6, 8
rubisco 79
runners in plants 152

S
saprotrophs 41, 43
sarcomeres 101
satellite DNA 31

saturated fatty acids 115, 170
SCID 28
secondary structure (proteins) 66, 67
seed banks 159
seed dispersal 110
seedless fruit 148
seedlings 110
seeds 110
segregation 23
selective re-absorption 106
semen 92
seminiferous tubules 90
sense/antisense technology 150
sensory neurones 119
sensory receptors 135
sewage 162
sex chromosomes 25
sex determination 25
sex linkage 25, 84
sexual reproduction (human) 54
sickle cell anemia 28, 131
sieve tubes 108
sigmoid growth curves 37
significance tests 86
simple reflexes 137
Simpson diversity index 157
sinusoids 168
size 3
skeleton 118
skin colour 85
Skinner 138
shulls (hominid) 129
social behaviour 139
sodium 12
soft tissue injuries 121
soil fertility 160
special creation 124
speciation 132
species 33, 132
species extinction 158
specificity of enzymes 14
sperm 90, 92
spermatogenesis 90, 92
spermatogenesis 93
spinal reflexes 137
spontaneous generation 124
sprains 121
standard deviation 36
statistical tests 86, 157
stems 109
stomach 47, 166
stop codons 65
substrate concentration 14
succession 155
surface areas 3
sweat 53
swimming (fish) 102
sympathetic system 140
synapses 100, 119, 141
synovial joints 102, 118

T
taxis 136
T-cells 97
tertiary structure (proteins) 67
test crosses 27, 88
testes 54, 90
thylakoids 80
thyroxin 164
tissue fluid 170
tissues 1

training 121
transcription 19, 61, 62
transfer RNA 19, 63, 64
transgenic organisms 29, 150
translation 19, 63, 65
translocation (phloem) 108
transpiration 108, 109
trisomy 22
tRNA 19, 63, 64
trophic levels 40, 154
T-test 157
tuberculosis 49
tumours 9

U
ultra violet light 161
ultrafiltration 105
unicellular organisms 1
units 3
urea 52
urine 52, 106

V
vaccination 96, 98, 146
vacuoles 6
vagina 54
variation 22, 36, 38, 88
vasopressin (ADH) 106
vegan diets 116
vegetarian diets 116
veins 48
ventilation 51
ventilation rate 120, 172
vesicles 7, 8, 68
veterinary techniques 146
villi 47, 167
viruses 4, 49, 62
vision 135
vitamins 113

W
Wallace 38, 125, 126
warming up (sports) 121
water 11
water uptake in plants 109
weedkillers 148
wheat breeding 149
wildlife reserves 159
wind pollination 151
withdrawal reflex 137
WWF 159

X
X chromosomes 25
xerophytes 107

Y
Y chromosomes 25

Z
zygotes 55